THE CONSTRUCTION OF TIME IN ANTIQUITY

Time has always held a fascination for human beings, who have attempted to relate to it and to make sense of it, constructing and deconstructing it through its various prisms, since time cannot be experienced in an unmediated way. This book answers the needs of a growing community of scholars and readers who are interested in this interaction. It offers a series of innovative studies by both senior and younger experts on various aspects of the construction of time in antiquity. Some articles in this book contain visual material published for the first time, while other studies update the field with new theories or apply new approaches to relevant sources. Within the study of antiquity, the book covers the disciplines of Classics and Ancient History, Assyriology, Egyptology, Ancient Judaism, and Early Christianity, with thematic contributions on rituals, festivals, astronomy, calendars, medicine, art, and narrative.

Jonathan Ben-Dov is an expert in ancient Jewish literature and ancient astronomy and calendars. He took part in the official publication of calendrical texts from the Dead Sea Scrolls and is now co-leader of a new project for the digitization of the scrolls. He has been a research fellow at New York University and at Durham University and is a member of the Israeli Young Academy of Sciences.

Lutz Doering is a scholar of ancient Judaism and early Christianity, specializing in the study of festivals, Jewish law, and letter writing. He has held research fellowships from the British Arts and Humanities Research Council and the Israel Institute for Advanced Studies, and he currently leads a project in the Cluster of Excellence "Religion and Politics" at Münster, Germany.

The Lysippan Kairos, 3D digital reconstruction (© SeungJung Kim and Dave Cortes)

THE CONSTRUCTION OF TIME IN ANTIQUITY

Ritual, Art, and Identity

Edited by

JONATHAN BEN-DOV

University of Haifa

LUTZ DOERING

University of Münster

CAMBRIDGE
UNIVERSITY PRESS

CAMBRIDGE
UNIVERSITY PRESS

University Printing House, Cambridge CB2 8BS, United Kingdom

One Liberty Plaza, 20th Floor, New York, NY 10006, USA

477 Williamstown Road, Port Melbourne, VIC 3207, Australia

314-321, 3rd Floor, Plot 3, Splendor Forum, Jasola District Centre, New Delhi - 110025, India

79 Anson Road, #06-04/06, Singapore 079906

Cambridge University Press is part of the University of Cambridge.

It furthers the University's mission by disseminating knowledge in the pursuit of education, learning and research at the highest international levels of excellence.

www.cambridge.org
Information on this title: www.cambridge.org/9781107108967
DOI: 10.1017/9781316266199

First published 2017

A catalogue record for this publication is available from the British Library

Library of Congress Cataloging in Publication data
Ben-Dov, Jonathan, editor.
The construction of time in antiquity : ritual, art, and identity / [edited by] Jonathan Ben-Dov, University of Haifa, Lutz Doering, University of Münster.
New York : Cambridge University Press, 2017.
LCCN 2017009661 | ISBN 9781107108967
LCSH: Time.
LCC BD638 .C665 2017 | DDC 115.093–dc23
LC record available at https://lccn.loc.gov/2017009661

ISBN 978-1-107-10896-7 Hardback
ISBN 978-1-107-52095-0 Paperback

CONTENTS

ILLUSTRATIONS

CONTRIBUTORS

Jonathan Ben-Dov is Associate Professor of Biblical Studies and George and Florence Wise Chair of Judaism, University of Haifa.

Lutz Doering is Professor of New Testament and Ancient Judaism, and Head of the Institutum Judaicum Delitzschianum, Westfälische Wilhelms-Universität Münster.

Robert Hannah is Emeritus Professor, University of Waikato, New Zealand.

Robert Hayward is Professor emeritus of Hebrew, Durham University, UK.

Sarit Kattan Gribetz is Assistant Professor of Classical and Medieval Judaism, Fordham University.

SeungJung Kim is Assistant Professor of (Ancient Greek) History of Art, University of Toronto.

Clemens Leonhard is Professor of Liturgical Studies, Faculty of Catholic Theology, Westfälische Wilhelms-Universität Münster.

Alexandra von Lieven is Außerplanmäßige Professorin (adjunct professor), Freie Universität Berlin.

Jörg Rüpke is Professor of Comparative Religion and Fellow in History of Religion at the Max Weber Centre, University of Erfurt.

John Steele is Professor of Egyptology and Assyriology, Brown University.

Sacha Stern is Professor of Rabbinic Judaism and Head of the Department of Hebrew and Jewish Studies, University College London.

Daniel Stökl Ben Ezra is Directeur d'Études (research professor), Ancient Hebrew and Aramaic, École Pratique des Hautes Études (EPHE), Paris. Jewish and Christian Lectionary database ThALES: www.lectionary.eu

Lorenzo Verderame is Assistant Professor (ricercatore) of Assyriology, "Sapienza" Università di Roma.

ACKNOWLEDGMENTS

In the academic year 2012–2013, the Institute of Advanced Study (IAS) at Durham University, where one of the editors of the present volume (Lutz Doering) then taught, chose "Time" as the annual theme for its interdisciplinary activities. The IAS was happy to support a series of activities led by this editor under the title, "Calendar and Festivals: Identity, Culture, and Experience." Among other things, this included an invitation to Jonathan Ben-Dov for a fellowship at the IAS during Epiphany term 2013. This became a fruitful period of time, not least because Robert Hannah, who is a contributor to the volume, concomitantly held a fellowship at the IAS. It was during this period that the editors organized an international conference, "The Construction of Time in Antiquity," held in the historical premises of University College Durham on March 3–4, 2013. We wish to thank the Institute of Advanced Study at Durham, and particularly its executive director, Professor Veronica Strang, as well as the Department of Theology and Religion at Durham University, and here especially its head of department at the time, Professor Alec Ryrie, for supporting the organization of the conference. The present book contains both the thoroughly revised contributions to this conference and further chapters specifically commissioned from additional specialists. We are delighted to be able to include in this collection not only studies of texts but also chapters discussing artworks from antiquity as well as other material finds, as seems appropriate for the rich and variegated field of "Time."

The volume was prepared for publication at Münster and Haifa. We are grateful to those who assisted us in the process. On the Münster team, we wish to thank Dr. Kim Czajkowski, now at Edinburgh, for comments on contents and language; Daniel Schumann, *wissenschaftlicher Mitarbeiter* at the Institutum Judaicum, for conforming the format of the submitted chapters to Cambridge University Press style; and research assistants Laura Bramkamp and Andreas Knöll for preparing the indices and checking the bibliographies for consistency. At Haifa, we offer thanks to Dr. Sonia Klinger for expert advice. At Cambridge University Press in New York, Beatrice Rehl and Joshua Penney steered the volume patiently through the publication process. Thanks are also due to Sathish Kumar Rajendran and his production team for their efficient work.

1

INTRODUCTION

LUTZ DOERING AND JONATHAN BEN-DOV

TO APPROACH THE THEME OF TIME AND TEMPORALITY IN THE twenty-first century requires us to think in multiple new directions and attend to a burgeoning body of literature. First of all, there is the context of the natural sciences: uniform, independent Newtonian time has become questionable under the public influence of Einstein's theories of special and general relativity, while physicists continue to debate, and to disagree about, the nature of time (Smolin 2013). Moreover, art and literature have presented us with ever evolving explorations of time as experienced on the scale of human life in the century past, from Marcel Proust's project of searching for the lost time to the switches of time and space in the Wachowskis' The Matrix – explorations that have begun to be analyzed by cultural theorists such as Ronald Schleifer (2000, for modernism) or Ursula Heise (1997, for postmodernism).

In fact, "time" has been a fertile ground for reflection in the humanities in general. Across various disciplines, the notion of time has been problematized and, as a result, enormously enriched – from philosophy to literary theory and art history, from historiography through the history of science to theology and the study of religion. Scholars, readers, and writers today are trained from their undergraduate years to diagnose the type of time used in a given text and, moreover, to fashion and mold it with the new tools available today. To mention only a few examples that are also indirectly relevant for the present volume: several studies over the past few years have been devoted to philosophical concepts of time – classical, continental, and analytical (e.g., McCumber 2011; Dyke and Bardon 2013). In literary theory, the concept of "chronotope," developed by Mikhail Bakhtin (1981, Russian 1975, drafted 1937/1938), has proven to be a powerful tool allowing readers to assess the configuration of space and time in a given narrative. Anthropologists have reflected, *inter alia*, on temporal experience as socialized and internalized by way of concepts and rituals (Gingrich 1994) or pointed to the constraining effects of time-telling devices (Birth 2012).

Closer to the temporal focus of the present volume, "time" has seen a recent upsurge also in the various disciplines concerned with the study of antiquity. The shadow of "the plurality of times" made its debut into the field with a groundbreaking study by Donald Wilcox (1987), who has shown the extent to which pre-Newtonian chronologies were devoid of the "Western" sense of uniform time. Theories of temporality in literature are by no means irrelevant for ancient texts, as demonstrated for the classics by Kennedy (2013) (and earlier Csapo and Miller 1998; Dunn 2007), as well as for biblical narrative by Sternberg (1987). This new awareness has also been applied to ancient philosophy, initially in the monumental work by Sorabji (1983), and more recently, for example, by Roark (2011), who reassessed the role of time in Aristotle's physics. A cultural history of Jewish temporality with systematic ambitions in two volumes has been published by Sophie Anne Goldberg (2000, 2004). A strong emphasis in research has been on calendars, both Greco-Roman (Hannah 2005) and Jewish (Stern 2001). Important studies have focused on the political and ideological dimensions of calendars, particularly Rüpke (2011, German 1995) on Rome, and Stern (2012) on empires more widely. This implies a certain shift in approach: as Stern (2012: 2) writes,

> Far more than a technical device, the calendar was at the heart of ancient societies and cultures, as an organizing principle of social life and as constitutive of ideologies and world views. The calendar should not be confined, as it sometimes is, to the history of science or to a marginal aspect of the history of religions. It firmly belongs to the core of social history.

Other contributions to calendar studies address specific issues such as the indeterminacy or the contrast between observation and calculation in lunar calendars (Ben-Dov, Horowitz, and Steele 2012). Feeney (2007) provides a wide-ranging study of Roman synchronism and chronology, myth and history, as well as calendar ideology.

Of course, time in antiquity remains closely related to the history of science, as far as the technology is concerned with which time could be measured, marked, or told (Hannah 2009). But the history of science also has important implications for social and intellectual history. An important advance has been the decipherment and explanation of the Antikythera Mechanism (Freeth et al. 2008); this early computer based on gears and wheels perfected the ancient ability to synchronize various calendrical and astronomical systems. This brings us to the practice and intellectual background of ancient astronomy and astrology more widely, which has been the focus of several studies, from Babylonian astronomical-astrological texts (Brack-Bernsen and Steele 2004), through the study of parapegmata and astrometereology (Lehoux 2007) and of the role of time and astronomy in Roman public architecture (Hannah and Magli 2011) to Roman astronomical poetry (Volk 2009).

The present volume fits within this recent interest in the study of time in antiquity and intersects with several of the lines of inquiry outlined earlier, while contributing a specific focus and angle. It proceeds from the observation that some of the fundamental questions of human society and culture have to do with the relationship between time and *human agency*. Time as experienced by human beings is constructed, and the chapters of this book amply demonstrate this for different historical periods, cultures, and expressions of human life. Ritual has a particular role in the construction of time by human agents and constitutes a powerful bridge between the foundational beliefs of a society and their materialization in real life (Durkheim 1915; Bell 1997). Other means of constructing time include visual representations of time or its divisions, the metaphorization of the human body in relation to time and temporality, or the deployment of propaganda steering both the properties of time and public memory. The overarching theme of the book is thus the interface of human agency and ancient time reckoning in their historical context.

Many of the following chapters focus on cyclical time. In part, this has to do with the basic cycles of nature and of the human body that relate to human perception of time: the phases generated by the sun and the moon, the sequence of the seasons, the cycle of illnesses brought about by seasonal weather, or the female menstruation cycle. It is by observing them that human beings are able to enforce a sense of order onto their activities. Although seemingly rigid and uniform, these cycles are mediated to every individual by a powerful social matrix that sets the parameters for how they are experienced and what they convey. The present volume addresses these themes in a variety of sources from the cuneiform tradition and ancient Egypt through to Jewish and Christian texts from late antiquity.

Developing an important line of recent research further, the first three chapters are concerned with time, ideology, and identity in the Hellenistic-Roman world. *Jonathan Ben-Dov* investigates the relationship between time and natural law in Jewish sources from the Hellenistic-Roman period and compares the results with the Greco-Egyptian Decree of Canopus of 238 BCE. He concludes that all of these sources found ways integral to their own national culture of connecting time with the idea of nature, with some interesting similarities in approach. *Sacha Stern* demonstrates the close relationship between politics, power relations, and the history of calendars in the Roman Empire. He suggests that the institution of the calendar of Asia in 8 BCE, Jewish retention of a lunar calendar, lunar dating practices in late antique Italy, and the display of the lunar Coligny calendar in Gaul express different ways of relating to the Roman rulers and the Julian calendar, ranging from political loyalty to autonomy and subtle dissidence. *Jörg Rüpke* argues that the instantaneous success of Caesar's calendar reform, despite the conservatism of contemporary society, can be explained by a twofold contextualization: first, by the process of "rationalizing" religion in the late republic and, second, by the Augustan development of "doubling religion," that is, superimposing the concept of *feriae* as devoted to a single god with a more open concept, in which imperial agency could dedicate a day to all, or unspecified, gods. The calendar thereby became a medium for the diffusion of imperial propaganda.

The next two chapters investigate the interplay between time, science, and ideology. *John Steele* examines the ways in which Babylonian texts concerned with astral medicine differentiate between "real" time, in the sense of durations of illnesses, the timing of harvesting medical ingredients and the performance of therapeutic rituals, and "constructed" time, by which schemes for the appropriate treatment of a patient are inferred from a given calendar date. Building on the recent study of the Antikythera Mechanism, *Robert Hannah* explores the philosophical background to this and related devices. Hannah argues that such instruments ultimately served to provide a theoretical basis into which the observable phenomena, especially regarding the planetary system, could be fitted, thereby exemplifying a worldview aligning well with Platonic cosmology.

Four chapters explore the fields of myth, metaphor, and visual art. For ancient Egypt, *Alexandra von Lieven* presents different types of divine figurations of time, such as personifications of the year, the months, the decades, the days, and the hours, as well as the so-called chronocrats, some of which are published here for the first time. These are found in both textual and pictorial sources, especially as decoration of temple walls and ceilings or in manuals of priestly knowledge from temple libraries. Such knowledge enabled the priests to officiate within rituals linked to cosmological cycles and gave them power over time. While much of this will have served the needs of the king, personifications of time are also attested in private manuscripts, suggesting that other members of the elite also had access to such practices. *Lorenzo Verderame* discusses the ancient Mesopotamian concept of time, from its mythological foundation to calendrical festivals. On a ritual and mythological level, renewal of kingship is associated with the Babylonian New Year, during which Marduk's absence and return are enacted by the king's disposal and reinstallation. In addition to the better-known association of royalty with the sun, the connection with the heavenly bodies as the driving forces of the calendar is upheld in the identification of the king with the moon god Nanna(r)/Sîn.

SeungJung Kim brings us forward to representations of time and moment in classical Greek art. Her chapter is based on her new 3D reconstruction of Lysippus's famous sculpture of Kairos (third quarter of the fourth century BCE) – realized together with graphic artist and sculptor Dave Cortes. Kim argues that the so-called Pioneer Group of vase painters at the end of the sixth century BCE began an engagement of the viewer that can be called phenomenological, in the sense that the vase decorations of this group exploit the struggle against gravity for an experience of the moment passed that is fundamentally *felt*. The Lysippan Kairos at the end of the classical period, then, providing a *spatial* solution to the gravitational equilibrium as the epitome of the elusive *temporal* concept is but the culmination of this move. The contribution by *Sarit Kattan Gribetz* examines the use of metaphors of women's bodies relative to time from the Hebrew Bible through Second Temple and rabbinic sources. Metaphors of pregnancy, labor, and birth are often invoked in relation to the anticipation of a nearing redemptive or eschatological time. Metaphors of women's bodies are also used when discussing calendrical time, especially in early rabbinic texts. Kattan Gribetz also engages the paradox that, the gendered nature of these metaphors notwithstanding, women themselves were barred by rabbinic doctrine from time-related commandments and from an active role in redemption.

The final set of four chapters deals with time in Jewish and Christian ritual and calendrical practice. *Lutz Doering* focuses on the ambiguities surrounding the beginning of the Sabbath and festivals in ancient Jewish sources, assumed to take place sometime in the evening but difficult to pinpoint. Jews up to the early rabbinic period responded in two ways, thereby "bracketing" the start of day: some required an early start of rest, whereby time prior to the day was invested with appropriate behavior, while others spent the beginning of the Sabbath and festivals at common meals, transitioning into the holy day under the controlled inertia of being seated at table and performing short rituals on account of the sanctity of the day. *Daniel Stökl Ben Ezra* compares the ways Judaism and Christianity impregnated popular cyclical time with sophisticated narratives. Both "biblified time" through the creation of a cycle of public liturgical readings of scriptural passages that had not necessarily been connected before. On the other hand, both "seasoned the Bible" by contextualizing selected passages in cyclical rather than linear times creating a multidimensional reenactment of its religious ideals. Inspired by the approach of Yuval (2006), Stökl Ben Ezra argues that part of the Jewish selection of annual readings in Palestine (where the main lectionary was triennial) is a response to the challenge posed by the powerful, ritualized narrative present in the Christian lectionaries from the late fourth century onward. Continuing the relationship between Jewish and Christian festivals, *Robert Hayward* explores the similarities between the texts and customs associated with the Jewish New Year, which falls in Tishri (September/October), and the texts of the Mass for the Wednesday of the Ember days of September in the rites of the Roman Church. Though proclaimed as a fast day, the Mass of the Wednesday emphasizes various themes of rejoicing and enthronement, all of which are shared with the Jewish New Year. Hayward suggests that both Jews and Christians may have had an interest in defending the Almighty and his creation, thus presenting a common front against Gnostic teachers active in Rome. In the final contribution to this volume, *Clemens Leonhard* argues that Christians naturally would have celebrated the festivals of their city and that this would initially have *inhibited* the development of "Christian" festivals: both Christians integrating into their municipal organization (as Tertullian claims they did) and those promoting citizenship in a heavenly city (as suggested by Gal. 4:26, Hebrews, and Revelation) would have rejected such festivals on similar grounds. According to Leonhard, Sunday was celebrated as a day of regular meetings only from the second century. Hence, there is no basis for claiming continuity of festal celebration between the first and the late fourth centuries. He provocatively claims that the suggestion that festivals and the construction of sacred time would continuously build up group identity by shaping collective memory is a simplistic cliché. This final chapter therefore reintegrates the early Christian construction of time into the wider context of the Roman Empire.

Bibliography

Bakhtin, M. M. 1981. "Forms of Time and of the Chronotope in the Novel: Notes toward a Historical Poetics." In M. Holquist (ed.), *The Dialogic Imagination: Four Essays*. Austin: University of Texas Press, 84–258 (Russian 1975).

Bell, C. 1997. *Ritual: Perspectives and Dimensions*. Oxford: Oxford University Press.

Ben-Dov, J., W. Horowitz, and J. Steele (eds.). 2012. *Living the Lunar Calendar*. Oxford: Oxbow.

Birth, K. 2012. *Objects of Time: How Things Shape Temporality.* London: Palgrave Macmillan.

Brack-Bernsen, L. and J. Steele. 2004. "Babylonian Mathemagics: Two Mathematical Astronomical-Astrological Texts." In C. Burnett et al. (eds.), *Studies in the History of Exact Sciences in Honour of David Pingree.* Leiden: Brill, 95–125.

Csapo, E. and M. C. Miller. 1998. "Democracy, Empire, and Art: Toward a Politics of Time and Narrative." In D. Boedecker and K. A. Raaflaub (eds.), *Democracy, Empire and the Arts in Fifth-Century Athens.* Cambridge, MA: Harvard University Press, 87–125.

Dunn, F. 2007. *Present Shock in Late Fifth-Century Greece.* Ann Arbor: University of Michigan Press.

Durkheim, E. 1915. *The Elementary Forms of Religious Life.* Trans. J. Ward Swain. New York: Free Press.

Dyke, H. and A. Bardon. 2013. *A Companion to the Philosophy of Time.* Blackwell Companions to Philosophy. Malden, MA: Wiley-Blackwell.

Feeney, D. 2007. *Caesar's Calendar: Ancient Time and the Beginnings of History.* Sather Classical Lectures 65. Berkeley: University of California Press.

Freeth, T. et al. 2008. "Calendars with Olympiad Display and Eclipse Prediction on the Antikythera Mechanism." *Nature* 454: 614–617.

Gingrich, A. 1994. "Time, Ritual and Social Experience." In K. Hastrup and P. Hervik (eds.), *Social Experience and Anthropological Knowledge.* London: Routledge, 125–134.

Goldberg, S. A. 2000. *La Clepsydre. Essai sur la pluralité des temps dans le judaïsme.* Paris: Albin Michel.

Goldberg, S. A. 2004. *La Clepsydre II. Temps de Jérusalem, temps de Babylone.* Paris: Albin Michel.

Hannah, R. 2005. *Greek and Roman Calendars: Constructions of Time in the Classical World.* London: Duckworth.

Hannah, R. 2009. *Time in Antiquity: Sciences of Antiquity.* London: Routledge.

Hannah, R. and G. Magli. 2011. "The Role of the Sun in the Pantheon's Design and Meaning." *Numen* 58: 486–513.

Heise, U. 1997. *Chronoschisms: Time, Narrative, and Postmodernism.* Cambridge: Cambridge University Press.

Kennedy, D. 2013. *Antiquity and the Meanings of Time: A Philosophy of Ancient and Modern Literature.* London: I.B. Tauris.

Lehoux, D. 2007. *Astronomy, Weather, and Calendars in the Ancient World: Parapegmata and Related Texts in Classical and Near Eastern Societies.* Cambridge: Cambridge University Press.

McCumber, J. 2011. *Time and Philosophy: A History of Continental Thought.* Montreal: McGill-Queen's University Press.

Roark, T. 2011. *Aristotle on Time. A Study of the Physics.* Cambridge: Cambridge University Press.

Rüpke, J. 2011. *The Roman Calendar from Numa to Constantine: Time, History and the Fasti.* Trans. D. Richardson. Malden, MA: Wiley Blackwell (German 1995).

Schleifer, R. 2000. *Modernism and Time: The Logic of Abundance in Literature, Science, and Culture, 1880–1930.* Cambridge: Cambridge University Press.

Smolin, L. 2013. *Time Reborn: From the Crisis in Physics to the Future of the Universe.* Toronto: Knopf Canada.

Sorabji, R. 1983. *Time, Creation and the Continuum: Theories in Antiquity and the Early Middle Ages*. Ithaca, NY: Cornell University Press.
Stern, S. 2001. *Calendar and Community. A History of the Jewish Calendar 2nd Century BCE – 10th Century CE*. Oxford: Oxford University Press.
Stern, S. 2012. *Calendars in Antiquity: Empires, States and Societies*. Oxford: Oxford University Press.
Sternberg, M. 1987. *The Poetics of Biblical Narrative: Ideological Literature and the Drama of Reading*. Bloomington: Indiana University Press.
Volk, K. 2009. *Manilius and His Intellectual Background*. Oxford: Oxford University Press.
Wilcox, D. 1987. *The Measure of Times Past: Pre-Newtonian Chronology and the Rhetoric of Relative Time*. Chicago: University of Chicago Press.
Yuval, I. 2006. *Two Nations in Your Womb: Perceptions of Jews and Christians in Late Antiquity and the Middle Ages*. Trans. B. Harshav and J. Chipman. Berkeley: University of California Press.

2

TIME AND NATURAL LAW IN JEWISH-HELLENISTIC WRITINGS

Jonathan Ben-Dov

1 Introduction

This chapter aims to read Hebrew and Greek reflections on the calendar as ideological statements about natural law. In Jewish sources, various constructions of the calendar are ultimately anchored in scriptural statements that relate to agriculture and the march of the seasons. As part of the calendar polemics in Second Temple Judaism, the correspondence of religious times with nature became a central bone of contention. Two dominant elements appear in all Jewish calendars, both of which can be considered as corresponding with nature, although they may and often do disagree between themselves: (1) the lunar cycle and (2) the sabbatical structure of the year. The present chapter aims to demonstrate that these two elements were depicted as part of natural law, and further to detect the implicit competition between them.

The definition of "natural law" to be used here is a slightly modified formulation of that used by Bockmuehl: "those cases where universal religious injunctions are derived from perceptions of the created order" (Bockmuehl 1995: 18). More specifically, this chapter examines the idea that specific ritual laws ascribed to Moses correspond to the order of nature and are thus authoritative not only by the voice of Moses but also by the authority of nature itself. While the formulation of the philosophical term "natural law" depends on the Hellenistic, particularly Stoic, *paideia*, the correspondence of Torah and nature is attested in "a well-documented and long-standing tradition within Second Temple Judaism itself."[1] In the

[1] Bockmuehl 1995: 44. Bockmuehl 2000 gives a wide-ranging survey of natural law in Judaism and Early Christianity. For the philosophical nuance of "natural law," see the provocative proposal by Koester 1968; and more recently Najman 1999, 2003b; Niehoff 2001; Sterling 2003.

absence of a clear concept of nature, some early Jewish sources subscribe to the same idea, albeit in a less rigorous formulation. The full thrust of the argument was put forth by Philo of Alexandria, in his long deliberations on the Sabbath and the number Seven, about the virtues of the Patriarchs, about the value of Jewish dietary laws, and so on. The Jewish-Greek philosopher Aristobulus had anticipated the philosophical argument. Josephus, Pseudo-Phocylides (Sterling 2003), and 4 Maccabees (Bockmuehl 1995) represent a less formal branch. However, such a concept was by no means limited to Jews writing in Greek; it was also expressed in Palestinian sources written in Hebrew: in the Book of Jubilees and in Qumran literature, both slightly underestimated in this respect.[2]

The association of natural law with the calendar exposes a distinctive trait of Second Temple Judaism, namely the aim to define the calendar as a mark of Jewish identity. Such an effort had not existed previously, as Jews were perfectly content with administering their religious life with the calendar of the empire.[3] The idea of a national calendar is a novelty, harnessing the calendar to the effort of identity formation. Interestingly enough, this effort took place at the same time as the efforts to establish the Torah as a natural law. The present chapter therefore correlates these two distinct trends: the search for a national calendar and the correlation of Jewish laws with nature. As such, the argument straddles the universal-particular continuum, a central axis for Jewish thinkers, mainly those in the Hellenistic-Roman Diaspora.

The motivation to align one's national calendar year with nature, while pointing at the inadequacy of other calendrical systems in that regard, was not present in the Hebrew Bible. In an earlier study, I claimed that this drive arose as part of the Greek ethnological and scientific discourse (Ben-Dov 2013). Thus in the writings of the first-century BCE astronomer Geminus (*Introduction to Phenomena* 8.7):[4] "For the command, by the laws as well as the oracles, to sacrifice in the manner of the fathers, was taken by all the Greeks to mean reckoning the years in accordance with the sun and the days and months with the moon." Geminus's treatise represents the current knowledge of Greek astronomy in the period between Hipparchus and Ptolemy. In his discussion "On Months" (chap. 8) Geminus provides the most comprehensive extant ancient account of calendars and cycles. In the quotation noted earlier, Geminus maintains a Greek trope according to which the calendar is intended to perpetuate the correct performance of the temple

[2] The present chapter thus to some extent modifies Bockmuehl's skepticism about natural law in the Dead Sea Scrolls. See Bockmuehl 2000: 88–111 and the criticism by Doering 2004.

[3] Stern 2001: 27–31; Ben-Dov 2013, 2014.

[4] Evans and Berggren 2006: 176.

rituals "in the manner of the fathers."[5] At the same time, he assiduously links the calendrical practices with national definition: the practice observed is followed by "all the Greeks"; the Egyptians, as we read later on (8.16–24), pursue different goals and thus adopt a different calendar. As in the Jewish passages quoted in § 2, a nation is identified, *inter alia*, by the way it constructs its calendar in alignment with the heavenly luminaries.

A similar outcome of the encounter between Greek concepts of nature and the ancient cultures of the East is the Greco-Egyptian hybrid embodied in the Canopus Decree of 238 BCE. In this extraordinary document, the Egyptian priests declare – under conspicuous royal patronage – that the famous traditional Egyptian year is faulty and in need of periodic correction:[6]

> That it may not happen that some of the public feasts held in the winter are ever held in the summer, the star changing by one day every four years, and that others of those now held in the summer are held in the winter in future times as has happened in the past and as would be happening now, if the arrangement of the year (τῆς συντάξεως τοῦ ἐνιαυτοῦ) remained of 360 days plus the five days later brought into usage (be it resolved) for a one-day feast of the Benefactor Gods to be added every four years to the five additional days before the new year, in order that all may know that the former defect in the arrangement of the seasons and the year and in the beliefs about the whole ordering of the heavens (τὸ ἐλλεῖπον πρότερον περὶ τὴν σύνταξιν τῶν ὡρῶν καὶ τοῦ ἐνιατοῦ καὶ τῶν νομιζομένων περὶ τὴν ὅλην διακόσμησιν τοῦ πόλου) has come to be corrected and made good by the Benefactor Gods.[7]

After millennia of perfect satisfaction with the Egyptian civil year of 365 days, the insight suddenly arose that it does not agree with the seasons and must thus be corrected. This new awareness surfaced as part of the immense confusion with regard to calendars in Ptolemaic Egypt, as the indigenous calendars of ancient Egypt had to correlate themselves with the newly introduced Macedonian systems.[8]

The writer of the Canopus Decree is worried that rites in the temples will not be productive because they do not properly correspond with the "order

[5] For this theme, see Herodotus, *Histories* 4.2.1; Aristophanes, *Clouds* 615–626 (quoted in Evans and Berggren 2006: 176).

[6] For the cultural background of the Canopus Decree in terms of the tension or symbiosis between Greek and Egyptian cultural elements, see mainly Clarysse 1999; Gorre 2009: 285–309; Stern 2013: 137–142.

[7] Lines 41–46 of the Greek text. Translation follows Bagnall and Derow 2004.

[8] The ways for solving this confusion remain highly disputed; see Grzybek 1990; Depuydt 1997; Bennett 2011.

of the year." Note the terms "the arrangement of the year/seasons" and the "ordering of the heavens," as well as the term διακόσμησις ("ordering") and particularly νομιζόμενα ("beliefs") – the exact opposite of the Greek φύσις ("nature"). These are quasi-philosophical terms, close to their later usage in Stoic texts yet innocent of professional philosophizing. The Canopus Decree is thus an example of a new mode of awareness for nature and its regularities, even at this early period of Greek presence in Egypt, in a text that is not particularly philosophical in orientation. This text shows how strongly rooted this new *Zeitgeist* was. That at least in some circles the suggested Canopus reform was considered of foreign character may be seen from the fact that, two centuries later, when Augustus reinstituted an additional day every fourth year, Egyptian administrative documents continued to distinguish between dates καθ' Ἕλληνας ("in the manner of the Greeks") and κατ' Αἰγυπτίους ("in the manner of the Egyptians") – the corrected year versus the old, uncorrected 365-day year.[9]

As the Egyptians struggled with their year and its correspondence with nature, Egyptian Jews were faced with a similar question, and it was not long before Palestinian Jews were concerned with similar problems too. At such turbulent times, the Sabbath was chosen to serve as the foundation stone of Jewish time as well as a foundation stone of the natural order, being the most characteristic Jewish marker of time reckoning. Gradually later, the Sabbath became characteristic not only of the weekly count but also of the entire structure of the year, or "the order of the seasons," what Philo more philosophically calls "nature."

2 The Calendar as a Marker of Jewish Identity

Hebrew authors in the Iron Age called to distinguish the Israelites from their neighbors not only by means of their un-polytheistic cult but also by means of their sexual morality and a variety of ritual practices that they rejected (see esp. Lev 18, Deut 18:9–14). Israelites did not, however, differentiate themselves by means of their time reckoning. In contrast, the calendar is for present-day Jews a dominant divider between them and their non-Jewish compatriots; while the general public uses the Gregorian calendar, Jews embrace a distinctive Jewish luni-solar calendar. This contrast has been explicitly stated in early rabbinic literature (i.e., late second to early third centuries CE), as in the following quote (Mek. *Pisḥa* 1, my translation):

[9] Lippert 2009: 187.

> Thus we learn that Israel reckons according to the moon, while the gentiles reckon according to the sun. Furthermore, once every thirty days the Israelites elevate their gaze to see their Father in heaven. When the sun eclipses, it is a bad sign for the gentiles, who reckon by the sun; and when the moon eclipses it is a bad sign for Israel [literally: for the foes of Israel – a euphemism, JBD], who reckon by the moon.

This statement was written during the rule of the Roman Empire, at a time in which the entire known world was following the solar Julian calendar, after the robust enforcement of that calendar throughout the empire. The Jews declared their subversiveness by means of upholding a different calendrical system.[10] An earlier Jewish statement representing a more sectarian stance conveys a similar idea, although the actual calendrical practice promoted in it is the opposite (Jub. 6:32–36, trans. VanderKam 1989):

> (32) Now you command the Israelites to keep the years in this number – 364 days. Then the year will be complete and it will not disturb its time from its days or from its festivals because everything will happen in harmony with their testimonies. They will neither omit a day nor disturb a festival. (33) If they transgress and do not celebrate them in accord with his command, then all of them will disturb their times. The years will be moved from this; they will disturb the times and the years will be moved. They will transgress their prescribed pattern . . . (35) . . . lest they forget the covenantal festivals and walk in the festivals of the nations, after their error and after their ignorance. (36) There will be people who carefully observe the moon with lunar observations because it is corrupt (with respect to) the seasons and is early from year to year by ten days.

In this second-century BCE statement, the author of the Book of Jubilees exhorts Jews to practice a schematic year of 364 days. This practice too opposes the empire, albeit a different empire with a different calendar: it was written under the luni-solar calendar of the Seleucids. Although Jubilees and the Mekhilta disagree as to the identity of the right calendar, they both agree that a Jewish mode of time reckoning should be different from that of the empire.[11] The Jubilees calendar is praised as standing "in harmony with the testimonies," although as we know that calendar would result quite rapidly in a large gap from the actual path of the seasons.

Close to the spirit of Jubilees is a statement in the Dead Sea Scrolls, which accuses nonsectarian Jews, following the Seleucid calendar in the Jerusalem

[10] Stern 2013: 331–353.

[11] In addition, both texts also express an inner-Jewish polemic against those who hold a similar calendar to that of the empire. See Ben-Dov 2013.

Temple, of "walking in the festivals of Gentiles" (4Q166 = 4QpHos[a] ii 15–16).[12]

A similar idea is raised by Philo of Alexandria, located in time between Jubilees and the early rabbis, as part of his *QE* 12:2:

> But not all (peoples) treat the months and years alike, but some in one way and some in another. Some reckon by the sun, others by the moon. And because of this the initiators of the divine festivals have expressed divergent views about the beginnings of the year, setting divergent beginnings to the revolutions of the seasons suitable to the beginnings of the cycles. Wherefore (Scripture) has added, "This month (shall be) to you the beginning," making clear a determined and distinct number of seasons, lest they follow the Egyptians, with whom they are mixed, and be seduced by the customs of the land in which they dwell.[13]

Similar to Jubilees and the Mekhilta, Philo praises the distinct Jewish year and calls for separation from the nations. In calling for a distinction between Jews and Egyptians, Philo clearly builds an analogy between the conflict of biblical Israel with the Egyptians and the reflection of that conflict in his own times, a millennium or so later. One might also say that Philo sees himself in a similar role to that of Moses, when proclaiming Jewish time to the Egyptians.[14]

It was described earlier how the confusion of calendars in Ptolemaic Egypt gave rise to a new awareness about them. Jews did not remain unaffected by this new conflict. Sacha Stern (2001: 32–34) has demonstrated how Jews were first prompted into reflecting about their national calendar after their encounter with the various calendars in Egypt, especially the civil calendar of 365 days. Before that, Jews had been perfectly content to reckon their time, even celebrate their festivals, according to the luni-solar calendar of the Achaemenid (or earlier, the Neo-Babylonian) Empire, and no distinct national reckoning was required. Thus, Philo and earlier Jewish-Alexandrian authors respond to the same set of historical circumstances that brought about the Canopus Decree.

Not all Jews under Roman rule were anxious like Jubilees and the rabbis to underscore a national calendar and its difference from that of the empire. Flavius Josephus, a prominent writer in Flavian Rome and earlier a priest in Jerusalem, wrote extensively on Jewish history and culture but did not note a distinctive Jewish time reckoning. For him, the Jewish months simply

[12] See Bernstein 1991.
[13] Philo, *QE* 1:1 (Marcus 1953: 4–5).
[14] For Philo's placement of himself as Moses, see Bloch 2013.

coincide with their corresponding Macedonian months, although in reality they could never have perfectly overlapped.[15] It seems that Jubilees was the first to contrast the true Jewish time reckoning with that of the gentiles. Its mode of thought, conceived under the Seleucid Empire, regained popularity among wider Judean circles two hundred years later, under the new calendar ideology of the Roman Empire.

3 Sabbath and the Lunar Calendar as Natural Law: The Second Century BCE

In the Hellenistic cultural furnace, at a stage when all ancient nations sought to reinforce their identity vis-à-vis the new dominant Greeks, each one of them sought characteristic traits to flag as a proof of its antiquity.[16] In the realm of time reckoning, it was only natural for Jews to boast the Sabbath and the seven-day week as the Jewish time mechanisms *par excellence*. Thus, the Book of Jubilees 2:19–20 (the original Hebrew is partially preserved in the Qumran scroll 4Q216 = 4QJub[a]) places the institution of Jewish identity with the Sabbath already at the moment of Creation:

> [He said to us: "I will now separate for myself] a people among my nations. And [they will keep sabbath. I will sanctify them as my people, and I will bless them. They will be my people and I will be their God.] And he chose the descendants of Jacob among [all of those whom I have seen. I have recorded them as my firstborn son and have sanctified them for myself] for all the age(s) of eternity.

According to Jubilees, the preeminent markers of Jewish identity – the Sabbath, circumcision, and the festival of Pentecost (Shavuʿot) celebrating the public announcement of the Torah and God's covenant with his people – were established already at the foundation of the world rather than at any of the other formative stages of biblical history.[17] The long oration following

[15] Josephus, *A.J.* 2:311; Stern 2001: 35. For Josephus's overlapping months, see Stern 2013: 255–259. Bennett 2011 analyzed the dates in the Zenon Papyri, recording Zenon's travels in Ptolemaic Palestine. He showed that while Zenon had been careful at first to specifically align the Egyptian dates with the Macedonian ones, he later gave up on this precision and settled for a general correspondence of the respective months.

[16] Gruen 2005, 2011.

[17] Segal 2007: 279–282. A more recent theory by Kugel (2012) focuses more closely on the concept of natural law in Jubilees. According to Kugel, all references in Jubilees to a metaphysical law engraved on the heavenly tablets and already extant at creation belong to the hand of a consistent interpolator. That interpolator seeks to modify the message of the original author, who promotes a Stoic-like idea, namely, that the patriarchs had discovered the laws independently, based on their moral conduct. According to Kugel, these two trends

the commandment of Pentecost in Jub. 6 also forms the occasion for an admonition on preserving the proper calendar, as quoted earlier. Thus, while the author of Jubilees clearly utilizes biblical categories, he does so to buttress his metaphysical notion of the Jewish definition. In the universal-particularistic continuum, Jubilees tends strongly toward the latter. One may say that this tendency is the main guideline in his reworking of the earlier creation story from Gen 1.[18]

Similarly to Gen 1, Jubilees anchors the sabbatical order in the moment of creation. Chapter 2 of Jubilees demonstrates how the order of seven days and the seven-based structure in general are inherent in world order. As no word for "nature" exists in Hebrew,[19] the idea is expressed by the flow of the narrative and by the laws resulting from it in the priestly sources of the Pentateuch: the Sabbath, the count of seven weeks, of seven years, and of 7 × 7 years for land emancipation. However, Jubilees adds several emphases to the biblical presentation to reinforce a novel perspective on the Sabbath and Creation, one that fits the contemporary thought of the second century BCE. These emphases have been conveniently drawn by VanderKam 2013 and will be partly recounted here.

First, Jubilees reinforces the importance of the number Seven in the creation narrative by inserting an additional count, unattested in Genesis, of the number of objects created on each of the days. In day 1 of the creation, seven "great works" were created, a number that is not matched in any of the other days. Furthermore, the middle item of this list – the spirits – is itself split into seven branches of angels and spirits. Altogether, says Jubilees, twenty-two items were created throughout the creation week, similarly to the twenty-two generations from Adam to Jacob. The awkward mention of seven creation acts on day 1 – unattested in the biblical account – must have

cannot be reconciled and must be the result of literary growth. As concerns the present chapter, however, the two trends merge beautifully to form a combined message: virtuous life can lead to the independent discovery of metaphysically ordained laws. Kugel's analysis is helpful in comparing Jubilees with Hellenistic literature, such as Philo's works (Kugel (2012: 391–405), and will be consulted again later.

18 VanderKam 2013: 26–27; earlier Steck 1977.

19 Cursory formulations of the idea of nature appear in the Dead Sea Scrolls in the framework of halakhic legal discussions. In these discussions, sectarian authors tend toward a "realistic" point of view vis-à-vis the more nominalistic views of their opponents. Thus one encounters the phrase משפט בריאתם (CD-A 12:15), literally "the habit (or: precept) of their creation," which in fact conveys the sense of "their nature." The phrase, however, refers to the nature of the locusts discussed in that particular passage and does not support a more abstract sense of nature. For the distinction between "realistic" and "nominalistic" views, see Schwartz 1992. Schwartz's ideas raised a lively debate, which cannot be fully covered here; see a recent summary in Amihay 2016.

resulted from a deliberation by the author to connect the first day with the unit of seven, placing the Seven already at the very beginning of the creation narrative. The result is a kind of balance and symmetry between One and Seven of the creation week. A connection, even unification, of the One and the Seven is a Greek idea, specifically Pythagorean, which also occurs in the writings of the Jewish-Hellenistic philosophers Philo and Aristobulus.[20] Jubilees does not go as far as using explicit arithmological terminology, but the assignment of seven acts to day 1 may reflect a similar concept. Finding the connection between One and Seven in Jubilees is rather surprising and calls for an evaluation of the author's background in Greek philosophy. Yet for the purposes of the present chapter, it shows how a second-century BCE Jewish writer in Palestine felt the need to inscribe the heptad into the world order.

Second, while the Pentateuch in Gen 1 does not yet dictate the Sabbath as a *commandment* demanding fulfillment but rather only first makes this claim in Exod 20:8, Jubilees merges the commandment of Exod 31 with the wording of Gen 1 and thus places the commandment already in earliest times.[21]

Third, the story of creation is brought in Jubilees only in chap. 2, being subordinated to retrospective narrative frame of Jubilees, contained in chap. 1. That chapter recounts the centrality of Israel's nationhood, and the Sabbath – and Creation – is then recorded as a means to buttress that centrality. The Book of Jubilees thus clarifies that the Sabbath is *not* a universal decree, possibly against other readers who would see it this

20 The similarity between Jubilees and the Greek heptad was recently pointed out by Werman 2007: 156–157 and VanderKam 2013: 27, who also quotes earlier references by Epstein and Büchler. Werman claims that Jubilees annuls the value of the heptad by laying the emphasis on the number 22 rather than on 7. According to her, Jubilees is thus an anti-philosophical text that seeks to dissolve Hellenistic numerology. In my opinion, however, this explanation is not compelling.

Philo later expanded significantly on the connection of the One (monad) and the Seven (*Op.* 16–35, esp. 29; *Decal.* 102–103; see Filler 2008b: 14–18), and in addition accorded special importance to the first day, distinguishing it from the other days of creation (*Opif.* 15). For Philo's arithmology, see Doering 1999: 366–370; Runia 2000; and in detail Filler 2008a, 2008b with earlier literature cited there.

21 See a nuanced presentation of the distinction between narrative and commandment within the sources of the Pentateuch in Stackert 2011 and forthcoming. The actual statement about time measuring in Gen 1:14 does not include the Sabbath in the list of key time periods determined by the luminaries. In Jub. 2:9, however, the "signs" of Gen. 1:14 are explicitly interpreted as Sabbaths, in the wake of Exod 31:12–17. This concept is later anchored in rabbinic midrash (e.g., Gen Rab. 6). The introduction of Exod 31 into Jub. 2 was also discussed by Steck 1977.

way.[22] In addition, while the Sabbath does correspond to nature, it is not *based* on it but rather originated in a divine decree.

The Jewish-Hellenistic philosopher Aristobulus, probably a contemporary of the Palestinian Book of Jubilees, also addresses the relation between the Sabbath and the order of nature.[23] Since his work survives in fragments only, we cannot be sure about its entire scope. Yet a passage in one of the extant fragments reveals his awareness of the question by assigning a cosmic role to the heptad. When trying to account for the strange concept that God "rested" on the seventh day, Aristobulus explains the true meaning of that day of rest:

> For it signifies that "in six days he made both the heaven, the earth, and everything in them" (Exod 20:11), that he might show the times and proclaim the order by which one thing precedes another. For, once he arranged all things, he thus holds them together and presides over their movement. Now, he has clearly shown it [sc. the seventh day] to us as legally binding, to be a symbol of the principle of the seven established all around us, through which we have knowledge of things both human and divine.[24]

Aristobulus then continues to show that "all the world comprising all animal and plant life as well revolves through periods of seven," and that "not only the Hebrews but also the Greeks recognize the seventh day as sacred, as the day around which the whole world of all animal and plant life revolve." Then follows a long anthology of excerpts from various Greek writers – with problems as to attribution and wording – who acknowledged the centrality of the heptad.

In a clever work of combined citation, Aristobulus links Exod 20:11 with Gen 1:14 and thus acts similarly to Jubilees, augmenting Gen 1 with a more practical and legal perspective from later on in the Pentateuch. This link

[22] See Doering 1999: 51–53. The opposite conclusion about the place of the story of creation in Genesis was drawn by Philo, *Opif.* 3; Philo stresses the universal role of the Torah and its correspondence with nature. Similarly, the universal role of the Sabbath and the Passover is emphasized in the fragments of Aristobulus; see Riaud 2006.

[23] For the person, his date, and work, see conveniently Holladay 1995, and cf. the *magnum opus* by Walter 1964. While Walter dates Aristobulus to the mid–second century BCE, Holladay and others date him some twenty years earlier.

[24] Aristobulus frg 5.12 (Euseb. *P.E* 13.12.12). Translation follows Doering 2005: 9. Extensive discussions of this notoriously elusive paragraph may be found in Walter 1964: 150–171. Holladay 1995: 185, 230 employs a more radical translation: "Our law code has clearly shown us that the seventh day *is an inherent law of nature*" etc. (italics mine). This reading however has been contested, not the least because it is not expected to find the term "law of nature" in the early second century BCE; see especially Doering 1999: 313–314.

serves to inscribe the seven-based framework not only into the path of the luminaries (as implied in Gen 1:14) but also to plants and animals and everything living, defining the heptad as an inherent order of the entire cosmos.

Aristobulus links the Seventh with the First:

> [God gave us] the seventh day, but which, in the real sense, might also be called first, that is, the beginning of light through which all things are seen together in their entirety (fr. 5.9; Eusebius, *Strom.* 138.1).[25]

In contrast to Jubilees, Aristobulus's argument prefers the universal to the particular. He does not use the word "Sabbath" but rather the term "the heptad," which sounds better to Greek ears and carries a less particularistic connotation. While Aristobulus shows the use of the heptad by various Greek authors, he is unable to point out how the seven is applied by them in any known comprehensive system of time reckoning. It is one thing to say that the heptad is a constitutive principle, but it is a totally different task to show that it constitutes the basis for an entire calendrical ephemeris. In fact, Aristobulus presents no claim for a comprehensive Jewish time reckoning on the basis of the heptad.

Both Aristobulus and Jubilees address here, perhaps unintentionally, a question that has dominated much of the subsequent Jewish calendrical discourse: can the structure of the year be aligned with the week? On the one hand, the Sabbath and the heptad are acknowledged as a typically Jewish component of time reckoning and as a foundation of world order. On the other hand, neither the lunar nor the solar year yields a perfectly heptad number. Thus, for example, it is impossible to tie the dates in the luni-solar year to the days of the week, a fact that caused serious trouble for the rabbinic calendar several centuries later. If both the Sabbath and the astronomical structure of the year are salient for its definition, and both are anchored in world order, how is it that the two factors cannot be reconciled? Both Jubilees and Aristobulus related to these competing definitions of Jewish time and yielded different results.

Although advocating the centrality of the heptad, Aristobulus did not posit a heptadic calendar year. On the contrary, when attending to the character of the Israelite year, he points rather to the centrality of the moon:

> But Aristobulus adds that, at the time of the feast of Passover, of necessity this would be when not only the sun is passing through an equinoctial sector, but also the moon as well.[26]

[25] Translation follows Doering 2005: 4–8.

[26] Aristobulus frag. 1 (translation follows Holladay 1995: 131). This text was preserved by Anatolius in his Easter treatise, in turn also lost but preserved by Eusebius, *Hist. Eccl.*

In this fragment, Aristobulus presents the Jewish "Rule of the Equinox": since Passover occurs on the day of the full moon, the moon will rise that day in an equinoctial section of the horizon; but in order that the sun rise in an equinoctial sector too on that day, the Passover should take place precisely within thirty days after the equinox. While Anatolius must have exaggerated the content of this rule for his own agenda,[27] an authentic core can be pointed out, in which Aristobulus boasted the Jewish insistence that the Passover falls within one month after the spring equinox.[28] In this fragment, the emphasis is on the luni-solar harmony of the Jewish year, with no mention of the heptad, so dear to Aristobulus elsewhere. Indeed there is no sign that Aristobulus conceived of the whole year as constructed by heptads, or pentecontads, or any other arithmological scheme.

In contrast to Aristobulus, the Book of Jubilees is more committed to sabbatical time reckoning. Since this principle is incommensurate with either solar or lunar observations, the author of Jubilees is required to introduce a novel year, previously unattested: a heptad year of exactly fifty-two weeks. While a year of 364 days was previously suggested in the Astronomical Book of Enoch, that year is never described as constructed of any number of full weeks.[29] In contrast, in Jubilees 6 this is its central trait:

> All the days of the commandments will be 52 weeks of days; (they will make the entire year complete). So it has been ordained and engraved on the heavenly tablets. One is not allowed to transgress a single year, year by year. Now you command the Israelites to keep the years in this number – 364 days. Then the year will be complete and it will not disturb its time from its days or from its festivals because everything will happen in harmony with their testimonies. They will neither omit a day nor disturb a festival.
>
> (6:30–32)

The year depicted here is *defined* as a complete number of weeks: fifty-two exactly, not one day more or less. If the year is preserved this way, it will remain in harmony with the testimonies. Note that here the harmony is not with nature but rather with the testimonies engraved on the tablets of

7.32.14–19. The section quoted here is #17. For this fragment, see Holladay 1995: 129–133; Riaud 2006.

27 This agenda has to do with the computus tables produced by Anatolius in the third century CE. See McCarthy and Breen 1993.

28 Stern 2001: 50–53.

29 The heptadic year of Jub. 6 is not entirely identical with other sectarian accounts of the 364-day year; see Ben-Dov 2009. The Book of Enoch does note, however, a lunar theory that operates on a heptadic basis; see the Aramaic fragment 4Q210 1 iii 6 "and it guides the lunar months by halves of sevenths" (Drawnel 2011: 223–224).

heaven.[30] To a certain extent, we may say that these heavenly tablets are the apocalyptic equivalent of the Stoic term "nature."[31]

The Book of Jubilees in chaps. 2 and 6 is concerned with the same basic questions that also concerned Aristobulus: the correspondence of the calendar with nature and the primacy of the typical Jewish component of the Sabbath. Both of them accord special importance to the number Seven as a window to the natural order. However, Jubilees carries the Sabbath forward in terms of actual calendar practice and is thus required to make the radical move of constructing a whole new calendar year based exclusively on the heptad.

Comparable logic is reflected in calendrical texts from Qumran, which endorse a similar calendar year to that of Jubilees. These texts assign each week of the year to the service of one particular priestly family in the Jerusalem Temple, and in turn they tie the order of the year to the lunar cycle. The authors of these calendar texts made sure to anchor their sabbatical order already in the day of creation:[32]

> ... [ ]to its being seen (or: appearance) from the east ...]to[sh]ine [in]the middle of the heavens at the foundation of [Creatio]n from evening until morning on the 4th (day) of the week (of service) [of Ga]mul in the first month in [the fir]st year.

This passage depicts the moment of creation, when the sun and the moon first rose. For the authors, it was clear that the sectarian calendar, with its sabbatical structure and the priestly families, was enacted already at the moment of creation. This instrumental convention becomes the warp and the woof of the very fabric of nature.

Aristobulus and Jubilees are thus two more or less contemporary authors, one in Alexandria and the other in Palestine, who struggle with the same set of problems and assumptions but find divergent ways to solve them. Both authors acknowledge the force of the heptad as both a typically Jewish mechanism and an inherent mechanism of nature. Both would like to bestow prestige on the Jewish way of counting time. Both are a product of a new Hellenistic *Zeitgeist*, which encouraged reflection on these aspects of time reckoning. Several astonishing similarities can be found between the two,

[30] For this concept in the Book of Jubilees, see García-Martínez 1997; Segal 2007: 273–313; Werman 2002.

[31] Jubilees' thought is thus akin to that of Philo, in the sense that he envisions a written law of nature, something that would have been conceived an oxymoron in Greek philosophical writing.

[32] 4Q320 1 i 1–5 (Talmon, Ben-Dov, and Glessmer 2001: 42–43).

but there are differences too. While Jubilees tends more to the particularistic end of the continuum, Aristobulus is keen to maintain a universalistic mode. And while Aristobulus promotes the heptad as a law of nature, Jubilees uses it more concretely to construct a novel calendar year.

4 Philo of Alexandria

The Jewish philosopher Philo was active in Alexandria in the second quarter of the first century CE, at least 150 years after Aristobulus. Neither Philo's treatment of time nor his ideas on nature can be covered here in an exhaustive way. Instead, I focus on several points pertinent to the present discussion.

Philo expended much effort in developing an innovative concept of natural law and describing the relation of the Torah to that norm.[33] According to him, the Patriarchs prior to Sinai were men of virtue, who were able to follow the laws of the Torah not according to revelation but on the basis of observing the natural world.[34] This constellation underscored Philo's Stoic ideal of living according to nature. As part of this effort, Philo sought to demonstrate how various branches of Jewish law correspond to the demands of nature and thus exist independently of their status in the Mosaic revelation: "they seek to attain to the harmony of the universe and are in agreement with the principles of the eternal nature" (*Mos.* 2:52), and "Moses is alone in that his laws, firm, unshaken, immovable, stamped, as it were, with the seals of nature herself" (*Mos* 2.14). Philo seeks to demonstrate that Mosaic Law promotes healthiness and hygiene, in contrast to the foul health of the Egyptians, and thus reflects a true image of nature.[35] He similarly highlights the correspondence between the Jewish calendar and festivals to astronomical notions of the sun's position and the heavenly sphere in general.[36] In addition, he points out the double role of Jewish festivals, both as particular Jewish institutions and as universal occasions: "Thus [unleavened bread] may be regarded from two points of view, one peculiar to the nation, referring to the migration just mentioned, the other universal, following the lead of nature and in agreement with the general cosmic order."[37]

[33] The first to draw attention to the philosophical background of Philo's treatment of nature, in a provocative yet problematic way, was Koester 1968. Further landmark studies are Horsley 1978; and more recently Najman 1999, 2003a, 2003b; Niehoff 2001: 247–266; Sterling 2003.
[34] See especially Najman 2003b: 60–61.
[35] Niehoff 2001: 254–258.
[36] Niehoff 2001: 258–266.
[37] *Spec.* 2.150; see Weitzman 1999.

Philo was a keen promoter of the heptad as a both Jewish and universal principle of time reckoning. In this respect, he is similar to Aristobulus and even uses similar ideas and arguments, quoting similar authorities but expanding more on the arithmological exegesis of the Seven.[38] He dedicated two long sections to the discussion of the numerical value of the Seven, both as part of his *Exposition of the Law*: a long section (§§ 89–128) in his treatise *On the Creation of the World*, and a shorter piece (§§ 1.8–16) in his *On the Special Laws*.[39] The contents of the two sections closely resemble each other, which might imply that the long section in *Opif.* is an expansion of the earlier one in *Spec.*[40] I quote here the concluding paragraph of the passage on the Seven (*Opif.* 128a) to highlight Philo's motivation in producing that section:

> These and yet more than these are the statements and philosophical insights of men on the number Seven, showing the reasons for the very high honor which that number has attained in nature, the honor in which it is held by the most approved investigators of the mathematical science among Greeks and barbarians, and the special honor accorded to it by that lover of virtue, Moses.

Despite his lengthy praise of the number Seven, Philo – like Aristobulus – does not inscribe that number into the structure of the year in his arithmological treatises. It is therefore most revealing that elsewhere Philo does collect the threads, tying the heptad with the year. In book 2 of *Spec.*, in the demonstration of the Ten Commandments, the account of the Sabbath as the fourth commandment (*Spec.* 2.39–69) takes the form of a lengthy essay on the Jewish festival year. At the end of his discussion of the annual festivals, Philo claims that the entire order of the year springs forth from the heptad:

> All this long exposition is due to my regard for the sacred seventh day, and my wish to show that all the yearly feasts prove to be as it were the children of that number which stands as a mother. (§ 214)

[38] On the relation between Aristobulus and Philo, see Doering 1999: 367; Holladay 1995: 229, and bibliography cited there.

[39] The fact that both sections on the numerical value of Seven are included in the *Exposition* may be significant. Niehoff (2013) has argued that Philo first developed an articulate notion of creation theology after his prolonged stay in Rome (38–41 CE), and that this turn in his thought was caused by his exposure to Stoic philosophy while in Rome. Niehoff's new ideas on Philo's intellectual development are rather attractive but need yet to be considered by experts for his *oeuvre*. It might be worth noting here that Philo's thoughts on the Seven resemble those of his predecessor Aristobulus, although admittedly the Stoic tint in the latter's work is less pronounced.

[40] See Runia 2000; Runia 2001: 260–309; Doering 1999: 367.

And again at the very end (§ 223):

> I have now completed the discussion of the number Seven and of matters connected with days and months and years that have reference to that number, and also of the feasts which are associated with it.

The seventh day is thus not merely the basis for the count of weeks but rather the foundational principle of the entire festival year and the seasonal order (compare § 57).[41] This insight gains arithmological force as Philo connects the ten annual festivals with the heptadic principle:[42]

> Now the part played by seven among the numbers has been described at length in an earlier place, where we have discussed the properties which it possesses within the decad, and its close connection with ten itself, and with four, which is the origin and source of ten. (*Spec.* 2.40)

Finally, this idea is yet further expanded as Philo finds a way to align the Seven with the phases of the moon, thus installing harmony in the two most contrasted elements of Jewish time. Thus in explaining the seven lambs in the festival sacrifice, he states:

> Seven lambs because the complete changes of form to which she (=the moon) is subject are measured in sevens. In the first seven from the conjunction we have the half moon, in the second the full moon, and when she is reversing her course she passes first into the half moon and then dies away into the conjunction. (*Spec.* 1.178; compare *Op.* 101)

Elsewhere (*Spec.* 2.177; compare *Opif.* 95), Philo applies his arithmology to the number Fifty, another prominent Jewish number as the day (or year) after the interval of $7 \times 7 = 49$ days or years. Fifty is the origin of the festival of Pentecost. In Philo's Pythagorean logic, however, fifty is the sum of $3^2 + 4^2 + 5^2$, together representing the squares of Pythagoras's famous right triangle.[43]

41 In stating this principle, it seems that Philo bases himself on the scriptural sequence in Leviticus 23, a proof text for the structure of the festival year. In that chapter, the Sabbath is declared in 23:3, just after the solemn announcement "These are the festivals of the Lord" (23:2) and before the enumeration of the actual festivals by a repetition of that same declaration (23:4). Verse 23:3 is often considered by Bible scholars as a later addition in the chapter, inserting the Sabbath – which is not a "festival" – in the framework of the festival year. Philo read the entire chapter and deduced from it an expanded message by way of a rabbinic-like syllogism: it is not only that "the calendrically determined sacred times are to be observed *in addition to* the already-commanded weekly Sabbaths" (Schwartz 2012: 21, italics added), but rather that the Sabbath is the constructive principle of the entire festival year.

42 For the connection of the Seven with the Decad, see further *Opif.* 95–105; see Filler 2008b.

43 Philo builds on the holiness of the number Fifty when recording the sacred times of the sect of the Thrapeutae (*Contempl.* 65). That sect, he reports, "assemble after seven sevens/weeks

Furthermore, even the structure of the equinoxes – commonly considered to be six months apart – also validates the Jewish reckoning. Put differently, each equinox is actually placed in the *seventh* month after the previous one![44]

Philo thus builds on the same rhetoric as his predecessors. He picks up from second-century BCE writers the need to establish a nationally distinct calendar and align it with nature. Like Aristobulus, he underscores the importance of the heptad on both the national and the universal levels. However, he exceeds Aristobulus in assigning the Sabbath a constructive role in the formation of the year and even relates it to the lunar calendar. In this, Philo comes closer to Jubilees but remains distinct in that the year of Jubilees is a purely heptad year. Philo thus stands conceptually midway between Jubilees and Aristobulus: the Sabbath plays a part in the construction of the year but does not create a strictly sabbatical calendar.[45]

5 Conclusion

Priestly literature in the Pentateuch points out how the heptad is inherent in the natural order. The Sabbath is at once a universal mechanism anchored in the creation of the world and a sign for a particular covenant with the Israelites. However, the priestly mode of thought does not fully flesh out the tension between the particular and the universal that grows out of this situation. In addition, it does not purport to produce any particular calendar that is characteristically Jewish.

The idea that one particular calendar corresponds more perfectly with nature arose, on entirely different grounds, among Egyptians and Greeks in Ptolemaic Egypt. A prominent example of such an idea is the Canopus Decree, when the old Egyptian year was for the first time announced as defective because of its disharmony with the seasons. The seed planted in 238 BCE led to a full-fledged discourse about time in Hellenistic and Roman Egypt. In turn, this discourse eventually led to the development of the Julian calendar.

(δι' ἑπτὰ ἑβδομάδων), holding in awe not only the simple Seven but its square too, because they know it is pure and perpetually virginal. This is the preliminary festival of the very great festival, which the Fifty has taken on (ἔστι δὲ προέορτος μεγίστης ἑορτῆς, ἣν πεντηκοντὰς ἔλαχεν), the most sacred and natural number."

44 *Spec.* 1.172, 182; cf. 2:151–154.

45 Kugel (2012: 391–405) has lucidly explored both the striking resemblance and the profound differences between Jubilees and Philo, especially with regard to the meaning of the patriarchal narratives in the Book of Genesis. The present discussion, I believe, applies a similar mode of thought to elements of the creation narrative.

The situation for Jewish writers in Ptolemaic and Roman Egypt was even more complicated. The confusion around local Egyptian calendars gave rise to the Jews' contemplation of their own national calendar. Thus already Aristobulus, and probably others before him, promoted the Sabbath as the Jewish mechanism *par excellence*: typically Mosaic and at the same time universally acknowledged as the driving force of reality. Yet Aristobulus did not find a way to merge the Sabbath with what he knew as the constitutive principle of the Jewish calendar: the moon.

The Palestinian Book of Jubilees, despite its pronounced particularistic emphasis, used a mode of argumentation not dissimilar to that of the Jewish-Hellenistic writers. Less philosophically oriented, this book develops the biblical heptadic scheme in a reflective way, sometimes reaching similar results to those of Philo later in Alexandria. This mode of thought is comparable to the sectarian legalistic view, which tends toward realism and stresses the status of the created objects at the time of creation.[46] Looking specifically into the Sabbath, the Book of Jubilees developed an indigenous notion of nature, which is identified with the content of the heavenly tablets. The metaphor of writing on the tablets serves in Jubilees in a similar way to the Greek philosophical concept of natural law. In addition, Jubilees reveals several surprising correspondences with his contemporary Aristobulus, including similarities in characteristic philosophical concepts. Taking seriously his commitment to the Sabbath, the author of Jubilees constructed a novel concept of a year made of exactly fifty-two weeks.

Philo carried the concepts of his predecessors further, making use of a full-fledged Stoic concept of *physis* ("nature"). Not only did he perfect the analogy between the law of Moses and natural law, he also used the Sabbath as the primary demonstration of that analogy. Philo's philosophical edifice rests, on the one hand, on the solid ground of Jewish priestly thought and, on the other, on current Egyptian-Greek discourse.

While Aristobulus, at least in the extant fragments, did not apply his heptadic ideas to the structure of the Jewish year, Philo certainly did so. However, even he did not go as far as to construct a wholly septenary Jewish year as in Jubilees. He remains loyal to the – by then, and at his place – particularly Jewish luni-solar year, but he finds ways to weave the heptad into the fine structure of that edifice.

The three cultures here surveyed all found indigenous ways to convey the idea of nature. The Canopus Decree adhered to nature as the order of the seasons and the ordering of heavens. The Book of Jubilees uses the concept of

[46] See note 19.

the law inscribed on the heavenly tablets. In contrast, Philo chose to define it with the Greek *physis*, quite a bold choice taking into account the history of Greek philosophy, but probably not without anchoring in the philosophy of his age. Based on these three sources, I suggest that the development of Jewish calendars has been connected with the birth of nature as an object for observation and as a model for imitation. This notion was demonstrated here not only in explicitly philosophical Greek treatises but also in less philosophically oriented texts originating from Judea.

Bibliography

Amihay, A. 2016. *Theory and Practice in Essene Law*. Oxford: Oxford University Press.

Bagnall, R. and P. Derow 2004. *The Hellenistic Period: Historical Sources in Translation*. Malden, MA: Blackwell.

Ben-Dov, J. 2009. "Tradition and Innovation in the Calendar of *Jubilees*." In G. Boccaccini and G. Ibba (eds.), *Enoch and the Mosaic Torah: The Evidence of Jubilees*. Grand Rapids, MI: Eerdmans, 276–293.

Ben-Dov, J. 2013. "Time and Identity: The Hellenistic Background of the Calendar Treatise in *Jubilees* 6." *Meghillot* 10: 31–56 (in Hebrew).

Ben-Dov, J. 2014. "Time and Culture: Mesopotamian Calendars in Jewish Sources from the Bible to the Mishnah." In U. Gabbai and S. Secunda (eds.), *Encounters by the Rivers of Babylon. Scholarly Conversations between Jews, Iranians and Babylonians in Antiquity*. Tübingen: Mohr Siebeck, 218–252.

Bennett, C. 2011. *Alexandria and the Moon: An Investigation into the Lunar Macedonian Calendar of Ptolemaic Egypt*. Studia Hellenistica 52. Leuven: Peeters.

Bernstein, M. J. 1991. "'Walking in the Festivals of the Gentiles': 4QpHosea[a] 2.15–17 and *Jubilees* 6:34–38." *Journal for the Study of the Pseudepigrapha* 9: 21–34.

Bloch, R. 2013. "Moses im Spiegel Philons: Autobiographisches in Philons Moses-Biographie." In idem, *Jüdische Drehbühnen: Biblische Variationen im antiken Judentum*. Tria Corda 7. Tübingen: Mohr Siebeck, 29–52.

Bockmuehl, M. 1995. "Natural Law in Second Temple Judaism." *Vetus Testamentum* 45: 17–44.

Bockmuehl, M. 2000. *Jewish Law in Gentile Churches: Halakhah and the Beginning of Christian Public Ethics*. Edinburgh: T&T Clark.

Clarysse, W. 1999. "Ptolémées et Temples." In D. Valbelle and J. Leclant (eds.), *Le décret de Memphis: Colloque de la foundation Singer-Polignac à l'occasion de la celebration du bicentenaire de la découverte de la Pierre de Rosette*. Paris: De Boccard, 41–65.

Depuydt, L. 1997. *Civil Calendar and Lunar Calendar in Ancient Egypt*. Leuven: Peeters.

Doering, L. 1999. *Schabbat. Sabbathalacha und -praxis im antiken Judentum und Urchristentum*. Texts and Studies in Ancient Judaism 78. Tübingen: Mohr Siebeck.

Doering, L. 2004. Review of Markus Bockmuehl, *Jewish Law in Gentile Churches: Halakhah and the Beginning of Christian Public Ethics*. *Review of Biblical Literature* 10/2004. www.bookreviews.org/pdf/3930_3802.pdf (accessed 15 November 2015).

Doering, L. 2005. "Excerpted Texts in Second Temple Judaism: A Survey of the Evidence." In R. M. Piccione and M. Perkams (eds.), *Selecta colligere II: Beiträge zur*

Technik des Sammelns und Kompilierens griechischer Texte von der Antike bis zum Humanismus. Alessandria: Edizioni Dell'Orso, 1–38.

Drawnel, H. 2011. *The Aramaic Astronomical Book (4Q208–4Q211) from Qumran. Text, Translation, and Commentary*. Oxford: Oxford University Press.

Evans, J. and L. Berggren 2006. *Geminos' Introduction to the Phenomena: A Translation and Study of a Hellenistic Survey of Astronomy*. Princeton: Princeton University Press.

Filler, E. 2008a. "Description of the Creation by Philo in the Light of Neopythagorean Theory of Numbers." *Da'at* 62: 5–25 (in Hebrew).

Filler, E. 2008b. "From Inception to Perfection: The Nature of Number Seven in Propensity towards Number One in Philo." *Da'at* 63: 5–18 (in Hebrew).

García-Martínez, F. 1997. "The Heavenly Tablets in the Book of Jubilees." In M. Albani, J. Frey, and A. Lange (eds.), *Studies in the Book of Jubilees*. Texte und Studien zum antiken Judentum 65. Tübingen: Mohr Siebeck, 243–260.

Gorre, G. 2009. *Les relations du clergé égyptian et des Lagides d'après les sources privées*. Studia Hellenistica 45. Leuven: Peeters.

Gruen, E. 2005 (ed.). *Cultural Borrowings and Ethnic Appropriations in Antiquity*. Stuttgart: Franz Steiner.

Gruen, E. 2011 (ed.). *Cultural Identity in the Ancient Mediterranean*. Los Angeles: Getty Research Institute.

Grzybek, E. 1990. *Du calendrier macédonien au calendrier ptolémaïque: Problèmes de chronologie hellénistique*. Basel: Friedrich Reinhardt Verlag.

Holladay, C. R. 1995. *Fragments from Hellenistic Jewish Authors*, vol. 3: *Aristobolus*. Chico, CA: Scholars Press.

Horsley, R. 1978. "The Law of Nature in Philo and Cicero." *Harvard Theological Review* 71: 35–59.

Koester, H. 1968. "ΝΟΜΟΣ ΦΥΣΕΩΣ: The Concept of Natural Law in Greek Thought." In J. Neusner (ed.), *Religions in Antiquity: Essays in Memory of Erwin R. Goodenough*. Numen Supplements 14. Leiden: Brill, 521–541.

Kugel, J. L. 2012. *A Walk through Jubilees: Studies in the Book of Jubilees and the World of its Creation*. Journal for the Study of Judaism in the Persian, Hellenistic, and Roman Periods, Supplement Series 156. Leiden: Brill.

Lippert, S. 2009. "*Au clair de la lune* – The Organisation of Cultic Service by Moon Calendar in Soknopaiou Nesos." In G. Widmer and D. Devauchelle (eds.), *Actes du IXe congrès international des etudes démotiques*. Cairo: IFAO, 183–194.

Marcus, R. 1953. *Philo Supplement II: Questions and Answers on Exodus*. Loeb Classical Library. Cambridge, MA: Harvard University Press; London: Heinemann.

McCarthy, D. P. and A. Breen 1993. *The Ante-Nicean Christian Pasch: De ratione paschali, the Paschal Tract of Anatolius, Bishop of Laodicea*. Dublin: Four Courts Press.

Najman, H. 1999. "The Law of Nature and the Authority of Mosaic Law." *Studia Philonica Annual* 11: 55–73.

Najman, H. 2003a. *Seconding Sinai. The Development of Mosaic Discourse in Second Temple Judaism*. Journal for the Study of Judaism in the Persian, Hellenistic, and Roman Periods, Supplement Series 77. Leiden: Brill.

Najman, H. 2003b. "A Written Copy of the Law of Nature: An Unthinkable Paradox?" *Studia Philonica Annual* 15: 54–63.

Niehoff, M. 2001. *Philo on Jewish Identity and Culture*. Texts and Studies in Ancient Judaism 86. Tübingen: Mohr Siebeck.

Niehoff, M. 2013. "The Emergence of Monotheistic Creation Theology in Hellenistic Judaism." In L. Jenott and S. Kattan Gribetz (eds.), *Jewish and Christian Cosmogony in Late Antiquity*. Texts and Studies in Ancient Judaism 155. Tübingen: Mohr Siebeck, 85–106.

Riaud, J. 2006. "Pâque et sabbat dans les Fragments I et V d'Aristobule." In C. Grappe and J. C. Ingelaere (eds.), *Le Temps et les Temps dans les littératures juives et chrétiennes au tournant de notre ère*. Journal for the Study of Judaism in the Persian, Hellenistic, and Roman Periods, Supplement Series 112. Leiden: Brill, 107–123.

Runia, D. T. 2000. "Philo's Longest Arithmological Passage: *De opificio mundi 89–128*." In L.-J. Bord and D. Hamidovic (eds.), *De Jérusalem à Rome: mélanges offerts à Jean Riaud*. Paris: Geuthner, 155–174.

Runia, D. T. 2001. *Philo of Alexandria. On the Creation of the Cosmos according to Moses. Introduction, Translation and Commentary*. Philo of Alexandria Commentary Series. Leiden: Brill.

Schwartz, B. J. 2012. "*Miqra' Qodesh* and the Structure of Leviticus 23." In C. S. Ehrlich, A. Runesson, and E. M. Schuller (eds.), *Purity and Holiness in Ancient Israel, Judaism, and Christianity: Essays in Memory of Susan Haber*. Wissenschaftliche Untersuchungen zum Neuen Testament 305. Tübingen: Mohr Siebeck, 11–24.

Schwartz, D. R. 1992. "Law and Truth: On Qumran-Sadducean and Rabbinic Views of Law." In D. Dimant and U. Rappaport (eds.), *The Dead Sea Scrolls: Forty Years of Research*. Studies on the Texts of the Desert of Judah 10. Leiden: Brill, 229–240.

Segal, M. 2007. *The Book of Jubilees: Rewritten Bible, Redaction, Ideology and Theology*. Journal for the Study of Judaism in the Persian, Hellenistic, and Roman Periods, Supplement Series 117. Leiden: Brill.

Stackert, J. 2011. "Compositional Strata in the Priestly Sabbath: Exodus 31: 12–17 and 35: 1–3." *Journal of Hebrew Scriptures* 11. www.jhsonline.org/Articles/article_162.pdf (accessed 15 November 2015).

Stackert, J. forthcoming. "From Sabbath Day to Sabbath Year: Continuity and Innovation." Lecture given at the University of St. Andrews, July 2013.

Steck, O. H. 1977. "Die Aufnahme von Genesis 1 in Jubiläen 2 und 4. Esra 6." *Journal for the Study of Judaism* 8: 154–182.

Sterling, G. 2003. "Universalizing the Particular: Natural Law in Second Temple Jewish Texts." *Studia Philonica Annual* 15: 64–80.

Stern, S. 2001. *Calendar and Community. A History of the Jewish Calendar 2nd Century BCE – 10th Century CE*. Oxford: Oxford University Press.

Stern, S. 2013. *Calendars in Antiquity. Empires, States, and Societies*. Oxford: Oxford University Press.

Talmon, S., J. Ben-Dov, and U. Glessmer 2001. *Qumran Cave 4 XVI. Calendrical Texts*. Discoveries in the Judaean Desert 21. Oxford: Clarendon Press.

VanderKam, J., trans. 1989. *The Book of Jubilees*. Corpus Scriptorum Christianorum Orientalium 511. Leuven: Peeters.

VanderKam, J. 2013. "Made to Order: Creation in *Jubilees*." In L. Jenott and S. Kattan Gribetz (eds.), *Jewish and Christian Cosmogony in Late Antiquity*. Texts and Studies in Ancient Judaism 155. Tübingen: Mohr Siebeck, 23–38.

Walter, N. 1964. *Der Thoraausleger Aristobulos: Untersuchungen zu seinen Fragmenten und zu pseudepigraphischen Resten der jüdisch-hellenistischen Literatur*. Texte und Untersuchungen 86. Berlin: Akademie Verlag.

Weitzman, S. 1999. "From Feasts into Mourning: The Violence of Early Jewish Festivals." *Journal of Religion* 79: 545–565.

Werman, C. 2002. "The Torah and Te'udah Engraved on the Tablets." *Dead Sea Discoveries* 9: 75–103.

Werman, C. 2007. "Jubilees in the Hellenistic Context." In L. LiDonnici and A. Lieber (eds.), *Heavenly Tablets: Interpretation, Tradition and Identity in Ancient Judaism*. Journal for the Study of Judaism in the Persian, Hellenistic, and Roman Periods, Supplement Series 119. Leiden: Brill: 133–158.

3

CALENDARS, POLITICS, AND POWER RELATIONS IN THE ROMAN EMPIRE

SACHA STERN

1 Introduction

Most ancient calendars were controlled and managed by political rulers. In Greek city-states, the city councils and magistrates decided when to intercalate the year with an additional month, and when to end and begin the months. In the Roman Republic, intercalations were initiated by the *pontifex maximus* and the pontifical college – a priestly body, but elected from the senatorial class and in a high position of political influence. And in monarchical states such as the neo-Assyrian and Achaemenid Empires, the king controlled the calendar. Control of the calendar, and more specifically of the length of the months and years, gave political rulers the means of controlling economic activity, state administration, religious cult, and in some political systems their own tenures of office – often to their personal advantage. In most ancient societies, the calendar was thus fundamentally political. The relationship between calendar and political authority is critical to our understanding as much of ancient calendars as of political forces in ancient society.[1]

[1] This chapter is based on parts of my recent book *Calendars in Antiquity: States, Empires, and Societies* (Stern 2012), in particular chaps. 5–6, where full references to primary sources and scholarly literature can be found, and to which the reader will be referred. The passage on Gen. Rab. 6:3–4, however, is original to this chapter and an additional contribution to my argument. My approach in this chapter is generally informed by postcolonial theory, although I shall not refer to this explicitly. The spread of the Julian calendar in the early decades of the Roman imperial period, which much of this chapter reflects on, contributed significantly toward a more general process that characterized the whole of antiquity, which was the gradual standardization of calendars and their evolution from flexible, typically lunar calendars to fixed, typically solar schemes. This general, macro-historical process is one of the main themes of my book and of the ERC Advanced Grant project at University College London on "Calendars in Antiquity and the Middle Ages: Standardization and Fixation," as part of which the present chapter was written.

In this chapter, I consider the relationship between calendars and politics, or more precisely, power relations, in the Roman Empire. The Roman Empire was characterized by the large-scale adoption of the Julian calendar in the provinces of the Latin West, and in the Greek East, by the adaptation of local provincial calendars to the Julian calendar's year length of 365¼ days. In Egypt, this adaptation was minimal, although revolutionary enough, given the historical antiquity of the ancient Egyptian calendar of 365 days: all it involved was the addition of one day every four years (the "leap year"). Elsewhere in the Eastern Mediterranean, local calendars were transformed, much more radically, from lunar calendars based on the phases of the moon – which had dominated until then the territories of the former Seleucid Empire – to solar calendars based on the Julian year.

The spread of the Julian calendar in the Roman Empire was not the result of cultural osmosis or what is often thoughtlessly called "Romanization," but rather the result of deliberate political decisions. Switching to the Julian calendar was a political act, usually one of subservience toward the emperor and the Roman Empire, which as we shall see was often promoted by the Roman authorities. When calendars of provinces and cities of the Greek East were retained but adapted to the Julian year – in contrast to the Latin West, where local calendars were entirely abandoned in favor of the Julian calendar – this was the result of a political compromise between subservience to the Roman Empire and a contrastive, somewhat illusory sense of civic autonomy and independence. The Julian calendar itself was rarely used in the eastern part of the Roman Empire before the end of antiquity, but it is found in public inscriptions recording imperial rescripts and edicts or honoring Roman officials;[2] this special usage is hardly surprising, but it highlights the political meaning that provincial people in the East would have attached to the Julian calendar.

A few peoples and provinces of the empire resisted the Julian calendar completely and retained their ancestral lunar calendars: the Jews and

[2] The earliest example is a Roman date in the Greek text of a *senatus consultum* inscribed in Priene, Asia Minor, in 135 BCE (thus prior to the institution of the Julian calendar): *I. Priene* 41 = *AE* 2007, no. 1428. Julian dates appear for example in the texts of the following imperial rescripts: *AE* 2006, no. 1403b, from Alexandria in Troas, 134 CE; *AE* 2001, no. 1949, from second- or third-century Salamis in Cyprus. The monumental inscription from the theater of Ephesus in honor of Gaius Vibius Salutarius of 104 CE is dated πρὸ η΄ Καλανδῶν Μαρτίων (i.e., *a. d. VIII kal. Mart.*, or February 22) and μηνὸς Ἀνθεστηριῶνος β΄ Σεβαστῇ (second of the month of Anthesterion, in the provincial calendar of Asia): *IEphesus* 27G, ll. 447–450 (Wankel 1979). Gaius Vibius Salutarius was an equestrian of Italian origin who had held offices in various provinces of the empire before his retirement in Ephesus (Spaul 1994: 238–239).

Samaritans of Palestine and the Diaspora, and much of the Greek peninsula (notably the city of Athens), Macedonia, and the northern regions from there down to the Danube. Although their retention of lunar calendars could be the result of cultural factors such as local patriotism (e.g., perhaps, in the case of the city of Athens) or religious conservatism (Jews and Samaritans), political motivations cannot be ruled out. The extent to which provinces, cities, and peoples in the Roman Empire adopted the Julian calendar, adapted their own calendars to it, or rejected it entirely varied widely in form and in degree, reflecting different power relationships between rulers and ruled and different political attitudes. The calendar served in this way as a means of asserting variously imperial power, provincial loyalty and subservience, political autonomy and independence, and sometimes even as a means of what might be identified as a subtle form of political and cultural dissidence. In this chapter, examples of these different political stances are briefly considered in turn.

The spread of the Julian calendar in the Roman provinces also served other purposes that were directly relevant to the empire and the efficiency of its administration. In the East, where calendars remained diverse, the adaptation of provincial calendars to the Julian calendar and hence to a common denominator – the year of 365¼ days – made it possible and easy to convert dates from Julian to local calendars, or between different local calendars. The stable and constant relationship between these calendars enabled, for example, the creation of multi-column conversion tables such as those attested in the late antique *hemerologia*.[3] The convenience of a single calendar, or of easily convertible calendars, would have been helpful to provincial administration but also to trade and commerce. But more than offering a common time-reckoning system, the Julian calendar, together with the calendars adapted to its fixed year of 365¼ days, was inherently well suited for the administration of a vast territorial empire such as the Roman Empire. Unlike lunar calendars that were usually dependent on unpredictable, locally variable factors such as lunar phases and the visibility and sighting of the new moon, and often also dependent on the whims of political rulers for decisions about when exactly to begin the months and the intercalation of years, the Julian calendar was fixed, unchanging, and therefore completely predictable. This made it possible for anyone in the Roman Empire to reckon time in an identical way, and hence for emperors, governors, officials, and military

[3] Kubitschek 1915. These so-called *hemerologia*, preserved in early medieval astronomical manuscripts, consist of twelve tables for each month of the Julian year, with the Julian calendar in the first column and more than a dozen Roman Eastern provincial calendars in the other columns.

commanders to synchronize their activities and movements at a high level of precision. This single, common time frame also contributed to the cultural cohesion of the empire, and perhaps to a growing sense of shared *romanitas* throughout it.

And yet, looking more closely at the evidence (as we presently do), it becomes evident that decisions about whether or how to adapt to the Julian calendar were not primarily governed by such administrative and cultural considerations. As has often been noted, Roman emperors were not interested in developing efficient systems of imperial administration or in Romanizing their provinces. Administrative efficiency and commercial convenience were similarly not high up in the agenda of the provincial cities that were responsible for adapting their local calendars to the Julian. For both rulers and ruled, the promotion of the Julian calendar had far more to do with imperial propaganda and the assertion of, or allegiance to, Roman political hegemony.[4]

2 Political Allegiance: The Calendar of Asia, 8 BCE

My first example is the institution of the calendar of Asia, which was adapted in 8 BCE from a lunar calendar to a solar, Julian-type scheme. This new calendar was instigated primarily as a ritual act of subservience to Augustus, although the cities of the province still found ways, in this context, of asserting their claim of political autonomy.[5]

Among the lunar calendars of Asia Minor and the Near East that adapted to the Julian calendar in the early Roman period, the calendar later known as "Asian" was perhaps the first. The decree that led to its institution has been uniquely preserved in more than one copy, in inscriptions from several Asian cities.[6] This epigraphic record confirms that the decree was widely diffused in the Roman province of Asia, as required by the text of the decree. There is abundant evidence, moreover, that its calendar soon became the official calendar of the province and remained so throughout the Roman period.

The inscriptions consist of an edict of the proconsul (Roman governor of the province of Asia), partly in Latin and in Greek, followed by a decree of the *koinon* (association, confederation, or council in Greek) of the cities of the province of Asia. In his edict (διάταγμα), the proconsul puts forward the proposal – to be read as a virtual directive – that special honors be given to

[4] On all these, see Stern 2012, chaps. 5–6; also Rüpke 1995; Feeney 2007.

[5] For a more detailed account and full references, see Stern 2012: 274–284.

[6] The full edition is Laffi 1967; an abbreviated edition (with only a composite text) is in Sherk 1969: 328–337 (no. 65).

the emperor Augustus by reforming the calendar of the province in such a way that the New Year occurs on September 23, the birthday of Augustus. In the Latin appendix, the proconsul also lays out the lengths of the months of the new calendar. The decree that follows, issued by the *koinon*, expresses an enthusiastic endorsement of this proposal. The New Year is to occur on September 23, Augustus's birthday; the first month, moreover, will be renamed *Kaisar* ("Caesar"). The months of the year and their lengths are then laid out in full; every month will begin on the 9th day before the Kalends of the Julian month (as on September 23). In a leap year, an intercalary day will be added to day 1 of Xandikos, which will thus have thirty-two days. Precise arrangements are made for the transition from the old to the new calendar: the current month of Peritios will run its course until the 14th, which will be January 23; the next day (i.e., January 24) will be renamed Dystros 1 of the new calendar.

The sequence of months laid down in the *koinon*'s decree differs from what the proconsul had originally proposed. The fragmentary passage from the Latin section of the proconsul's edict indicates that the last four months of the year, according to him, should have followed the sequence of 30–31–30–31 (days); whereas according to the *koinon*'s decree, the sequence of the last four months was 30–31–31–30. There is no need to assume a scribal (or other) error. The scheme of the proconsul was simply not well thought out; and this is what justified the *koinon*'s decision to modify it. Indeed, the month lengths proposed by the *koinon* had the distinctive advantage of corresponding exactly to those of the Julian calendar (from October to September), with the result that all Asian months began on the same day of the Roman, Julian month (the 9th before the Kalends).

By modifying the proconsul's scheme and improving its synchronization to the Julian calendar, the *koinon* were paradoxically asserting on the one hand their commitment to conform to the calendar of Rome, and on the other hand, their autonomy with regard to the proconsul's authority. The *koinon*'s decree represents, in this respect, a manifestation of the complex (albeit unequal) political relationship between provincial cities and the Roman Empire. This complex relationship is also evident in the proconsul's letter, which blends the authoritarian status of an "edict" with the conciliatory tone of a recommendation.

But these small concessions to provincial autonomy should not obscure the fact that the institution of the calendar of Asia was, fundamentally, an act of submission to the imperial rulers. It is clear from the contents, tone, and emphasis of both the proconsul's edict and the *koinon*'s decree that the main and perhaps sole purpose of this calendar reform was for the province to demonstrate its loyalty to the emperor Augustus. In the Greek version of his edict, indeed, the proconsul makes no mention at all of a change of the calendar from lunar to Julian, or to a new sequence of month lengths; the only calendar change that he proposes is the institution of the New Year on Augustus's birthday. Although this New Year necessitated the conversion of the lunar calendar to a 365-day year (because otherwise, the lunar New Year could not have remained permanently on September 23), conversion of the calendar was only a technical and incidental aspect of the decree's implementation, which did not interest the proconsul and which he did not bother to mention explicitly. A list of month lengths does appear in the Latin text of his edict, but its mediocre design (which we have noted earlier) betrays his lack of attention to technical, calendrical details. The point of the proconsul's edict was only that the province should observe Augustus's birthday as a New Year's day, and this point was not lost on the *koinon*, who added for good measure the renaming of the first month as *Kaisar*.

The motivation for this calendar reform, on both sides, was thus essentially political. The political gains that both the proconsul and the province stood to make in the process were, indeed, potentially considerable. In contrast, neither the proconsul nor the *koinon* referred in their decrees to the practical, administrative advantages of synchronizing the calendar of Asia to the Roman calendar. That this argument was ignored may appear surprising, because it should have been evident that synchronization of the calendars would have greatly facilitated the local administration of the province as well as its wider integration into the Roman Empire. This omission has much to teach us about how calendars were perceived in ancient society; it also calls us to revise the common modern assumption that calendar reforms were motivated in antiquity by arguments such as administrative convenience or efficiency. In the context of this decree, at least, the proconsul and the *koinon* did not perceive the calendar as an administrative instrument, as we would tend to look at it today, but rather only a statement of political loyalty.

The timing of this decree is also highly significant. In early 8 BCE, at the time of the decree, Augustus was about to suspend temporarily the leap years, and the month Sextilis, in the Julian calendar, was about to be renamed "Augustus." Calendar reform was in the air. It is likely that the proconsul of Asia, on whose directive the decree of the calendar of Asia was issued, did not act on a purely spontaneous initiative. At the very least, the calendar change he was proposing – largely honorific, as was to be the renaming of the month of August in the Julian calendar – reflected the current policies and ambitions of the emperor in Rome. This sheds light on how the imperial authorities, both central and in the provinces, could have been responsible, directly or indirectly, for the spread of the Julian calendar (or adaptations of it) in the eastern provinces of the Roman Empire.

The precedent set by the province of Asia in 8 BCE did not take long to be followed by the other provinces of the Roman East. Nearly all the lunar calendars of the former Seleucid Empire under Roman rule converted, in the following decades, to various forms of Julian-type calendars; not all began the year on Augustus's birthday, but they all shared the common, fundamental structure of a 365-day year, with an additional day every four years. This widespread adaptation of provincial calendars to the Julian calendar was a clear manifestation of Rome's expanding hegemony in this region. Indeed, as the Roman Empire expanded further into the Near East in the centuries that followed, the local lunar calendars of the new territories were almost simultaneously converted into solar, Julian schemes: for example, in the province of Arabia after its annexation in 106 CE, and in the city of Dura-Europos after its annexation in the last decades of the second century. The expansion of the Roman Empire in the Near East was thus conterminous, and often simultaneous, to the adoption of the Julian year.

The political significance of the Julian calendar and its adoption in the Roman Empire is all the more evident when we consider the specific elements of the Julian calendar that were being adopted by these provincial calendars. As we have seen in the case of the province of Asia, but similarly throughout the eastern and western provinces of the empire, what was adopted was not the entire *Roman* calendar with its religious festivals, its special events, its days qualified as *fas* and *nefas*, and so on but rather only the bare structure of the *Julian* calendar, that is, twelve months of various lengths (in the provinces of the East, not necessarily conterminous to the months of the Julian calendar) adding up to a year of 365 days, with an extra day every four years (for the leap year). I emphasize "Julian" as opposed to "Roman" because this calendar structure and year length were not essentially Roman – they did not characterize in any way the Roman calendar of the Republican period, which did not have a 365¼-day year – but only instituted by Julius Caesar in his major calendar reform of 46 BCE. In the period of Augustus, when the Julian calendar or adaptations of it were widely adopted in the western and eastern provinces of the empire, this calendar structure was still associated specifically with Julius Caesar and his adoptive heirs, rather than with Rome, and justifiably so, since Augustus was putting his own mark on the calendar with the renaming of its eighth month as "Augustus" and with the adjustments he made, in the decade following 8 BCE, to the schedule of the leap years. The adoption of the Julian calendar therefore did not represent an act of "Romanization" as much as one of personal loyalty to Julius Caesar, Augustus, and their heirs – it was, in other words, a primarily political act of allegiance to the ruling family of the empire.[7]

3 Political Opposition: The Jewish Lunar Calendar in Rabbinic Sources

However, there were some notable exceptions. In Jewish and Samaritan Palestine and the Jewish communities in the Diaspora, for example, lunar calendars were maintained, somewhat subversively, throughout the Roman imperial period. I say "subversively" because we have evidence that the preservation of a lunar calendar, at least for Jews, represented a conscious statement of identity and difference within the Roman Empire, which in some circumstances could have come close to political dissidence. A particularly good example from rabbinic literature is a passage in the Midrash Genesis Rabbah, which comments on Gen. 1:16 (the creation of the sun, moon, and stars), and of which this is my translation:[8]

> (3) Rabbi Levi (said) in the name of Rabbi Yossi son of Lai: It is the way of the world that the big should reckon by the big, and the small by the small. Esau reckons by the sun which is big, and Jacob by the moon which is small. Rav Naḥman said: That is a good sign. Esau reckons by the sun which is big: just as the sun rules by day[9] and does not rule by night, so Esau has a share in this world, but does not have a share in the world-to-come. Jacob reckons by the moon which is small: just as the moon rules by night and by day, so Jacob has a share in this world and in the world-to-come. Rav Naḥman said another saying: Rav Naḥman said:[10] As long as the light of the big (luminary) is there, the light of the small one is not known; but when the light of the big one sets, the light of the small one becomes known. So, as long as the light of Esau is there, the light of Jacob is not known; but when the light of Esau sets, the light of Jacob becomes known. As it is

[7] An extreme example of this phenomenon may be identified in the case of Cappadocia, a client kingdom whose calendar was probably adapted to the Julian calendar (with the addition of a leap year to its Persian-Egyptian year of 365 days) immediately after the institution of the Julian calendar in 46 BCE. This calendar reform, the first of its kind in the East, could only have been intended as a gesture of personal loyalty to Julius Caesar by the Cappadocian king Ariobarzanes III, who was in great need of finding favor with him because of his earlier allegiance to Caesar's rival, Pompey. See Stern 2012: 269–271.

[8] Gen. Rab. 6:3–4 (42–43 Theodor and Albeck). Although one of the sages cited, Rav Naḥman, is Babylonian, the text as a whole is a Palestinian composition and the political context it refers to is evidently late Roman Palestine.

[9] Alternatively: "over the day."

[10] This duplication is possibly a textual error.

written: "Arise, shine, for your light is coming . . . for behold, darkness shall cover the earth." (Isa 60:1–2).

(4) ". . . and the stars" (Gen 1:16). Rabbi Aḥa said: (It is) like a king who had two officials,[11] one ruled a city (*'ir*) and the other ruled a province (*medinah*). Said the king: "Since this one has reduced himself to rule a city, I decree upon him that whenever he goes out, the council (*boule*) and assembly (*demos*) shall go out with him; and whenever he goes in, the council and assembly shall go in with him." So said the Holy One Blessed is He: "Since the moon has reduced[12] herself to rule the night, I decree upon her that whenever she comes out, the stars shall go out with her; and when she goes in,[13] the stars shall go in with her.

Genesis Rabbah is usually dated to the fifth century CE, although the rabbis cited in this passage are of the third to fourth centuries. As everywhere else in Genesis Rabbah, Esau is symbolic of Rome, whereas Jacob is Israel. This passage not only draws the contrast between Rome, whose calendar is solar, and Israel, whose calendar is lunar (the saying of R. Levi) but also relates this contrast of calendars to a political, eschatological claim (the saying of Rav Naḥman, in its two alternative versions). According to Rav Naḥman, the lunar calendar embodies Israel's ultimate triumph over Rome: just as the moon shines in the night after the sun has set, so Israel will rise after the Roman Empire will have set. This makes observance of a lunar calendar, for the Jews in the Roman Empire, not only the marker of a distinct identity, but also – through the medium of an eschatological promise – the embodiment of a politically dissident, even oppositional stance toward the Roman Empire.[14]

In the saying that immediately follows Rav Naḥman's, Rabbi Aḥa compares the sun to an official who has been appointed over a province, and the moon to an official appointed over a city; to compensate for this lower appointment, or as a reward for his humility, the official of the city is given the privilege of being escorted everywhere by the *boule* (city council)

[11] The term used for this is the Greek *epitropos*, which is versatile and has no specific technical meaning; my translation "official" is intended to reflect this. In contrast, the terms for council (*boule*) and assembly (*demos*) in the text that follows are technical and specific. The term *medinah* can sometimes mean "city," but in the context of this passage, only the meaning of "province" seems possible.

[12] In the context of the moon, "reduced" may have several meanings: it means, as in the parable of the king, that she has humbled herself by accepting to rule over the night (or "at night"), which is inferior to ruling over the day, but it also may mean that she has reduced the strength of her light and subjected herself to the regime of lunar phases.

[13] I.e., sets.

[14] Cf. Ben-Dov, Chapter 2, in this volume.

and *demos* (popular assembly), just like the moon is escorted by the stars. The Midrash does not explain, but maybe this is obvious, that this escort is actually appropriate to the city official's position, as *boule* and *demos* were the main political institutions of Hellenistic cities (but not of Roman provinces).

It is possible to read this saying of Rabbi Aḥa as completely separate from what precedes it; but if it is linked to the foregoing passage, as the textual juxtaposition suggests, and if Rabbi Aḥa's parable is related to the contents of the foregoing passage, then the sun or governor of the province must be identified again with Esau or Rome, and the moon or governor of the city with Israel. If so, Rabbi Aḥa's parable implicitly provides a further political dimension to the contrast between the solar and lunar calendars of Rome and Israel. The interpretation of this parable requires some caution, however, because as in any parable, a certain distance can be expected between signifier and signified. The association of the moon, identified with Israel, with the institutions of *boule* and *demos*, is only parabolic and does not need to imply a real-life connection between Jews and these civic institutions – even if a connection of this kind is likely, inasmuch as in many late Roman Palestinian cities the members of city councils are likely to have been predominantly Jewish.[15] Likewise, the association of the sun, identified with Rome, with the governor of the province is not necessarily meant to reflect the fact that provincial governors, in Palestine and elsewhere, were generally not Jewish. What is certain, however, is that this parable takes up a real-life contrast in the Roman Empire between the governors of provinces and cities, and likens it to the opposition between sun and moon, that is, Rome and Israel.

The oppositional, sometimes competitive and evenly balanced relationship between provincial governors and autonomous cities characterized the whole of the Roman imperial period – for example, in the case of the institution of the calendar of Asia in 8 BCE, as we have seen earlier – but may have become more pronounced in the later Roman period. In the early empire, provinces were governed by proconsuls or other high officials who were appointed by the emperor and in command of one or two legions, which clearly gave them the upper hand; whereas cities with their territories were governed autonomously by locally elected magistrates and

[15] In early rabbinic literature, at least, city councilors (*bulevtaya*) are frequently assumed to be Jews: see Goodblatt 2006: 413–415, with references for example to y. Pe'ah 1:1, 16a (the city councilors of Sepphoris who transgressed the prohibition on *lashon hara*, slander, against their absconding colleague Yoḥanan), y. Hor. 3:8, 48c and y. Šabb. 12:3, 13c (city councilors making daily *salutatio* to the Jewish Patriarch), and y. Pesaḥ 4:1, 30c (a Rabbi Shimon the *bulevta*).

councils with no military command, which meant subservience to the provincial governor. In the later Roman Empire, however, provincial governors declined in status as they ruled over much smaller provinces and no longer commanded any legions. The city councils also declined in political status, but they gave way to other forms of local, oligarchic leadership. The rise of these new oligarchies is evident in our passage of Genesis Rabbah, where the emperor appoints an official to govern a city: in the fourth to fifth centuries, when this text was composed, emperors and their administration were becoming increasingly involved in the government of cities, and the autonomy and authority of city councils was increasingly bypassed by strong, individual leaders. But however governed, the cities and their aristocracies remained an important locus of political power in late antiquity, in a good position to compete against the declining authority of provincial governors. It is thus not surprising that this late Roman-period rabbinic parable presents the governor of the city as formally inferior to the governor of the province, yet able to consider himself, in a certain way, his equal.[16]

By comparing the sun and moon, that is, Rome and Israel, to officials appointed by the king over a province and a city, the parable conveys a message that is politically subversive in two ways. First, it suggests that the true emperor (or "king") of the world is not the Roman emperor, but God, to whom Rome is only like the governor of a province. Second, it suggests that although Israel (or the moon) is *prima facie* inferior to Rome (or the sun – as assumed in the first part of our Genesis Rabbah passage), this inferiority can be challenged and negated, similarly to the governor of a city who can claim sociopolitical parity with the provincial governor, on the strength of his prestigious escort of *boule* and *demos*. In both these ways, this parable about the sun and the moon thus draws on the complex relationship between province and city in the later Roman Empire to challenge the hegemony of Rome and make an implicit claim of parity between Rome and Israel.

In contrast to the province of Asia in 8 BCE, where the *koinon* of cities surrendered to the Roman imperial rulers by abandoning their lunar calendar

[16] For a variety of perspectives on the distribution of power in the late Roman cities and provinces, including discussion of the decline of provincial governors, the decline of city councils, and the rise of new oligarchies, see for example Liebeschuetz 1972: esp. 101–114, 167–192, 208–219; Whittow 1990; Lendon 2001; Liebeschuetz 2001: 104–136; Laniado 2002; Rapp 2005. Laniado's study, in particular, suggests that cities retained a considerable amount of political autonomy in late antiquity, to the extent of becoming involved in the appointment of provincial governors in the late sixth century (see esp. 225–254). I am grateful to Fergus Millar and Benet Salway for referring me to some of these readings.

tradition in favor of a Julian-type solar year, the Jews maintained a lunar calendar throughout the period of Roman rule. This rabbinic passage suggests that the identification of the Jews with the moon and their adherence to a lunar calendar could be perceived as a statement of political strength in relation to the imperial rulers. Although this perception of political strength was only illusory and a reflection, perhaps, of provincial wishful thinking, it is significant that adherence to a lunar calendar could be interpreted and experienced in this manner by Jews living in Palestine in the later Roman period.

4 Lunar Subculture in Roman Sicily: The Catania Inscription of 383 CE

The contrast between the Jewish lunar calendar and the Julian solar calendar, which in this Palestinian rabbinic parable is completely taken for granted, is likely in reality to have been considerably more complex, especially in the context of the Jewish Diaspora where Jews were implicated in a generally more complex web of variant identities. This is evident from a funerary inscription from Catania (Sicily), which was erected in 383 CE by Aurelius Samuel, a Jew, in memory of his wife, and where the date of her death is given as follows (ll. 3–6):[17]

> . . . *fatum conplebit XII Kal(endas) Novebr|es, diae Veneris, luna octaba, Mero| | baudes iterum et Satornino con|sulibus*. . .

> ... (she) completed her allotted life on the 12th (before) the Kalends of November, Friday, lunar day 8, when Merobaudes for the second time and Saturninus were consuls.

The inclusion of a lunar day, which refers most likely to the date in the Jewish calendar (this would have been the 8th of Marḥeshwan), may be interpreted, at first sight, as a statement of Aurelius Samuel's Jewish identity. This would tie in with the other, conspicuously Jewish elements of this inscription: a line of Hebrew at the top of the inscription, the biblical name (Samuel) of the dedicator, the reference farther down the inscription to "the law which the Lord gave the Jews" (*licem quem Dominus dedit Iudeis*), and finally the Jewish symbol of *menorot* in the bottom corners of the epitaph. Yet in other ways, this inscription, written almost entirely in Latin, belongs firmly to the Greco-Roman epigraphic tradition. Its date, in particular – if we ignore temporarily the lunar element – conforms to a thoroughly Roman

[17] Ursino Museum, Catania, no. 540; *CIJ* i. 650; *AE* 1984, no. 439; *JIWE* i. 145; Korhonen 2004, no. 228. For a more detailed account and full references, see Stern 2012: 340–341.

dating formula, consisting of the day of the month (down-counted inclusively from the following Kalends), the Julian month, and the consular year. The day of the week, by the fourth century, was also not uncommon in Roman datings; in this inscription, it is designated without apparent compunction by the name of a Roman goddess, Venus.[18] More generally, the practice of recording a detailed date of death was a distinctly Roman funerary epigraphic tradition.[19]

The lunar date itself should also be viewed as part of Roman traditional dating practices, even though, as mentioned earlier, it refers in this inscription most likely to Aurelius Samuel's Jewish calendar, that is, the calendar that regulated for him the dates of the Jewish festivals. The *formula* employed for this lunar date – *luna* in the ablative, followed by the ordinal number – was not at all unique to the Jews, nor indeed was the practice of dating according to the moon. A good number of Latin inscriptions from the third century and later include a lunar date together with the Julian date, and they all employ this same formula; most of these are from Italy, many are Christian, and some are also pagan. This lunar formula can be traced back to an inscription from Pompeii from 60 CE, and earlier still, to an inscription from Etruria from 67 BCE, still in the Republican period. Lunar dating was clearly an ancient Italian tradition, going back to a period when Italian calendars – including even the calendar of the city of Rome – were lunar, before the Roman hegemony over the Italian peninsula led to the diffusion and adoption of the Republican Roman calendar and, eventually, in the late first century BCE, of the Julian calendar. The late antique survival of lunar dating practices, usually in conjunction with Julian dates, is evident not only from epigraphic sources but also from lunar *parapegmata*, counting devices that enabled the continuous count of lunar month days – many lunar *parapegmata* are attested in late antique Italy and have been discussed at length elsewhere. The calendar codex of Philocalus (354 CE), furthermore, includes a lunar column, alongside the normal Roman dating system. The formula *luna* + ordinal is also pervasive in late antique Christian treatises on the date of Easter, which had to be lunar like the Jewish Passover (Easter occurred on the

[18] I owe this observation to Jörg Rüpke. It could be argued, however, that in late Roman society this was generally understood in this context as the name of the planet, and not of a goddess.

[19] Carletti 2004, arguing for the pagan, Roman origins of this practice and refuting Shaw's 1996 claim that this tradition was specifically Christian, which I followed in Stern 2001: 134–135 and 2012: 340. Detailed dates of death are also extensively represented in the late antique, Christian and Jewish cemetery of Zoar (southern Palestine), but this is exceptional for the Roman East (see Meimaris and Kritikakou-Nikolaropoulou 2005; 2008).

Sunday following *luna* XIV of the first lunar month); again, this formula was not specifically Christian but the heir of an older Latin tradition.[20]

The survival of lunar calendar reckoning, and more specifically of lunar dating with the *luna* formula, in Italian and Sicilian inscriptions long after the institution of the Julian calendar and its wide diffusion in the Italian peninsula and the Latin West, may be interpreted as an expression of popular, covert resistance to the imposition of the non-lunar, Julian calendar in the Roman Empire. It is significant to note that *luna* dates only appear in private inscriptions and are not known to have ever been used for official purposes. Perhaps this usage expressed no more than a conservative disposition, among ordinary people in Rome and elsewhere in the Latin West, to preserve their local, ancient calendrical traditions that had been lunar. But even if not politically dissident or subversive, this deviation from the dominant Julian calendar should be regarded as part of an unofficial subculture of the Roman world.

In this light, the lunar date in the Catania inscription was not simply a statement of Aurelius Samuel's Jewish identity, even if it belonged to the lunar calendar that he is likely to have used as his Jewish calendar. This lunar date was embedded, indeed, in a thoroughly Roman dating formula, including the Julian date and consular year; even the lunar formula, *luna octaba,* was an ancient Latin tradition of dating that others, pagans and Christians, similarly used. Aurelius Samuel's use of this lunar formula was a way of participating in a broader subculture or, perhaps, of merging the Jewish, somewhat dissident adherence to a lunar calendar with a similar subculture of the Latin West. Although the survival of lunar dating in Italy was much less likely an expression of politically dissidence than it was among the Jews (at least some Jews, as we have seen), it is interesting to note that in this inscription, Jewish and Latin traditions converged because they shared common, if subtle, subversive objectives in relation to the dominant calendar of the Roman Empire.

5 Dissidence and Subversiveness: The Gallic Calendar of Coligny

The connection between lunar calendars and political dissidence is more evident in the Latin West, in the case of the Coligny calendar. This lunar calendar was discovered in the French Jura on the fragments of a monumental bronze tablet; its language is Celtic, and it attests the persistence of native Gallic calendars in the Roman period. The Coligny calendar consists basically of a five-year lunar cycle, which some scholars

[20] On all the foregoing, see Stern 2012: 313–330.

have sought to expand into thirty-year or twenty-five-year cycles, for the sake of improving its calendrical accuracy. I shall not discuss these theories in any detail; instead, I focus on the cultural and political context of this remarkable inscription.[21]

The Coligny calendar has often been presented as "very ancient," mainly on the basis of its archaic Celtic names of months and days; in spite of these archaisms, there is a strong argument for viewing this calendar as a product of the Roman period. On paleographic grounds, to begin with, the inscription can be dated to the late second century CE. The mere fact that it is a written text (as opposed to oral), and that its writing is Latin, indicates a distinctly Gallo-Roman cultural context. Some other physical features of the inscription betray Roman influence: for example, the peg holes that are aligned along each day of the calendar may have been drawn from Greek and Roman *parapegmata*. Most importantly, the tabular concept and design of the Coligny inscription clearly imitates the Roman tradition of monumental calendar displays, the so-called *Fasti*.

The structure of the calendar itself, albeit lunar, shows further evidence of Roman influence. Unlike the lunar calendars that generally prevailed in earlier antiquity, from Mesopotamia to the Levant, Asia Minor, Greece, Italy, and presumably other parts of Europe including Celtic Gaul, where the months were irregular and determined by empirical new moon sightings or ad hoc decisions, the Coligny lunar calendar is schematic and fixed, with alternating twenty-nine- and thirty-day months. The concept of a lunar calendar as a fixed scheme, which only became normal among Jews and Christians in the Middle Ages, was still unusual in the second century CE, when Jews and other lunar calendar users in and outside the Roman Empire (with the only exception, perhaps, of lunar calendars in Egypt) were still relying on monthly lunar observations, and when Christians had not yet begun to develop their fixed Easter cycles. The Coligny scheme is likely to have arisen by imitation or emulation of the Julian calendar, which had the distinctive and unique characteristic of being schematic and fixed.

Proponents of pre-Roman origins, inspired by a touch of Celtic romanticism, have suggested that the Coligny calendar was the product of ancient Druidic astronomical lore, itself the result of age-old astronomical inquiry. But however much astronomy was really known to the Druids, the amount of astronomical knowledge that would have been necessary to construct the

[21] The definitive edition is Duval and Pinault 1986; a sound interpretation can be found in McCluskey 1998: 54–76. For a more detailed discussion and full references, see Stern 2012: 303–313.

Coligny calendar has been grossly exaggerated. This calendar is actually far more simple and rudimentary than has been made out. Its structure consists of a simple alternation of twenty-nine- and thirty-day months, with possibly only a few variations to the month of Equos that show signs of trial and error and do little to improve the calendar's lunar accuracy, and the intercalation of one month at two-and-a-half-year intervals, which is all rather elementary and inaccurate. Even if the five-year cycle of this calendar were corrected (somewhat, but not entirely) through the use of a hypothetical thirty-year or twenty-five-year cycle (as has been suggested by many scholars, but without any evidence in the inscription), this correction could have been determined empirically without much astronomical expertise. The dates of the solstices and equinoxes, which according to some scholars are annotated in the calendar, could have been entered in the Coligny calendar by simply tracking its days alongside those of the Julian calendar, where solstices and equinoxes could be easily identified. Nothing compels us to assume, therefore, that the Coligny calendar was the result of centuries of Druidic astronomical inquiry. Not only was the inscription designed and erected in the second century CE, in a format that clearly betrays Roman influence, but the schematic structure of its lunar calendar could have been similarly designed in the same period, when reference to the widely available Julian calendar would have greatly facilitated the calculation of its more complex features.[22]

The Romanization, or more precisely Julianization, of the Gallic lunar calendar, as embodied in the monumental, *Fasti*-type inscription of Coligny and in its design as a fixed, schematic calendar, is somewhat reminiscent of the Julianization of provincial calendars in the Roman East, such as the calendar of the province of Asia, where, as we have seen, the lunar calendar was recast as a fixed scheme compatible with the Julian calendar. In this respect, the calendar of Coligny was congruent with a general trend in the Roman Empire, as well as in other great empires in antiquity (for example, Achaemenid and Seleucid) for empirical, flexible calendars to become increasingly standardized and fixed. It was part of a much broader historical process, the outcome of the large-scale political changes that were brought about with the rise of great empires in the Near East and the Mediterranean from the middle of the first millennium BCE until the end of antiquity.

However, the Coligny calendar did not go as far in the Roman East: in spite of its overtly Romanizing features, it remained tenaciously lunar and retained a strong sense of Gallic identity. This is evident not only in its use of Celtic month and day names but also in its lunar (or lunar-like) structure of

[22] I am taking issue mainly with Olmsted 1992; for more details, see Stern 2012: 303–313.

twenty-nine- and thirty-day months and intercalary months, which bore no resemblance to the structure of the Julian calendar. Although fixed, the Coligny calendar was not synchronized in any way to the Julian calendar; even the dates of the solstices and equinoxes, if this is what they are, did not (indeed could not) consistently conform to those of the Julian calendar, at least over a long-term period. In contrast to calendars of the Roman East, which, as the calendar of Asia did, adopted instead various models of the 365-day year, the Coligny calendar remained at least formally lunar (even if poorly synchronized to the moon). In this respect, it is appropriate to categorize this calendar as "dissident." By publicly displaying a model of the Celtic calendar that bore no structural identity or synchronicity with the Julian calendar, this monumental inscription constituted at once a mimicry and a grand subversion of the Julian *Fasti*.

As a dissident calendar, the calendar of Coligny seems not to have had any official status. It does not appear in any dated inscription from Roman Gaul: all dated inscriptions use only the Julian calendar. This suggests that neither the Coligny calendar nor any other version of the Gallic lunar calendar was used as an official calendar in the Gallo-Roman cities. However, the Coligny calendar was not a socially marginal phenomenon. The monumental scale of the inscription, on a bronze plate that measured originally 1.48 m × 0.90 m, indicates that it could only have been produced by people of means. The involvement of the local aristocracy in its production seems evident from the high level of numeracy and literacy that the creation of this lunar cycle and its complex inscription would have required. Furthermore, this calendar appears to have been widely diffused. Fragments of a similar calendar have been discovered in the town of Villards d'Héria, 31 km east of Coligny, which shows that the Coligny calendar was not an isolated phenomenon. The towns of Coligny and Villards d'Héria are relatively close, but the topography is such that access between them is difficult; whereas Coligny was in the tribal territory of the Ambarri, Villards d'Héria was in that of the Sequani. This indicates that Gallic lunar calendar inscriptions spread to more than one region and tribe of Roman Gaul.

In the absence of evidence that both calendars were identical, it is impossible to determine the extent to which the Gallic fixed calendar was standardized, and, hence, to what extent calendar dissidence was structured and organized in the context of Roman Gallic society. But the hybrid, Gallo-Roman features that characterize this lunar calendar have much to say about power relations in the provincial context of Roman Gaul. Rather than interpreting this Gallic calendar in terms of a simplistic model of Romanization and resistance to it, as has been assumed by earlier scholars, I propose a more complex model of cultural and political interaction whereby Roman culture and the political power that it embodied were at once espoused by the Gallic aristocracies (the fixed calendar, the monumental inscription) and perverted, where it was appropriated but re-negotiated and reformulated in their own terms and to their own perceived political advantage. This monumentally inscribed Celtic, lunar calendar was a subversive mimicry of the Julian *Fasti* that challenged Roman cultural hegemony in Gaul and represented a political statement by the local aristocracies that erected it.

6 Summary

In this chapter, I have sought to demonstrate the close relationship between politics, power relations, and the history of calendars in antiquity. In the context of the Roman Empire, the institution of the calendar of Asia in 8 BCE, the Jews' contrasting retention of a lunar calendar, the survival of lunar dating practices in late antique Sicily and Italy, and the erection of the Coligny calendar inscription were all different ways of negotiating relationships with the Roman rulers and the dominant Julian calendar. These different stances ranged from political loyalty and subservience to autonomy, subversion, and subtle forms of dissidence. As I have argued elsewhere,[23] it is through these political processes, rather than as a result of scientific progress, that calendars and dating practices emerged and evolved throughout the course of ancient history.

Bibliography

Carletti, C. 2004. "*Dies mortis – depositio:* un modulo 'profano' nell'epigrafia tradoantica." *Vetera Christianorum* 41: 21–48.

Duval, P.-M. and G. Pinault. 1986. *Recueil des inscriptions gauloises*, Vol. 3: *Les Calendriers*, Paris: Éditions du CNRS.

Feeney, D. 2007. *Caesar's Calendar: Ancient Time and the Beginnings of History*. Berkeley and Los Angeles: University of California Press.

Goodblatt, D. 2006. "The Political and Social History of the Jewish Community in the Land of Israel, c. 235–638." In S. T. Katz (ed.), *The Cambridge History of Judaism*, Vol. 4: *The Late Roman-Rabbinic Period*. Cambridge: Cambridge University Press, 404–430.

Korhonen, K. 2004. *Le iscrizioni del museo civico di Catania*. Helsinki: Societas Scientiarum Fennica.

Kubitschek, W. 1915. *Die Kalenderbücher von Florenz, Rom und Leyden*. Denkschriften der Kaiserlichen Akademie der Wissenschaften in Wien, Philosophisch-historische Klasse 57, Abhandlung 3, Vienna.

[23] Stern 2012.

Laffi, U. 1967. "Le iscrizioni relative all'introduzione nel 9 a.C. del nuovo calendario della provincia d'Asia." *Studi Classici e Orientali* 16: 5–98.

Laniado, A. 2002. *Recherches sur les notables municipaux dans l'empire protobyzantin*. Paris: Association des Amis du Centre d'Histoire et Civilisation de Byzance.

Lendon, J. E. 2001. *Empire of Honour: The Art of Government in the Roman World*. Oxford: Oxford University Press.

Liebeschuetz, J. H. W. G. 1972. *Antioch: City and Imperial Administration in the Later Roman Empire*. Oxford: Clarendon Press.

Liebeschuetz, J. H. W. G. 2001, *The Decline and Fall of the Roman City*. Oxford: Oxford University Press.

McCluskey, S. 1998. *Astronomies and Cultures in Early Medieval Europe*. Cambridge: Cambridge University Press.

Meimaris, Y. E. and K. I. Kritikakou-Nikolaropoulou. 2005. *Inscriptions from Palaestina Tertia*, Vol. 1a: *The Greek Inscriptions from Ghor Es-Safi (Byzantine Zoora)*. Meletemata 41. Athens: National Hellenic Research Foundation; Paris: Diffusion de Boccard.

Meimaris, Y. E. and K. I. Kritikakou-Nikolaropoulou. 2008. *Inscriptions from Palaestina Tertia*, Vol. 1b: *The Greek Inscriptions from Ghor Es-Safi (Byzantine Zoora) (Supplement), Khirbet Qazone and Feinan*. Meletemata 57. Athens: National Hellenic Research Foundation; Paris: Diffusion de Boccard.

Olmsted, G. 1992. *The Gaulish Calendar*. Bonn: Rudolf Habelt.

Rapp, C. 2005. *Holy Bishops in Late Antiquity: The Nature of Christian Leadership in an Age of Transition*. Berkeley: University of California Press.

Rüpke, J. 1995. *Kalender und Öffentlichkeit: Die Geschichte der Repräsentation und religiösen Qualifikation von Zeit in Rom*. Berlin: de Gruyter.

Shaw, B. 1996. "Seasons of Death: Aspects of Mortality in Imperial Rome." *Journal of Roman Studies* 86: 100–138.

Sherk, R. K. 1969. *Roman Documents from the Greek East*. Baltimore: Johns Hopkins Press.

Spaul, J. E. H. 1994. "Governors of Tinginata." *Antiquités Africaines* 30: 235–260.

Stern, S. 2001. *Calendar and Community: A History of the Jewish Calendar, 2nd Century BCE–10th Century CE*. Oxford: Oxford University Press.

Stern, S. 2012. *Calendars in Antiquity: States, Empires, and Societies*. Oxford: Oxford University Press.

Wankel, H. 1979. *Die Inschriften von Ephesos*. Vol. 1a. Bonn: R. Habelt.

Whittow, M. 1990. "Ruling the Late Roman and Early Byzantine City: A Continuous History." *Past and Present* 129: 3–29.

4

DOUBLING RELIGION IN THE AUGUSTAN AGE: SHAPING TIME FOR AN EMPIRE

Jörg Rüpke

"CALENDAR" IS A NOTORIOUSLY COMPLEX CONCEPT. IT MAY denote a system of time reckoning, which could be judged in terms of scientific precision and astronomical adequacy. It denotes a constellation of festivals, expressing or teaching social identities. It denotes an administrative and political instrument, determining coordination of long distance exchanges within an empire as well as local rhythms. It denotes a graphic form of representing time – and so on. The Roman calendar of the late republican and early imperial Roman periods has figured prominently in most of these perspectives and the histories narrated on their bases.[1] The reason is that the "Julian reform" of 45 BCE and the shape of the calendar developed during the Augustan semi-century (44 BCE–14 CE) were of enormous and far-reaching consequences. Large parts of the world across many states, cultures, and religions use this calendar, its ridiculously differing lengths of months (28 to 31 days) and its funny names for the months, referring to second- and third-rank Roman deities (Ianus, Maia?) and giving numbers that are out of tune with their position in the year. November for instance is not the ninth month of the year, even if many languages developed from Latin or in contact with it make this etymology transparent and the inappropriateness of the name widely felt. Evidently, despite all these deficits, the revolutionary transformation of the Roman calendar in the first century BCE produced something suitable for an empire, if not a world. This observation is my starting point.

And yet, despite the revolutionary character, the steps taken and modifications implemented were achieved in a basically conservative society, which

[1] See e.g. the very different perspectives of Samuel 1972; York 1986; Samuel 1988; Aveni 1991; Rüpke 1995; Richards 1998; Rüpke 2001; Stern 2001; Rüpke 2006, 2011b; Stern 2012; Feeney 2007.

was certainly not the most advanced in terms of astronomic competence at the time. The lasting impact of the political and military formation of the *Imperium Romanum* might help explain the long-term success – as well as account for the resistance in the form of counter-calendars.[2] It hardly accounts for the shape and the thoroughness of the short-term development. This paradox is the problem addressed in this chapter. In a careful analysis and reconstruction of the calendar reforms, I claim that the explanation of the paradox lies in a twofold contextualization of these reforms. First, in the late republican process of "rationalizing" religion[3] or even the creation of a "scientific" discourse on cultural practices, and second in the specifically Augustan[4] development of doubling religious, political, and certain other cultural practices in the form of transportable representations that could be communicated throughout the empire. Taken together, this enabled a massive though short-termed acceleration of religious change. Constructing time is a business that is as pervasive as it is precarious.

1 Caesar's Two Calendar Reforms

Gaius Iulius Caesar's revolution of the calendar of the city of Rome had two components, which were practically combined but neither logically nor institutionally dependent on each other: (1) the realignment of the traditional calendar with the seasons and (2) the improvement of this calendar to a length equal with the solar year.

The first component, the realignment of the civic calendar with the solar year, had to bring back spring festivals to spring or harvest festivals to autumn and the like. The instrument used was a series of "intercalations" through the latter part of the year 46 BCE.[5] After the ordinary intercalation during the end of February, two months of greater length were added between November and December. By the end of December, that year had reached the length of 445 days.[6] Although an astronomical norm was the driving force behind the determination of a new starting date for the next year (the first regular one), it evidently left no traces in the few contemporary references or the more extensive later accounts. Given the circumstances, an observation of the summer solstice with sufficient preparatory time before the actual intercalation might have been the

[2] See the ever broadening analyses of Stern 2001, 2002, 2012; Stern and Burnett 2014.

[3] On which see Rüpke 2012a.

[4] See in general features Galinsky 1996, 2007.

[5] Cicero, *Fam.* 6.14.2.

[6] Censorinus 20.8; Dio 43.26.1.

touchstone used for better aligning a correct winter solstice in the years to come and a corresponding (new style) January 1, 45 BCE. Adding two intercalatory months with an overall length of 67 days must have been a conscious decision.

Even if many ancient calendars were interested in being somehow in correspondence with the solar year for reasons of agricultural and other practicalities, the "new year days" of these calendars or their mechanism of verifying the correspondence of a precise astronomical event with a civic date varied widely. The determination of an exact point of correspondence between the civic and the solar year is an arbitrary decision. In Rome, the discourse on a basic conformity of the average civil year with the solar year had left traces from the early second century BCE onward. The *lex Acilia* of 191 BCE interfered with the practices of intercalation in an unknown manner but evidently led to reducing the gap between the civic and solar years, at least for a few decades.[7] By the 50s BCE, Cicero had asked for a "careful handling of the intercalation" to align the natural production of animals and fruits with the sequence of sacrifices.[8] Cicero added no further argument or indicator. The same holds true for Varro. Here, the treatment of the calendar in *On Latin Language*, Book 6, is relevant, a text probably composed around 47BCE, directly before the reform.[9] The natural year is defined as running from the winter solstice to summer and again to the winter solstice.[10] For the civil year, as far as the annual festivals of the gods are concerned, no relationship to this natural year is established. Nor is any equation with an astronomical event given for any of the festivals. Agricultural associations are not prominent either. In his Verrine orations of 70 BCE, Cicero judged any corrections of the calendar with reference to a fixed astronomical date a Greek practice.[11]

The strong impetus for the intercalation of 46 BCE, unusual as it is in many respects, must have come from such Greek sources, most probably also from Egyptian astronomers, who were credited with the technical background of the reformed length of the year elsewhere.[12] This is more than a technical

[7] Macrobius, *Sat.* 1.13.21; Warrior 1992; Rüpke 2011b: 68–69.

[8] Cicero, *Leg.* 2.29, referring back to 2.20.

[9] The dedication of the twenty-five books was announced to Cicero in 47 (Cicero, *Att.* 13.12.3) and realized in the latter's lifetime; the probable lack of a final revision (Ax 1995) would put the writing of an early book for a quick writer as Varro (not much) before that announcement.

[10] Varro, *Ling.* 6.8.

[11] Cicero, *Verr.* 2.2.129. Feeney 2007: 196–211 has thoroughly analyzed the lack of fixed points of reference and the total change in the period after the reform.

[12] See e.g. Dio 43.26.2, who names Alexandria.

detail. Unlike the ethnographic gaze of Cicero and the historiographical[13] and antiquarian approach of Varro, Caesar used a non-Roman order of knowledge to formulate binding norms. He naturalized, as it were, these foreign concepts by giving the ritual role necessary for implementing the correction to the time-honored figure of the *pontifex maximus,* who headed the priesthood responsible for intercalation, announcements regarding the calendar, and the determination of the position of movable festivals.

In contrast to the massive intervention of the first component, the second reform was rather cautious, as cautious as the far-ranging goal of producing a civil year of a length equal with the solar year would allow. Caesar was careful to tamper as little as possible with the religious structure of the calendar. This structure was interpreted as being built around the Ides, known to be the days of the full moon in ages past. To this purpose, the days in the first half of the month, starting from the Kalends, the first day of the month, were left untouched. Likewise, the distance from the Ides to the rituals of the second half remained unchanged. Where could those new days that were necessary to prolong the regular year from 355 to 365 days be placed – to reduce the length of the intercalatory period to the absolute minimum of one day every four years? The placement (and the marking of these "new" days as being added ones) in the *Fasti Praenestini* allows deciphering their rationale. These days were not simply added at the end of the months. They were rather placed in the tiny space between the last festival of a month and the last day of the month. Obviously, the very last day was interpreted as directly leading into the new month and the new Kalends, the latter of which would enable the sighting of the new crescent.[14] The cost widely felt was a change in the counting of all the days after the Ides, leading to some irritation.[15]

The most important element, and doubtlessly the primary aim of the reform, was reducing the length of the intercalatory period. The Roman calendar of the preceding centuries implied the intercalation of a whole "month," which replaced the last days of February and thus resulted in an addition of 22 or 23 days every second year.[16] As the history of the second and first centuries BCE demonstrated, this biannual intercalation was not felt as binding by the relevant authorities (who, at least from the previously

[13] See Rüpke 2014a.
[14] See Rüpke 2011b: 112–113.
[15] The extent of the long-term consequences has been exaggerated by a tradition of research represented e.g. by Radke 1990 or Feeney 2007: 156–160. Cf. e.g. the divided reaction of the population described in Macrobius, *Sat.* 1.10.2.
[16] The mechanism had been reconstructed by Michels 1967.

mentioned *lex Acilia* onward, were the pontiffs) and could not be relied on as a precise rhythm. Only a few years before the Julian reform, the popular politician Clodius had proposed a law on intercalation, the details of which are unknown to us. Given the enormous consequences in terms of office holding or payments of interests, the lack of any period of prior announcement of the substantial lengthening of the year through intercalation was a considerable nuisance. Caesar must have addressed this problem in his extraordinary intercalation, producing the only contemporary dating "before the intercalatory month(s)" in our sources: people must have been forewarned.[17]

We can exclude the idea that Caesar was only interested in reducing the power of the pontiffs. He was the supreme pontiff and had started to fill the college with candidates of his choice.[18] The minor irregularities in intercalations after the reform demonstrate that the college retained the right to decide on intercalation: its members did not necessarily read the details of the astronomical arguments accompanying the change. By all probability, it was they who misinterpreted the norm to intercalate "in every fifth year," that is, after four years, as an imperative to make four years without any intercalation follow on each intercalation.[19]

Taking the whole range of administrative reforms into account, Caesar's new calendar was part of a larger project of reforms aimed at improving public finances and administration on a grand scale. Concerns with specificities of urban problems at Rome were as much a part of this as the enlargement of the number of magistrates serving throughout the empire.[20] Evidently, the calendar reform was planned against the background of an ever more "global" outreach of Roman activities.[21] Sacha Stern has convincingly placed it among similar processes in other empires to set up comprehensive frameworks of civic time, even if I would regard the translatability of local and supra-regional calendars rather than the predictability of calendars as concomitant to the formation of empires.[22] And yet, the Roman calendar had to undergo further changes before it could fulfill such functions.

17 See Cicero, *Att.* 6.14.2.

18 See Rüpke 2008: 130 = Rüpke 2005: 135.

19 See Macrobius, *Sat.* 1.14.13–15; Rüpke 2011b: 116–117.

20 Suetonius, *Jul.* 40–42; Dio 43.25–26; Suetonius gives the pride of place in his narrative to the calendar reform, Dio to the reform of the magistrates, finally adding the calendar reform to his naming of new maximum periods of office.

21 Rüpke 1995: 371.

22 Stern 2012: 223, paralleling the new calendar and Caesar's dictatorship. The permanent dictatorship lying at the bottom of Stern's argument is, however, a development of a slightly later period and Caesar's plans for a long absence on an expedition to the East. Stern rightly

These will be analyzed later, as the epistemological and communicative preconditions for such changes receive a closer look.

2 Calendar and Knowledge about the Calendar

As far as we can see, the written form of the Roman calendar, the *fasti*, was probably an invention of the late fourth century BCE. This type of calendar served as an instrument for rationalizing the political and especially the juridical use of time in the city of Rome. The main information given by the *fasti* from the early third century BCE onward was on the suitability of days for certain types of legal action and public assemblies.[23] Probably as a result of the initiative of Quintus Ennius, most likely the first annalistic historian, in the early 160s BCE, these *fasti* were reinterpreted as a grid for historical memory by way of introducing references to the founding of temples and consular lists.[24] Here, a discourse not only about political and historical identity but also about religious practices started. The introduction of deities into Rome and the foundation of annual festivals and temples were all noted in the calendar. This was not mere chronological documentation but rather a way of thinking historically about religion. A similar interest could also be seen in the contemporary translation into Latin of Euhemeros of Messene's *Sacred History* and its narratives about the genesis of gods from beneficial historical figures. Against this background, the new form of the calendar had a much wider significance. In that form, an incipient history of religion was produced.[25] The history of the Roman calendar system formed part of this argument. The calendar became self-reflexive.

Statements about origins are claims of an interested party rather than neutral "findings." A historical narrative is a critical appraisal of tradition and an indication of divergent positions and their growing explicitness. Thus, developments in the time of the Julian reform are relevant for our understanding of the degree of reflexivity of cultural and religious practices. Processes of explicit reflection on religion at Rome have been pointed out before.[26] For knowledge about the calendar (and for the moment I deliberately use this vague term), contemporaries would above all

stresses the absence of any evidence for Caesar's explicit intentions with regard to the empire (224).

23 Rüpke 2011b: 44–67; Rüpke 2012a: 100–110.

24 Rüpke 2012a: 152–171.

25 Rüpke 2012b: 156.

26 Rüpke 2012a.

participate in a discourse about "the year," *de anno.*[27] It is a discourse about the properties of the "natural year," about astronomical cycles and the measurement of time.[28] Treatises with this title, for instance by Suetonius writing at the beginning of the second century CE, have not been preserved, but most of our knowledge about Roman calendars comes from passages under this title, ranging from a section in Varro's *On Latin Language* (6.8–11) of the first century BCE through Censorinus's *On the Birthday* (18–21) of the early third century CE to Macrobius's dialogue *Saturnalia* (1.12–16, leading to a treatment of Sol/Apollo in 1.17) of the early fifth century CE. Inherently, this type of discourse had a comparative component: the "year of the Romans" was different from "the year of the Greeks" or Egyptians. This type of treatment also included the subdivisions of the year; "on months" was the main interest of Greek treatises, given the variety of names and types of months in the world of the Greek luni-solar year.

What were the categories used by contemporary thinkers? In Marcus Terentius Varro's treatise on Latin language, information on the festivals and days of larger religious rituals follow under the heading of *vocabula dierum* (6.12–32), followed briefly by the "names of the months" (*mensium nomina,* 6.33–34). It is to this type of information that Macrobius assigns the activities of Caesar's "scribe" (probably a member of the pontiffs as *pontifex minor*) Marcus Flavius, who "drew a list of the single days," which made it possible to establish a stable structure of the months and the year.[29] In Varro's treatise on religion, the "Antiquities of divine things," a group of three books (8–10) was dedicated to holidays (*feriae*) and games (*ludi circenses* and *scaenicae*). In Varro (and this is confirmed by later treatments), there is a clear sense of the constructive character of such religious qualifications. Against the background of the natural character of the seasons *(naturale discrimen),* the institutionalization of certain days is openly expressed, be it for religious or for social reasons (*dierum . . . qui deorum causa, tum qui hominum sunt instituti*).[30] One fragment of these books on festivals talks about the institution of *feriae*, "holidays," by "decree of the pontiffs."[31] This is a self-aware interpretation of all religious practices (not of the divine realm as such!) as historical human institutions.[32]

[27] For an exhaustive list, see Degrassi 1963, 1: xxv–xxvi; briefly Rüpke 1994: 128; Rüpke 1997: 201.

[28] See Elias 1988 for this metaphor; see also Zerubavel 1982.

[29] Macrobius, *Sat.* 1.14.2.

[30] Varro, *Ling.* 6.12.

[31] Varro, *Ant. rer. div.* fr. 78 Cardauns (Bk. 8).

[32] Rüpke 2014a.

Of course, Varro's interpretation is a historiographical claim rather than a historical finding. Romans of the republican period had been very reluctant to regulate the sphere of the divine by laws.[33] At the end of the republic, however, certain intellectuals not only claimed to know about religion (and other spheres of culture) and its origins, but also started to attempt to formulate religious norms on such a basis. Certainly, this type of antiquarian knowledge was of importance in Hellenistic Athens and in late-republican Rome. Mastery of knowledge of the past qualified for legitimatizing recent changes as a return to tradition.[34] Systematization produced new knowledge and inspired new, supposedly "forgotten," practice. Reflecting on the very character of historical change as "institutionalization"[35] offered even more space for conscious religious politics in general and calendar politics in particular. Such an epistemological climate could call for open reform rather than veiled change. Varro mused about the religion he would have installed if he had not felt obliged to the antiquity of the Roman people.[36] His contemporary, the philosopher and statesman Marcus Tullius Cicero, considered radically different options of public religion, even if, as a good skeptic, he withheld any final judgment.[37] Nevertheless, it required the personal union of a *pontifex maximus* and a *dictator* as well as the ruthlessness of a Gaius Iulius Caesar to pursue and realize a reform that would basically last for more than two millennia. And yet, the Julian calendar reform was the only precondition for the diffusion and longevity of this calendar. The calendar of January 1, 45 BCE was far from the "Julian calendar" of the Roman Empire and its globalized Gregorian variant.

3 Emptying the Calendar

Caesar's employment of Egyptian knowledge is a commonplace of later accounts.[38] Given his special relationship to Alexandria, Cleopatra, and Egypt, this must have been an element of boasting and propaganda for the new calendar rather than a result of ancient investigative journalism. Egypt was pointed out as the only society possessing a stable year.[39] In fact, the close assimilation of the year of 365 days of the Nile valley to the solar year had

[33] See the review of laws on priesthoods by Rüpke 2005: 1617–1650.
[34] Thus Wallace-Hadrill 2008: 239 (on the calendar, see 239–248); Rüpke 2012a: 172–185.
[35] Stressed by Cancik 2008: 28; see also Cancik 2006 on the spatial dimension of the argument.
[36] Varro, *Ant. rer. div.* fr. 12 Cardauns (Bk. 1).
[37] Cicero, *Nat. d.* 3.95.
[38] E.g., Marcobius, *Sat.* 1.16.39: *ab Aegyptiis disciplinis hausit.*
[39] E.g., Macrobius *Sat.* 1.12.2; cf. 1.14.3.

inspired other imperial formations.[40] The combination of such a year with a Roman technique of intercalation, which had not been related to the actual lunations and was now reduced to a single day, allowed for maximal correspondence with the solar year.[41]

The technical perfection of a year of 365¼ days instead of the Egyptian practice of a year of 365 days (and radically different from it insofar as it included intercalation) engendered a further element of the reformed calendar – that is, the integration of dating practices according to heliacal risings and settings. In other words, Roman *fasti*, the civic calendar, could be combined or read in a synopsis with lists of activities dated by astronomical events. These lists were called *parapegmata* based on the use of movable nails to mark days on such objects.[42] The integration must have been intentional, given Caesar's personal interest in astronomy testified by his lost treatise "On stars," probably written in the 40s BCE. The reform act in itself seems to have involved the presentation of this or some similar treatise on the whole matter, a text (or a source for the text) used by Pliny the Elder in the eighteenth book of *Natural History* about a century later.[43]

The inclusion of the *parapegma* tradition necessarily points to a reflection on the use of calendars in the countryside. Pliny the Elder was writing in the third quarter of the first century CE and trying to give advice for Italy. Being aware of the dependency of such dates of risings and settings on the geographical latitude (apart from local impediments to sight), Pliny opts above all for Caesarean dates to establish an Italian formula.[44] Pliny's explicit reasoning suggests that this reflection was not part of Caesar's text or at least of his public presentation of the data. The many dates in the ensuing tradition that proved misleading for Italian locations point in the same direction. If this interpretation is valid, Caesar did not only aim at the Italian countryside but rather integrated a truly global element into the calendar. He thereby freely added Alexandrian data to a Roman calendar, a place more than ten degrees to the north.

[40] See Stern 2012: 224.

[41] Macrobius, *Sat.* 1.14.3.

[42] Feeney 2007: 198–201. On the genre and the technical implications, see Rehm 1941, 1949; Eriksson 1956; Wenskus 1990; Lehoux 2007; see also Ben-Dov 2014 and Leitz 1995.

[43] See Pliny the Elder, *Nat.* 18 and the *index auctorum* for the same book. The description is given by the tenth-century *Commentarium Bernense* on Lucanus (10.187), which, however, frequently contains valuable ancient material. In this case, it concurs with the description of Caesar's extensive communication with the Senate, described by Dio immediately after his description of the calendar reform (Dio 43.27.1). Cf. Macrobius, *Sat.* 1.16.39.

[44] Pliny the Elder, *Nat.* 18.214: *haec* erit *Italiae ratio.*

The success and importance of this element of the reform are demonstrated by its reception long before Pliny. Cicero polemicized against a dictator who extends his commands to the movements of the stars.[45] Ovid, in his late Augustan commentary on the Roman calendar regularly inserts references to risings or settings of constellations, even if he remains reluctant to develop longer narratives on them.[46] Even before, *fasti* inscribed on stone started to show astronomical references, for instance the *Fasti Venusini* noted the solstice for June 26.[47]

The concentration on *fasti* is, however, misleading (and has misled me in earlier analyses). The epigraphic form was an urban one and spread only slowly, with serious modifications, which we are not able to trace, as they were hardly ever realized in the form of lasting inscriptions.[48] The more relevant developments must have taken place in the form of manuscripts, rolls, and codices, preserved only in the classicizing forms of late ancient didactic collections.[49] The reaction of groups in Gaul in the form of the creation of the counter-calendar of Coligny[50] demonstrates *e negativo* the spread and presence of certain structural elements of the *fasti*. The calendar of the city of Rome had been stripped of its urban shape and had become the framework for the reform and recreation of local calendars all over the empire.[51]

4 *Feriae quod eo die*: Refilling for an Empire

We do not know which material form the Julian calendars took in many cities, particularly in Western Europe and North Africa. We can only make surmises about the local festivals present in such documents. But we know about one element that must have been present – festivals of the emperor. The most detailed document is a law from the Flavian township of Irni in the center of the Hispania Baetica, probably documenting a law from Domitian's time:

45 Plutarch, *Caes.* 59.3.
46 See Rüpke 1996.
47 *Inscr. It.* 13.2.59.
48 One of the tantalizing exceptions are the so-called *Fasti Guidizzolenses*; see Rüpke 1995: 160–164. A possible earlier step are the very fragmentary *Fasti Sorrinenses minores*, ibid. 145–148.
49 See ibid., 90–94 and 151–160 on the Chronograph of 354 and the text collection of Polemius Silvius in Gaul. For the Chronograph, see also Salzman 1990; Wischmeyer 2002; and Burgess 2012.
50 See Stern, Chapter 3, in this volume.
51 See the analyses of Stern 2012.

> Item *(rubrica)*. The days *(diebus)* on which matters may not be brought before the court, and the days on which decisions may not be given for the third day. – Whoever delivers justice in the municipium must ensure that no iudex or arbiter, or the recuperatores, makes legal argument in a private matter, nor must he give decisions for the third day, on such days as are already or will in the future be festivals, including those regarded as being held in honour of the imperial house (*quos dies propter venerationem domus Augustae festos feriarumve numero esse haberique o[p]ortet oportebit*), or on days on which games are held by the decuriones or the conscripti, or on which a banquet or distribution of food to the citizens is held, or a dinner is given at the expense of the citizens for the decuriones or conscripti, or on days when an assembly of the municipium takes place, or on such days as are decided under the terms of this law to be days on which no court decisions may be handed down owing to the grain or wine harvest, unless the iudex or arbiter or the recuperatores and those whose case is to be heard wish unanimously that it be heard at that time, and it is not a day that is a festival or regarded as such, or a day set aside for venerating the imperial house. No iudex, arbiter, or recuperator may give a decision in a private case on the days determined here, or decide a case, or give consideration to a verdict or opinion on those days, unless the iudex or arbiter or the recuperatores and those whose case is to be heard wish unanimously that it be heard at that time, and it is not a day that is a festival or regarded as such, or a day set aside for venerating the imperial house. Whatever is undertaken in breach of these dispositions cannot be recognized as legal and valid.[52]

It is the imperial festivals that are declared imperative for this place of Latin law – and probably in like manner throughout the empire. The terminology is imprecise. The term *feriae*, which had been the technical term for "holidays," implying restrictions in legal and some types of agricultural business, was modified and in the long run supplanted by the term *dies festus*. The latter had lost some of those associations, which the term *feriae* implied, but which were valid only for institutions of the city of Rome (and its surroundings). Outside of the precise context of the *fasti*, *feriae* became a loose term for a sequence of days without public business, particularly related to periods of harvesting.

In the urban tradition of the Republican period, *feriae* had been narrowly defined as the temporal property of a god. The analogy of spatial property is not explicit in ancient texts, but it helps to understand the problems of

[52] *Lex Irnit.* 10 C 25–51, probably chap. 92 of the original law. The text repeats formulations of chap. 90 (10 B 26–41). English translation based on González 1986: 198, quoted from Rüpke 2014b: 127.

boundary marking – it was better to avoid the succession of two *feriae* dedicated to different gods by an intermittent day – as a means of making the property relation visible by means of bans on practices on both "areas."[53] Evidently, the festive days referred to in the Law of Irni do not presuppose that all members of the imperial family are already consecrated, that is, declared gods. Again, it was the context of the late republic that started a process, totally reshaping the appearance of the Roman *fasti* as well as the Roman "calendar" as an imperial framework.[54]

A careful analysis of the extant epigraphic *fasti* in general and the well-preserved *Fasti fratrum Arvalium* in particular allows a reconstruction of a development of formulas and concepts that were carefully preserved in their wording. The *feriae* defined in Caesar's lifetime and for a short time afterward (42 BCE) were *feriae C. Caesaris honoris causa,* "holidays on behalf of the honor of Gaius Caesar." For instance, on August 2 for the victory of Munda in 45 BCE we read: "Holidays on behalf of the honor of Gaius Caesar after the defeat of Hispania citerior and because he had totally defeated the king Pharnax in Pontus."[55]

Technically, these *feriae* were not dedicated to Caesar as a deified being; otherwise, they would have been expressed by a plain dative construct, such as in the formula *feriae*. The concept of holidays *honoris causa* is taken from the ritual of *supplicationes*, an opening of all urban temples organized as a ritual of thanksgiving after victories.[56] Such gratitude is directed to the gods, as the wording betrays: *ut dis immortalibus habeatur honos,* "in order that the immortal gods should receive honor." At the same time, the thanksgiving to the gods is "in the name of the commander" (*nomine imperatoris*). It is he and the importance of his victory that is indicated and acknowledged by the length of the ritual, reaching from the traditional event of one day to events of two, three, eventually even twenty or fifty days.[57]

In the middle of the first century BCE, *honos* still refers to the theological principle, the divine addressee. For the calendar, a form of words was found that specified Caesar as the addressee of undoubtedly godlike honors, while circumventing the full deification implied by the dative locution *feriae alicui*. Even the classification of Caesar's birthday (*dies natalis*) as *feriae* in 42 BCE did

[53] See Rüpke 1995: 492–515.

[54] The following is fully developed in Rüpke 2011b: 126–130, extensively used here.

[55] *Inscr. It.* 13.2.31 (–35 for the following entries). *[Feriae C. Caesa]ris h(onoris) c(ausa) Hisp(ania) [citerior]e devicta [et quod in P]onto regem [Pharnace]m dev[i]cit,* Degrassi integrates here: *ex s(enatus) c(onsulto).*

[56] For a general treatment, see Rüpke 1990: 215–217; Naiden 2006; cf. Février 2009.

[57] Rüpke 1990: 216.

not address him as *divus Iulius*, but, once again – by the testimony of the *fasti Amiterni* – as *C. Caesar.*[58]

After the defeat of Sextus Pompeius, new measures were taken in 36 BCE to honor him by the declaration of special *feriae.*[59] However, either the Senate or Caesar's heir *C. Iulius divi Iuli filius Caesar* shrank from directly adopting the Caesarean model. The theological conceit now chosen for September 3 – *feriae et supplicationes ad omnia pulvinaria, q(uod) e(o) d(ie) Caesar August(us) in Sicilia vicit,*[60] according to the *Fasti fratrum Arvalium* – rested on the same conceptual link to the *supplicationes* already mentioned.[61] As such *supplicationes* were addressed to all "immortal gods," the *feriae* implicitly took part in this definition. Against this indeterminate background, the actual, personal reference of the occasion for celebration was brought all the more clearly to the fore.

Although the formula for September 3 is maintained in later *fasti* (those from Amiternum in particular), the experimental form subsequently petered out. The formula used for the celebration of the victory at Actium on September 2, probably in 30 BCE, is the one that later became the norm: *Feriae Imp. Caesaris h(onoris) c(ausa), quod eo die vicit Actium.*[62] The four letters *ex s(enatus) c(onsulto)* – "*feriae* by decision of the Senate" – were added as a correction in superscript over the beginning of *Imp.* Dio reports that the *feriae* decision had been arrived at on the day the news of the victory arrived; this would be a further analogy for the institution of the *supplicatio.*[63] The theological reference, which had been standard for "normal" holidays fell entirely out of use, the resulting gap being filled by the entirely different expression *ex senatus consulto*: procedural legitimization took the place of indications of a divine proprietor. Thus, the problematic concept of *feriae* of "all" the gods, suggested by the formula used in 36 BCE, was also set aside.[64]

58 Dio 47.18.5–6; *Inscr. It.* 13.2.189.

59 See also Fraschetti 1990: 91–93.

60 "Holidays and thanksgiving at all 'cushions' (*pars pro toto* for temples) because on that day Caesar Augustus has won in Sicilia."

61 See *Inscr. It.* 13.2.33 (*fasti fratrum Arvalium* 3. 9.).

62 Cf. ibid., August 1, *feriae* since 30 BCE: *F(eriae) ex s(enatus) c(onsulto), [q(uod) e(o) d(ie) Imp. Caesar rem pu]blic(am) tristiss(imo) p[e]riculo [libera]vit,* "Holiday decreed by the Senate, because on that day Imperator Caesar had liberated the Republic from most severe danger." Or September 23, *feriae* since 30 BCE: *F(eriae) ex s(enatus) c(onsulto), q(uod) e(o) d(ie) Imp. Caesar Aug(ustus) pont(ifex) ma[x(imus)] natus est,* "the Imperator Caesar Augustus, supreme pontiff, was born."

63 See Degrassi 1963: 505; Dio 51.19.1–2.

64 The *ex senatus consulto*, the reference to the Senate's decision, is absent from the *fasti Amiterni* for Caesar's *dies natalis*, celebrated on July 12, from Caesar's victories in Spain in 49 and over

It is for these very *feriae* of September 2 that we find in the *fasti* in the Arvals' sanctuary a peculiar hybrid formula, where the words *ex senatus consulto* were subsequently added to *honoris causa*. Considering not only the doubts that had led to the search for another solution as early as 36 BCE but also the "normal" form provided on August 1 for the *feriae* of 30 BCE, there is probably no need to interpret this as intimating that Augustus or like-minded individuals at first decided in favor of the old *honos* formula, and so to assume that the correction was added later, perhaps in 27 BCE. At the time when this calendar, perhaps the earliest monumental marble calendar at Rome, was created, Caesarean entries were still in the clear majority,[65] so it is not surprising that the stonemason spontaneously adopted the *honoris causa* formula; but, as a grave material error (and a serious embarrassment), this had to be corrected: the festivals associated with Octavian and Augustus were – in official parlance – held not "in his honor," but merely "because he accomplished this or that on this day."[66]

By the end of the reign of the first Augustus – admittedly a long period – the Senate had voted on around thirty new *feriae*. The long wording of new entries did determine the appearance of the calendar. The clear layout of columns of numbers and letters and the ample space between the columns of the months were sprinkled by heaps of small letters, all pointing out some detail of imperial genealogy, emperors' biographies, or their achievements. The monarchy took shape even beyond reading distance. Discounting the Ides, which were traditional holidays (usually given to Iuppiter on account of his relationship to the bright sky of the full moon), imperial festivals started to outnumber the old festive dates even in the city of Rome. For reception throughout the empire, the latter could easily be filtered away. For the urban connoisseur, even those were tinged with the Principate. Restoration of temples was frequently combined with a change in the dedication day noted in the *fasti*.[67]

5 Epilogue

For the development reconstructed here, the poetic commentary on the Roman *fasti* composed by Ovid for the first six months around 4 CE and

Pharnaces in 47, both celebrated on August 2, and from his victory at Pharsalus on August 9 (*Inscr. It.* 13.2.189–191); the *fasti Verulani* list Caesar's capture of Alexandria on March 27 with the short version (*Inscr. It.* 13.2.169).

[65] Degrassi 1963: 369.

[66] Against Fraschetti 1988.

[67] See Rüpke 2011b: 124.

partly revised in exile for dedication to Germanicus after the death of Augustus in 14 CE offers a good test case. No doubt, this is a comment on an Augustan, that is, imperial, copy of the *fasti*. The ritual and architectural activities on the Roman ground were represented in the ever more popular form of such *fasti*, populating not only individuals' lives but also public or associations' architectural space in luxurious creations of inscriptions or wall paintings. If this is a doubling of religion, Ovid's *Libri fastorum* was a reflection of that, a tripling of religious, of calendrical, and in particular of those imperial practices that created the center and hence the space of an empire.[68] The two consecutive dedications in the beginnings of Books 1 and 2 unanimously stress the massive presence of imperial personage, altars, "domestic festivals" (Book 1) or names, and "labels" (Book 2).

The role of the deified Iulius is limited in Ovid's poem, as has frequently been noted. Given his entanglement in the civil wars, out of which the Augustan monarchy grew, such praise was not fully justified. For the development of the Roman calendar, however, Caesar's absence is even more interesting. His presence as a reformer of the calendar is nearly negligible. This is far from self-evident. In his report on the reform early in the second century CE, Suetonius did not use the concept of "year" (*annus*), but had Caesar "correct the calendar" (*fastos correxit*).[69] In 238 CE, Censorinus stated that the reformed "year" is remembered down to his own day as "Julian years" (*Iuliani [scil. anni]*).[70] Ovid's contemporary Verrius Flaccus noted details of the reform in his monumental copy displayed at Praeneste, the so-called *Fasti Praenestini*, which integrated *fasti* and a commentary on *fasti* down to details of the length of the year (on February 23 or 24, the day of intercalation) and of the precise days being added by Caesar (e.g., on April 26).[71]

Ovid avoids every occasion for offering information on the new calendar. The Julian reform is lacking from the short introductory history of the calendar at the beginning of his Book 1 (1.27–44). He neither mentions days added by Caesar (e.g., at the end of January) nor comments on the mechanism of intercalation (the *bisextilis*) on the "sixth day before the Kalends of March." The only reference given is at the end of the lengthy introduction to the month of March. After the second mention of the

[68] Rüpke 2011a.
[69] Suetonius, *Jul.* 40.1.
[70] Censorinus 20.11.
[71] See *Inscr. It.* 13.2.119 (February) and 131 (April).

addition of two months by Numa before the original start of the year in March, Caesar is acknowledged as an astronomer and credited with the improved length of the year, given as 365 and a fifth (!) of a day. This is hardly a mere misunderstanding,[72] but rather it is a discrediting account of Augustus's predecessor, praised for his astronomical knowledge immediately before.[73] Changes in the calendar are credited to power and merit, not mere knowledge. Change is legitimized by tradition rather than research. In the Golden Age of Augustus, the role of antiquarianism and historiography was changing. Veiled change had replaced open reform as exemplified by the hotly debated Julian one again. The emperor was a first, a *princeps*, not a monarch. But such niceties and details were part of an urban discourse on time. Time in the empire developed more quickly as the many instances of "Julianification," as Sacha Stern called it, show, rendering local calendars accountable in Julian terms.

Bibliography

Aveni, A. 1991. *Rhythmen des Lebens: Eine Kulturgeschichte der Zeit*. Stuttgart: Klett-Cotta.

Ax, W. 1995. "Disputare in utramque partem: Zum literarischen Plan und zur dialektischen Methode Varros in de lingua Latina 8–10." *Rheinisches Museum für Philologie* 138: 146–177.

Ben-Dov, J. 2014. "A Jewish Parapegma? Reading 1 Enoch 82 in Roman Egypt." In S. Stern and C. Burnett (eds.), *Time, Astronomy, and Calendars in the Jewish Tradition*. Time, Astronomies, and Calendars 3. Leiden: Brill, 1–25.

Bömer, F. 1957. *P. Ovidius Naso, Die Fasten 1: Einleitung, Text und Übersetzung*. Heidelberg: Winter.

Burgess, R. W. 2012. "The Chronograph of 354: Its Manuscripts, Contents and History." *Journal of Late Antiquity* 5/2: 345–396.

Cancik, H. 2006. "'Götter einführen': Ein myth-historisches Modell für die Diffusion von Religion in Vergils Aeneis." In D. Elm von der Osten et al. (eds.), *Texte als Medium und Reflexion von Religion im römischen Reich*. Stuttgart: Franz Steiner, 31–40.

Cancik, H. 2008. *Religionsgeschichten: Römer, Juden und Christen im römischen Reich. Gesammelte Aufsätze 2*. Tübingen: Mohr Siebeck.

Degrassi, A. (ed.). 1963. *Inscriptiones Italiae 13: Fasti et elogia. Fasciculus 2: Fasti anni Numani et Iuliani, accedunt ferialia, menologia rustica, parapegmata*. Rome: Libreria dello stato.

Elias, N. 1988. *Über die Zeit: Arbeiten zur Wissenssoziologie 2*. Frankfurt a. M.: Suhrkamp.

Eriksson, S. 1956. *Wochentagsgötter, Mond und Tierkreis: Laienastrologie in der römischen Kaiserzeit*. Studia Graeca et Latina Gothoburgensia 3. Stockholm: Almquist & Wiksell.

[72] Thus e.g. Bömer 1957: 16.

[73] Ovid, *Fast*. 3.155–166. *Iuncta* in part of the manuscripts (3.164) must be regarded as an attempt at rescuing Ovid; *quinta* is *lectio difficilior*, ascertained by *lustrum*, a five-year period, in the following line.

Feeney, D. 2007. *Caesar's Calendar: Ancient Time and the Beginnings of History*. Sather Classical Lectures 65. Berkeley: University of California Press.
Février, C. 2009. *Supplicare deis. La supplication expiatoire à Rome*. Recherches sur les rhétoriques religieuses 10. Turnhout: Brepols.
Fraschetti, A. 1988. "Commemorare il principe: Ricerche preliminari sui calendari di epoca augusteo-Tiberiana." In P. Gignoux (ed.), *La commémoration: Colloque du centenaire de la Section des Sciences Religieuses de l'École Pratique des Hautes Études*. Leuven: Peeters, 115–140.
Fraschetti, A. 1990. "'Cognata numina': Culti della città e culti della famiglia del principe in epoca augustea." In J. Andreau and H. Bruhns (eds.), *Parenté et stratégies familiales dans l'antiquité romaines*. Rome: École française, 85–119.
Galinsky, K. 1996. *Augustan Culture: An Interpretive Introduction*. Princeton: Princeton University Press.
Galinsky, K. 2007. "Continuity and Change: Religion in the Augustan Semi-Century." In J. Rüpke (ed.), *A Companion to the Roman Religion*. Malden, MA: Blackwell, 71–82.
González, J. 1986, "The Lex Irnitana: A New Copy of the Flavian Municipal Law." *Journal of Roman Studies* 76: 147–243.
Lehoux, D. 2007. *Astronomy, Weather, and Calendars in the Ancient World: Parapegmata and Related Texts in Classical and Near-Eastern Societies*. Cambridge: Cambridge University Press.
Leitz, C. 1995. *Altägyptische Sternuhren*. Orientalia Lovaniensia analecta 62. Leuven: Peeters.
Michels, A. K. 1967. *The Calendar of the Roman Republic*. Princeton: Princeton University Press.
Naiden, F. S. 2006. *Ancient Supplication*. Oxford: Oxford University Press.
Radke, G. 1990. *Fasti Romani: Betrachtungen zur Frühgeschichte des römischen Kalenders*. Orbis antiquus 31. Münster: Aschendorff.
Rehm, A. 1941. *Parapegmastudien: Mit einem Anhang Euktemon und das Buch De signis*. Abhandlungen Bayerische Akademie der Wissenschaften, phil.-hist. Klasse NS 19. Munich: Bayerische Akademie.
Rehm, A. 1949. "Parapegma." *Paulys Real-Encyclopädie der classischen Altertumswissenschaft* 18/4: 1295–1366.
Richards, E. G. 1998. *Mapping Time: The Calendar and Its History*. Oxford: Oxford University Press.
Rüpke, J. 1990. *Domi militiae: Die religiöse Konstruktion des Krieges in Rom*. Stuttgart: Steiner.
Rüpke, J. 1994. "Ovids Kalenderkommentar: Zur Gattung der libri fastorum." *Antike und Abendland* 40: 125–136.
Rüpke, J. 1995. *Kalender und Öffentlichkeit: Die Geschichte der Repräsentation und religiösen Qualifikation von Zeit in Rom*. Religionsgeschichtliche Versuche und Vorarbeiten 40. Berlin: de Gruyter.
Rüpke, J. 1996. "Quis vetat et stellas . . . ? Les levers des étoiles et la tradition calendaire chez Ovide." In B. Bakhouche et al. (eds.), *Les astres 1: Les astres et les mythes, la description du ciel*. Publications de la Recherche Université Paul Valéry. Montpellier: Séminaire d'Étude des Mentalités Antiques, 293–306.
Rüpke, J. 1997. "Kognitive Einheit ritueller Sequenzen? Zur kommunikativen Funktion kalendarischer Gattungen in Rom." In: G. Binder and K. Ehlich (eds.), *Religiöse*

Kommunikation: Formen und Praxis vor der Neuzeit. Stätten und Formen der Kommunikation im Altertum 6. Trier: Wissenschaftlicher Verlag Trier, 191–223.
Rüpke, J. 2001. "Ein neues Jahrtausend und noch immer der alte Kalender: Antike Konstanten in der europäischen Zeitrechnung." *Gymnasium* 108: 419–438.
Rüpke, J. 2005. *Fasti sacerdotum: Die Mitglieder der Priesterschaften und das sakrale Funktionspersonal römischer, griechischer, orientalischer und jüdisch-christlicher Kulte in der Stadt Rom von 300 v. Chr. bis 499 n. Chr.* Potsdamer altertumswissenschaftliche Beiträge 12/1–3. Stuttgart: Steiner.
Rüpke, J. 2006. *Zeit und Fest. Eine Kulturgeschichte des Kalenders*. München: Beck.
Rüpke, J. 2008. *Fasti sacerdotum: A Prosopography of Pagan, Jewish, and Christian Religious Officials in the City of Rome, 300 BC to AD 499*. Oxford: Oxford University Press.
Rüpke, J. 2011a. "Reichsreligion? Überlegungen zur Religionsgeschichte des antiken Mittelmeerraums in römischer Zeit." *Historische Zeitschrift* 292: 297–322.
Rüpke, J. 2011b. *The Roman Calendar from Numa to Constantine: Time, History and the Fasti*. Malden, MA: Wiley-Blackwell.
Rüpke, J. 2012a. *Religion in Republican Rome: Rationalization and Ritual Change*. Philadelphia: University of Pennsylvania Press.
Rüpke, J. 2012b. *Religiöse Erinnerungskulturen: Formen der Geschichtsschreibung in der römischen Antike*. Darmstadt: Wissenschaftliche Buchgesellschaft.
Rüpke, J. 2014a. "Historicizing Religion: Varro's Antiquitates and History of Religion in the Late Roman Republic." *History of Religions* 53/3: 246–268.
Rüpke, J. 2014b. *From Jupiter to Christ: The History of Religion in the Roman Imperial Period*. Oxford: Oxford University Press.
Salzman, M. R. 1990. *On Roman Time: The Codex-Calendar of 354 and the Rhythms of Urban Life in Late Antiquity*. The Transformation of the Classical Heritage 17. Berkeley: University of California Press.
Samuel, A. E. 1972. *Greek and Roman Chronology: Calendars and Years in Classical Antiquity*. Handbuch der Altertumswissenschaft 1,7. München: Beck.
Samuel, A. E. 1988. "Calendars and Time-Telling." In M. Grant and R. Kitzinger (eds.), *Civilization of the Ancient Mediterranean: Greece and Rome 1*. New York: Scribner, 389–395.
Stern, S. 2001. *Calendar and Community: A History of the Jewish Calendar, 2nd Century BCE–10th Century CE*. Oxford: Oxford University Press.
Stern, S. 2002. "Jewish Calendar Reckoning in the Graeco-Roman Cities." In Bartlett, J. R. (ed.), *Jews in the Hellenistic and Roman Cities*. London: Routledge, 107–116.
Stern, S. 2012. *Calendars in Antiquity: Empires, States, and Societies*. Oxford: Oxford University Press.
Stern, S. and C. Burnett (eds.). 2014. *Time, Astronomy, and Calendars in the Jewish Tradition*. Time, Astronomies, and Calendars 3. Leiden: Brill.
Wallace-Hadrill, A. 2008. *Rome's Cultural Revolution*. Cambridge: Cambridge University Press.
Warrior, V. M. 1992. "Intercalation and the Action of M. Acilius Glabrio (cos. 191 B. C.)." *Studies in Latin Literature and Roman History* 6: 119–144.
Wenskus, O. 1990. *Astronomische Zeitangaben von Homer bis Theophrast*. Hermes Einzelschriften 55. Stuttgart: Steiner.

Wischmeyer, W. 2002. "Die christlichen Texte im sogenannten Filocalus-Kalender." In A. Primmer, K. Smolak, and D. Weber (eds.), *Textsorten und Textkritik: Tagungsbeiträge*. Wien: Akademie, 45–67.
York, M. 1986. *The Roman Festival Calendar of Numa Pompilius*. American University Studies 17 (Classical Languages and Literature), 2. New York: Lang.
Zerubavel, E. 1982. "The Standardization of Time: A Sociohistorical Perspective." *American Journal of Sociology* 88: 1–23.

5

REAL AND CONSTRUCTED TIME IN BABYLONIAN ASTRAL MEDICINE

JOHN STEELE

ASTRAL MEDICINE REFERS TO THE INTERACTION OF ASTRONOMY and astrology with medicine and has a long history that can be traced from early antiquity to the eighteenth century.[1] In the Ancient Near East, references to astral influence on medical practice can be found as early as the Old Babylonian period (early second millennium BCE) within "prescriptive" texts that describe the production of medical remedies for treating a patient's ailment. As part of the preparation of the remedy, often the ingredients once mixed are to be left to spend the night under the stars.[2] For example, a Hittite text from Hattuša states, "you mix (the ingredients) in first-quality beer, you let it spend the night under the stars, in the morning you strain the first-quality beer and you give it to him to drink."[3] As pointed out by Erica Reiner, this process of "spending the night under the stars" refers not just to leaving the ingredients time to be absorbed into the carrier liquid (usually, as in this case, beer), but to the activation of the remedy by the light of the stars.[4] Some prescriptions state that the mixed ingredients should by placed in front of the Goat star. This star is associated with the goddess Gula, who is the goddess of healing.[5]

Several new forms of astral medicine appear in texts from the first millennium BCE including texts that associate particular illnesses with the signs of the zodiac,[6] texts that prescribe the preparation or application of remedies on specific dates of astronomical significance, and texts in which the ingredients

[1] For an overview of the history of astral medicine in the west, see Greenbaum 2015.
[2] Reiner 1995: 49.
[3] KUB 37 55 iv 15; trans. Reiner 1995: 49.
[4] Reiner 1985: 594.
[5] Reiner 1995: 52.
[6] E.g., BM 34731 edited and discussed by Leibovici 1956; Heeßel 2008: 8–9; Geller 2010b: 68–71.

that are to be used to make a remedy are determined by astrological schemes. Time and the calendar play important roles in both of the last two groups of texts, and it is these texts that I focus on in this chapter. I argue that time is used in two ways within these texts: first, time may be used in the straightforward sense of the duration of an illness or the specific days on which medicinal ingredients are harvested or healing rituals are performed; second, time may provide the basis from which the method of treating an individual is constructed.

1 Babylonian Calendars

Before discussing the role of time and the calendar in medicine, it is useful to briefly outline the basic features of Babylonian calendars.[7] Throughout Babylonian history, the civil calendar employed lunar months that began either on the evening of the first visibility of the new moon crescent or, if the moon could not be seen (e.g., because of bad weather), thirty days after the beginning of the previous month. Thus, months contained either twenty-nine or thirty days. A normal year contained twelve months; but to keep the calendar roughly in line with the seasons, an intercalary month was added in approximately every three years. Before about the fifth century BCE, intercalary months were added whenever they were deemed necessary according to royal decree; in later centuries, a cycle of seven intercalations in every nineteen years was followed.[8] A year, therefore, could contain either twelve lunar months, totaling about 354 days, or thirteen months, totaling about 384 days.

The variable length of the lunar month and the uncertainty over whether a year would contain twelve or thirteen months led to the use of a simplified schematic calendar in many administrative contexts. In this calendar, each month was taken to be thirty days in length and each year to contain twelve months, making a total of 360 days in a year. In addition to its use in administration, the schematic 360-day year was used in certain astronomical calculations, and it also appears in the account of the assignment of the sun and moon to the roles of defining the year in the Babylonian creation myth *Enūma Eliš*.[9] The 360-day schematic calendar never replaced the luni-solar

[7] For a fuller treatment, see Steele 2011a.

[8] Britton 2007, but see Stern 2012: 94–123.

[9] Brown 2000 has argued that the 360-day year is an "ideal" year, illustrating how the universe should behave, discrepancies from which (e.g., twenty-nine day months or intercalary years) could be interpreted as bad omens.

calendar in civil use, however. Its use was only to simplify calculation, the results of which could then be mapped onto the civil luni-solar calendar.

2 Time and Diseases

People can become sick at any time. Nevertheless, many illnesses occur frequently at certain times of the year, the most obvious examples being the common cold and influenza, both of which are more prevalent during the winter, and foodborne illnesses that are more frequent in the summer, as the bacteria that cause food poisoning thrive in the warm weather. Some seasonal illnesses are not directly related to the change in weather over the course of the year but rather to periods of time when people gather together, especially if those people come from different locations – witness the high occurrence of illnesses among students in the second and third week of term caused by exposure to viruses and bugs from different part of the country for which the carriers have built up a resistance.

Seasonal illnesses, designated as MU.AN.NA "(of) the year," are referred to in a small number of cuneiform texts.[10] They include the epidemic disease *di'u*, which may be malaria, referred to as a seasonal disease in two Neo-Assyrian letters from scholars to the king.[11] The same disease appears alongside two other seasonal diseases in a short list found in a *tamītu* tablet, both again designated as "(of) the year."[12]

The hemerology usually known in modern scholarship as the "Babylonian Almanac" mentions occasionally the possibility of catching diseases on certain days of the year. This text is preserved with some slight variations in more than sixty copies from sites across Mesopotamia.[13] From the copies of the Almanac found in the Sippar library, we read that on the ninth of Month II, you should not eat fish, and on the thirtieth of Month V, you should not eat the meat of a pig, as in both cases you may be seized by an illness.[14]

The length of time for which a patient suffers from a disease and the number of days after which particular symptoms manifest are mentioned in the medical diagnostic series SA.GIG. Indeed, a whole tablet of the series, tablet 16, is concerned with the duration of illnesses, and other examples are scattered throughout the series. For example, an entry on tablet 16 reads: "If on the third day of his illness his skin is healthy (but) the illness returns, he

[10] Stol 2007: 15–16.
[11] SAA X 236 and SAA VIII 1.
[12] Lambert 2007, no. 1: 245–246.
[13] A classification of the hemerologies may be found in Livingstone 2010.
[14] Isma'el and George 2002.

will die."[15] In this case, the length of time is part of the symptom. Apart from one day, a period of three days is the most commonly cited time frame. Less frequently we find periods of five days, six days, seven days, ten days, and thirty-one days. Other numbers of days appear only occasionally. Time can also appear in the prognosis. For example, we find entries such as "If ditto (sunheat, mentioned earlier) (and) blood in his nose has begun to flow on that same day, that man will be ill twenty-five days."[16] In both of these cases, it is a period of time that is mentioned, rather than a specific calendar date or range of dates.

3 Time in Medical Treatment

The calendar could also play a role in medical treatment, both in the timing of treatment and in determining what the treatment should be. The Babylonian Almanac, for example, warns that on the seventeenth of Month I, a doctor (*asû*) should not treat a patient;[17] it is quite possible that some of the general warnings concerning favorable and unfavorable days in the Almanac may also have been considered when deciding about the treatment of a sick person.

Other texts refer to specific times at which medical remedies should be prepared or administered and rituals performed. Instructions for the preparation of medical remedies made from herbs and roots sometimes state that the plant should be harvested at night.[18] Some texts specify that herbs should be gathered on moonless nights, that is, on the nights around conjunction. For example, a fragmentary text prescribes, "on a day when the moon disappears from the sky . . . you pull up? [. . .] the stars must not see (it?), on the 29th day [. . .] hair from his head."[19] A similar reference to the day of invisibility of the moon appears in a discussion of a childhood illness in a commentary to the diagnostic medical handbook SA.GIG that reads: "If he (Bēl-ūri) falls upon him in the third year, he may remain for a long time at the head of his bed. That he should not so remain, a *šakirû*-plant which has been uprooted on the thirtieth day you shall bray, mix with river water and anoint him regularly therewith."[20] The text continues with more or less the same prescription for if the child falls sick to this illness in his seventh year, this time offering as an

[15] SA.GIG 16 41'; ed. Heeßel 2000: 176.
[16] SA.GIG 21 9; translation adapted from Stol 2007: 29.
[17] Casaburi 2003: 33.
[18] Reiner 1995: 36–37.
[19] Reiner 1995: 135.
[20] Kinnier Wilson 2007: 63.

alternative to picking the herb on the thirtieth day the possibility to pick it on the day of an eclipse.

A text describing a ritual to be performed by an *āšipu* ("ritual expert," sometimes translated as "exorcist") to cure a particular sickness caused by a ghost says that the ritual should be performed "On the fifteenth, the day when Sîn and Šamaš stand together."[21] Sîn and Šamaš are the moon and the sun god, respectively. During the ritual, the patient is told to sit facing the north with the moon to the left in the west and the sun in the east, respectively setting and rising as daylight begins.

The limited evidence we have for the timing of healing activities seems to indicate that the days of opposition, when the moon is full and the moon and sun are opposite each other in the sky, and conjunction, when the moon cannot be seen, were considered of particular significance within the Babylonian medical tradition. Parallels may be drawn here with the broader ritual tradition in Babylonia, which places particular emphasis on the day of the full moon, said to be the fifteenth day, and the day of the disappearance of the moon, said to be the thirtieth day, along with other days such as the first and the seventh days, which are not (to my knowledge) attested in medical texts.[22] In both cases, the days associated with opposition and conjunction are taken from the schematic 360-day calendar. It remains an open question whether when medicine and rituals were performed in practice, they were always undertaken on the fifteenth and thirtieth days or whether the true days of opposition or conjunction were used if these differed.

Time could also be used directly in the treatment of a patient by determining what medical remedy should be used. For example, among a large archive of medical tablets, which probably came from Achaemenid-period Sippar, several tablets contain copies of lists of certain stones in oil that are to be applied to parts of the body, plants that are to be drunk in certain liquids, and certain colored wools to be tied to a part of the body as an amulet. The lists are given for each month of the year.[23] Two further tablets from the same archive contain copies of a text that lists a stone, a tree, and a plant for each day of the month.[24]

Several texts refer to the use of "stone, plant and wood" in the treatment of a sick patient.[25] For example, BM 34035, a mystical explanatory work, explains, "When you perform plant, stone and wood and the art of the

[21] Scurlock 2006: no. 91 5.
[22] On the use of time in rituals, see Livingstone 1999.
[23] Finkel 2000: no. 55.
[24] Finkel 2000: no. 56.
[25] Heeßel 2005. As Heeßel notes, the order stone-plant-wood is not standardized.

exorcist for a sick man – one performs (it) with its commentary."[26] This explanation indicates that the "stone, plant and wood" texts are to be used in the treatment of sick patients by the *āšipu*.[27]

The Late Babylonian astrological text LBAT 1593 explains how the "stone, plant and wood" are to be used: "the animals (*ú-ma-mu*) of 13 and 4,37 you take one with the other, you salve, feed, and fumigate the patient with the stone, plant and wood, the handbook (*bi-ib-lu*) of Month I from the 1st to the 30th" (see presently for the meaning of the numbers 13 and 4,37).[28] This text indicates that the stone, plant, and wood are to be used as ingredients for making a medical remedy that will be applied to the patient either by application to the skin or by ingesting the remedy orally or by inhalation.[29] What is interesting about the use of the "stone, plant and wood" material as described in this text, however, is that the ingredients for the remedy are not determined by the patient's symptoms but rather by the day of the month (in this case of Month I) on which the treatment is being made. A modern parallel might be holistic systems of medicine that treat the overall well-being of the patient, strengthening the patient's ability to fight off the disease, rather than focusing on treatment of the disease itself.

The "handbook" texts referred to in LBAT 1593 can be identified as the so-called *Kalendertext* tablets. Weidner 1967 published two more or less complete *Kalendertext* tablets, and several further examples have been identified and published since then.[30] The basic structure of these tablets is as follows: each line begins with a series of four numbers. The third and fourth numbers represent the month and the day and increase by one day each line. The first two numbers give a position in the zodiac in signs and degrees (the first number being the sign, the second the degree). This position increases by 9 signs plus 7 degrees each line; 9 signs plus 7 degrees equals 277 degrees or 4,37 degrees written sexagesimally. A short piece of text follows the four numbers. This text usually names certain things that are correlated with the numbers at the beginning of the line. Among these may be the names of cities, temples, animals or parts of animals, stones, plants, and different kinds of wood. The inclusion of stones, plants, and woods suggests a connection of some of these *Kalendertext* tablets with medicine; some other examples refer to "anointing," again suggesting that they are connected with medicine.

26 Livingstone 1986: 73.
27 On the *āšipu* and the *asû*, see e.g. Geller 2010a: 43–55.
28 Translation adapted from Reiner 2000: 424.
29 On the preparation of medical ingredients generally, see Böck 2011.
30 For a summary of known *Kalendertext* tablets, see Brack-Bernsen and Steele 2004.

The number 277 = 4,37, which is characteristic of the *Kalendertext* scheme, is mentioned in the passage in LBAT 1593 alongside the number 13. This latter number is characteristic of a second scheme in which the position in the zodiac increases by thirteen degrees per day. This scheme is customarily referred to as the *Dodecatemoria* scheme after its name in Greek astrology. The *Dodecatemoria* scheme is known from several Babylonian sources, including a text that lists entries from the scheme alongside entries from the *Kalendertext* scheme.[31]

The *Dodecatemoria* and *Kalendertext* schemes operate with the schematic 360-day calendar. In the schematic calendar, the mean motion of the sun is equal to one degree per day. Operating purely with mean motions, therefore, dates in the schematic calendar correspond directly with the position of the sun in the zodiac. It is easy to demonstrate that the *Dodecatemoria* scheme corresponds to the mean motion of the moon. Again, operating only with mean motions, in one month of 30 days the sun will have moved 30 degrees along the ecliptic. In the corresponding time, the moon will have traveled one complete circuit of the zodiac to get back to its starting point, plus another 30 degrees to catch up to the sun again. The total distance travelled by the moon in one month is therefore 360 + 30 degrees = 390 degrees. Thus the daily motion of the moon will be equal to 390 degrees divided by 30 days, which equals 13 degrees per day. The *Dodecatemoria* scheme assumes that we start with day zero (or day 360) of the year with the sun and moon at conjunction at a position of zero degrees (or 360 degrees) in the zodiac. Therefore, on day one, the sun will be at one degree in Aries (the first sign of the zodiac), and the moon will be at thirteen degrees in Aries. On day two, the sun will be at two degrees in Aries and the moon will be at twenty-six degrees in Aries, and so on.

The significance of the number 277, characteristic of the *Kalendertext* scheme, is less immediately obvious than that of the number 13 of the *Dodecatemoria* scheme and has only recently been discovered. Among the erroneous suggestions that have been made in the past is that 277 days is an approximation to the length of pregnancy in humans.[32] Lis Brack-Bernsen reconstructed the true origin of the *Kalendertext* scheme.[33] She showed that it is derived by a mathematical manipulation of the *Dodecatemoria* scheme. If we imagine the *Dodecatemoria* as a table of numbers with the headings sign – degree – month – day, and then swap the column headings sign and degree

[31] Hunger 1996.
[32] Reiner 1995: 115.
[33] Brack-Bernsen and Steele 2004.

for month and day, and then reorder the table by the dates in the new month and day columns, the result is the *Kalendertext* scheme. Thus, the *Kalendertext* scheme is a mathematical construction of the *Dodecatemoria* scheme. Recall that the *Dodecatemoria* scheme represents the mean position of the moon according to the schematic calendar. The *Kalendertext* scheme is therefore a mathematical construction from a schematic astronomical model to produce a calendrical-astrological scheme.

The *Kalendertext* tablets relating to medicine, that is, those that contain the lists of stones, plants, and woods, and those that refer to anointing, relate those entries to the numbers generated by the *Kalendertext* scheme. What this means is that the ingredients that are to be used to make a medical remedy are determined through a mathematical and astronomical construction from the date in the schematic calendar.

The well-known scribe Iqiša owned two particularly interesting *Kalendertext* tablets from Uruk. He was active at the end of the fourth century BCE and identified himself as a *mašmaššu*.[34] Iqiša owned a sizable collection of scholarly tablets including copies of tablets from the main omen series, lexical texts, astronomical and astrological texts, medical texts, and more than twenty commentary texts.[35] Iqiša's medical texts include copies of tablets 9 and 16 of SA.GIG. Among the astronomical and astrological tablets are several otherwise unattested astrological compositions, which incorporate the zodiac into existing traditions, such as the prediction of the rise and fall of the market, and which suggest that Iqiša himself may have been responsible for some of these innovations, or at the very least, that he was at the forefront of astrological practice of the time. Iqiša's two *Kalendertext* tablets (surely originally a series of twelve, one for each month) are also, as we shall see, innovative in their content.

Iqiša's two *Kalendertext* tablets are slightly different from most tablets of this kind in that they give the names of the months and the signs of the zodiac rather than the corresponding number from one to twelve.[36] Following the

[34] Hunger 1971.

[35] For a summary of tablets attributed to Iqiša, see Clancier 2009: 53. On Iqiša's commentary texts, see Frahm 2011: 292–296.

[36] Interestingly, these two *Kalendertext* tablets (SpTU III 104 and 105) replace the name of the first sign of the zodiac LU "Aries" with the name of the corresponding first month of the year BAR (not MAŠ as read by von Weiher); SpTU III 104 replaces the fourth sign of the zodiac ALLA "Cancer" with ŠU "Month IV"; SpTU III 105 replaces the fifth sign of the zodiac UR.A "Leo" with IZI "Month V" in the first line (see Reiner 1995: 115). These substitutions highlight the equivalence of the schematic calendar and the division of the zodiac into 360 degrees: twelve signs of the zodiac correspond to the twelve months of the year, and each sign of the zodiac is divided into thirty degrees corresponding to the

Kalendertext data, each line lists various items that are to be used to anoint someone. Generally, these items are the blood, fat, and (where appropriate) hair of an animal. The accompanying table is a schematic translation of the first twelve lines of the *Kalendertext* for Month IV.

Month IV	1	Aries	7	Sheep-blood, sheep-fat, and sheep-hair, you anoint.
Ditto	2	Capricorn	14	Goat-blood, goat-fat, and goat-hair, you anoint
Ditto	3	Libra	21	"Empty place," you anoint.
Ditto	4	Cancer	28	Crab-blood or crab-fat, ditto.
Ditto	5	Taurus	5	Bull-blood or bull-fat or bull-hair, ditto.
Ditto	6	Aquarius	12	Eagle-head, feathers, and blood, ditto.
Ditto	7	Scorpio	19	"Empty place," ditto.
Ditto	8	Leo	26	Lion-blood, lion-fat, or lion-hair, ditto.
Ditto	9	Gemini	3	Rooster-head, blood, and feathers, ditto.
Ditto	10	Pisces	10	Dove-head, blood, swallow-head, blood, ditto.
Ditto	11	Sagittarius	17	Anzu(-bird?)-head, Anzu(-bird?)-feathers, Anzu (-bird?)-blood, ditto.
Ditto	12	Virgo	24	*šigušu*-barley-flour, raven-head, and raven-feather, ditto.

For the remainder of the month, the text accompanying the *Kalendertext* data repeats with the sign of the zodiac given by the scheme. In other words, the sign of the zodiac generated by the *Kalendertext* scheme determines the animal that provides the source of the material with which the patient will be anointed; this rule is confirmed by the second of Iqiša's preserved *Kalendertexte*, where the same ingredients are listed for the same signs of the zodiac. The *Kalendertext* scheme therefore provides a mechanism for determining the ingredients that are to be used to make a medical remedy based on the date. Furthermore, in most cases the ingredients are directly related to the sign of the zodiac.[37] For example, in Cancer, the ingredients are taken from a crab, in Leo from a lion, and in Aries from a sheep. In other cases, especially where the name of the zodiacal sign is not an animal, a nearby animal constellation is used instead. For example, the zodiacal sign Taurus, named

division of the month into thirty days (see e.g. Steele [2007: 303]). Similar, although more systematic in extending to each sign, use of month names to represent zodiacal signs is found in other astronomical and astrological tablets, e.g. BM 36609+ (published by Roughton, Steele, and Walker 2004) and BM 36303+ (published by Steele 2015).

37 See Reiner 1995: 116–117.

after "The Stars" (i.e., the Pleiades), also contains the constellation of the "Bull of Heaven," and the ingredients are taken from a bull. The meaning of KI.KAL-*tim*, literally "empty/uncultivated place," is as yet unexplained.[38]

The direct association of the ingredient used to make the remedy with which to anoint the patient with the animal represented by the zodiacal sign (or a nearby constellation) is in effect another layer of construction based on time: a position in the zodiac is constructed for every date in the year through the astrological scheme, and the determination of the ingredients to be used in making the medical remedy is constructed from that zodiacal position.

There may be yet one further level of construction. Iqiša's *Kalendertext* tablets suggest that products made from the blood, fat, and hair of animals are used in the treatment of sick patients. Bearing in mind that these animals include a lion and an eagle, we must ask how likely it is that remedies made from parts of these animals were used in practice. I suggest that it is not very likely – perhaps the king or another important individual could pay to have a lion slaughtered to provide the ingredients for a medical remedy, but surely most patients would not be able to. One possibility is simply that Iqiša's *Kalendertext* tablets were never intended to be used in practice, that they were purely a scholarly invention. But I would like to suggest an alternative.[39] There exists in Babylonian medical texts a tradition of *Decknamen*: "secret names" (or, perhaps better, "nicknames") for medical ingredients.[40] Often the

[38] A possible connection can be made between the entry for Scorpio with "empty place." The sign *tim* in KI.KAL-*tim* is the same as the signs GÍR-TAB "scorpion" that are used for the name of the zodiacal sign. Scorpions, of course, also live in uncultivated areas. This does not explain, however, why KI.KAL-*tim* is also given for Libra, except on the tangential grounds that Scoprio and Libra are neighboring zodiacal signs, and it is not obvious what animal could be associated with Libra ("The Balance").

[39] See Steele 2011b: 336–338.

[40] The term *Decknamen* is due to Köcher 1995. In the third tablet of the pharmacological series URU.AN.NA, there appears a list of regular medical ingredients together with their associated *Decknamen* (Köcher 1995; Kinnier Wilson 2005: 48). The list is presented in two columns, separated by the cuneiform "colon" sign (a stack of two horizontal wedges with the corner to the left) as is common in commentary and lexical texts. Before the *Decknamen* in the second column appears the cuneiform sign AŠ (a single horizontal stroke), which Köcher has argued is to be read as a logogram for *pirištu* ("secret"), based on the fairly rarely attested logographic reading SAG.AŠ for this word. Köcher may well have been influenced in his use of the term *Decknamen* by the well-known Greek papyrus PGM 12.401–444, which gives "secret names" for herbs, etc.: "Because of the curiosity of the masses they [i.e., the scribes] inscribed the names of the herbs and other things which they employed on the statues of the gods, so they that? they [i.e., the masses], since they do not take precaution, / might nor not? practice magic, [being prevented] by the consequence of their misunderstanding" (Betz 1986: 167). However, the AŠ sign before is probably more likely simply to be used as a formatting mark rather than to be read as a word. In my opinion,

Decknamen are rather disgusting ingredients such as the excrement and blood of various animals that, however, turn out to be code names for common herbs and minerals.[41] *Decknamen* are attested in a variety of medical texts, especially in a number of Hellenistic-period commentary texts, and a list of *Decknamen* is included in the third tablet of the pharmacological series URU.AN.NA. The association between the plant and the *Decknamen* may result from similarities between the appearance of the plant and the named object, or through scholarly punning on the reading of cuneiform signs.

Among the most commonly found *Decknamen* is "lion's blood," which refers to the juice drawn from the middle of a tamarisk plant. Lion's blood is of course one of the ingredients mentioned in Iqiša's *Kalendertext* tablets. The therapeutic medical commentary tablet BRM 4 32, which was probably also owned by Iqiša, contains other *Decknamen* including an explanation that a "raven" refers to the *amilanu* plant and the dove to the "single" plant.[42] Another commentary from Uruk written around the same period or slightly earlier, SpTU I 49, refers to crab's blood in broken context and associates the egg of a raven with the *illuru* flower, while another, SpTU I 50, refers to the hair of a lion, wolf, fox, and dog as treatments for the "hand of Ištar." Although only a limited number of the parts of the animals referred to in Iqiša's *Kalendertext* tablets are attested as *Decknamen* for common herbs or minerals in such texts, the examples discussed make it quite plausible that all of the entries in the *Kalendertexte* were intended as *Decknamen*.[43] If this suggestion is correct, it has two consequences for our understanding of these texts. First, it suggests that, at least for Iqiša, the *Decknamen* are not simply code names. The association between the animals that are used in the *Decknamen* and the signs of the zodiac only makes any sense if there is also an underlying connection between the animal and the plant or mineral. Thus, the *Decknamen* are not simply made-up "secret names" but must reflect an underlying reality in which these animals are associated with the plants and minerals.

therefore, the *Decknamen* need not be truly "secret" names so much as alternative nicknames for the ingredients.

41 Köcher 1995.

42 MLC 1863 edited by Geller 2010a: 168–173.

43 At the Fourth Regensburg Workshop on Babylonian Astral Sciences held in Berlin on May 14–16, 2014, Marvin Schreiber discussed a tablet from Babylon that he has identified as containing a list of *Decknamen* that agrees with those found on Iqiša's *Kalendertext* tablets, providing strong confirmation of the interpretation I am proposing here. The tablet will be published and discussed in Schreiber's forthcoming Humboldt University PhD dissertation, "Die astrologische Medizin der spätbabylonischen Zeit."

The second consequence, which is more relevant to the present discussion, is that this association adds yet another layer of construction onto the way that time is used in medicine. From a given date, a position in the zodiac is determined mathematically using the *Kalendertext* scheme, an astrological construct created mathematically from a simple model of lunar motion. The position in the zodiac is then used to determine the animal that is going to be used in the treatment of a patient. This relationship represents a second level of construction, which associates zodiacal signs with animals. Finally, the animals are related to plants and minerals through the *Decknamen*, a third level of construction. The ultimate result is a way of determining from the date which common ingredients to use to make a medical remedy with which a patient is treated.

4 Conclusion

My aim in this chapter has been to discuss some of the ways in which time played a role in Babylonian medicine. It is clear that time is used in at least two ways: real and constructed. Real time, in the sense of durations of time for which symptoms persist, the timing of harvesting ingredients for medical remedies, and the performance of rituals of healing, can be either independent of or fixed within the calendar. Durations of illnesses, for example, are simply counted in days and months, but whether the period counted begins on, for example, the first day of a month or the seventeenth is irrelevant. The timing of harvesting ingredients and of rituals, however, is often linked through the calendar to the phase of the moon. Constructed time, however, takes a calendar date and constructs from that date schemes and associations that allow one to determine how to treat a patient. It remains an open question how extensively either of these two roles of time in medicine was used in regular medical practice. Both the constructed time of the *Kalendertext* scheme and the preference for certain durations of symptoms in the diagnostic texts, such as three days and ten days, fit comfortably within Babylonian scholarly traditions, but at the moment we do not have a clear answer to whether they were anything more than that. If my interpretation of Iqiša's *Kalendertext* tablets as containing *Decknamen* is correct, however, it would mean that the *Kalendertext* schemes were capable of being used in everyday medicine.

Bibliography

Betz, H. D. 1986. *The Greek Magical Papyri in Translation Including the Demotic Spells.* Chicago: University of Chicago Press.

Brack-Bernsen, L. and J. M. Steele. 2004. "Babylonian Mathemagics: Two Mathematical Astronomical-Astrological Texts." In C. Burnett et al. (eds.), *Studies in the History of the Exact Sciences in Honour of David Pingree.* Leiden: Brill, 95–125.

Britton, J. P. 2007. "Calendars, Intercalations and Year Lengths in Babylonian Astronomy." In J. M. Steele (ed.), *Calendars and Years: Astronomy and Time in the Ancient Near East.* Oxford: Oxbow Books, 115–132.

Brown, D. 2000. *Mesopotamian Planetary Astronomy-Astrology.* Groningen: Styx.

Böck, B. 2011. "Sourcing, Organizing, and Administrating Medicinal Ingredients." In K. Radner and E. Robson (eds.), *The Oxford Handbook of Cuneiform Culture.* Oxford: Oxford University Press, 690–705.

Casaburi, M. C. 2003. *Ūmē ṭābūti: "I giorni favorevoli."* History of the Ancient Near East, Studies 8. Padova: S.a.r.g.o.n.

Clancier, P. 2009. *Les bibliothèques en Babylonie dans la deuxième moitié du 1er millenaire av. J.-C.* Alter Orient und Altes Testament 363. Münster: Ugarit-Verlag.

Finkel, I. L. 2000. "On Late Babylonian Medical Training." In A. R. George and I. L. Finkel (eds.), *Wisdom, Gods, and Literature: Studies in Assyriology in Honour of W. G. Lambert.* Winona Lake, IN: Eisenbrauns, 137–224.

Frahm, E. 2011. *Babylonian and Assyrian Text Commentaries: Origins of Interpretation.* Guides to the Mesopotamian Textual Record 5. Münster: Ugarit-Verlag.

Geller, M. J. 2010a. *Ancient Babylonian Medicine: Theory and Practice.* Oxford: Wiley-Blackwell.

Geller, M. J. 2010b. "Look to the Stars: Babylonian Medicine, Magic, Astrology and *Melothesia*," *Max Planck-Institut für Wissenschaftsgeschichte Preprint* 401.

Greenbaum, D. G. 2015. "Astronomy, Astrology and Medicine." In C. Ruggles et al. (eds.), *Handbook of Archaeoastronomy and Ethnoastronomy.* New York: Springer, 117–132.

Heeßel, N. 2000. *Babylonische-assyrische Diagnostik.* Alter Orient und Altes Testament 43. Münster: Ugarit-Verlag.

Heeßel, N. 2005. "Stein, Pflanze und Holz. Ein neuer Text zur 'medizinischen Astrologie.'" *Orientalia* N.S. 74: 1–22.

Heeßel, N. 2008. "Astrological Medicine in Babylonia." In A. Akasoy, C. Burnett, and R. Yoeli-Tlalim (eds.), *Astro-Medicine: Astrology and Medicine, East and West.* Micrologus's Library 25. Florence: SISMEL, 1–16.

Hunger, H. 1971. "Die Tafeln des Iqīšā." *Die Welt des Orients* 6: 163–165.

Hunger, H. 1996. "Ein astrologisches Zahlenschema." *Wiener Zeitschrift für die Kunde des Morgenlandes* 86: 191–196.

Isma'el, K. S. and A. R. George. 2002. "Tablets from the Sippar Library XI: The Babylonian Almanac." *Iraq* 64: 249–258.

Kinnier Wilson, J. 2005. "Notes on the Assyrian Pharmaceutical Series URU.AN.NA = *maštakal.*" *Journal of Near Eastern Studies* 64: 45–52.

Kinnier Wilson, J. V. 2007. "Infantile and Childhood Convulsions, and SA.GIG XXIX." In I. L. Finkel and M. J. Geller (eds.), *Disease in Babylonia.* Cuneiform Monographs 36. Leiden: Brill, 62–66.

Köcher, F. 1995. "Ein Text medizinischen Inhalts aus dem neubabylonischen Grab." In R. M. Boehmer, F. Pedde, and B. Salje (eds.), *Uruk – die Gräber*. Mainz: von Zabern, 203–217.

Lambert, W. G. 2007. *Babylonian Oracle Questions*. Mesopotamian Civilizations 13. Winona Lake, IN: Eisenbrauns.

Leibovici, M. 1956. "Sur l'astrologie médicale Néo-Babylonienne." *Journal Asiatique* 244: 270–280.

Livingstone, A. 1986. *Mystical and Mythological Explanatory Works of Assyrian and Babylonian Scholars*. Oxford: Clarendon Press.

Livingstone, A. 1999. "The Magic of Time." In T. Abusch and K. van der Toorn (eds.), *Mesopotamian Magic: Textual, Historical and Interpretative Perspectives*. Ancient Magic and Divination 1. Groningen: Styx, 131–137.

Livingstone, A. 2010. "The Babylonian Almanac: A Text for Specialists?" In H. D. Baker, E. Robson, and G. Zólyomi (eds.), *Your Praise Is Sweet: A Memorial Volume for Jeremy Black from Students, Colleagues and Friends*. London: British Institute for the Study of Iraq, 85–101.

Reiner, E. 1985. "The Uses of Astrology." *Journal of the American Oriental Society* 105: 589–595.

Reiner, E. 1995. *Astral Magic in Babylonia*. Transactions of the American Philosophical Society 85/4. Philadelphia: American Philosophical Society.

Reiner, E. 2000. "Early Zodiologia and Related Matters." In A. R. George and I. L. Finkel (eds.), *Wisdom, Gods, and Literature: Studies in Assyriology in Honour of W. G. Lambert*. Winona Lake, IN: Eisenbrauns, 421–427.

Roughton, N. A., Steele, J. M., and Walker, C. B. F. 2004. "A Late Babylonian Normal and Ziqpu Star Text." *Archive for History of Exact Sciences* 58/6: 537–572.

Scurlock, J. 2006. *Magico-Medical Means of Treating Ghost-Induced Illnesses in Ancient Mesopotamia*. Leiden: Brill.

Stern, S. 2012. *Calendars in Antiquity: Empires, States, and Societies*. Oxford: Oxford University Press.

Steele, J. M. 2007. "Celestial Measurement in Babylonian Astronomy." *Annals of Science* 64: 293–325.

Steele, J. M. 2011a. "Making Sense of Time: Observational and Theoretical Calendars." In K. Radner and E. Robson (eds.), *The Oxford Handbook of Cuneiform Culture*. Oxford: Oxford University Press, 470–485.

Steele, J. M. 2011b. "Astronomy and Culture in Late Babylonian Uruk." In C. L. N. Ruggles (ed.), *Archaeoastronomy and Ethnoastronomy: Building Bridges between Cultures*. Cambridge: Cambridge University Press, 331–341.

Steele, J. M. 2015. "A Late Babylonian Compendium of Calendrical and Stellar Astrology." *Journal of Cuneiform Studies* 67: 187–215.

Stol, M. 2007. "Fevers in Babylonia." In I. L. Finkel and M. J. Geller (eds.), *Disease in Babylonia*. Cuneiform Monographs 36. Leiden: Brill, 1–39.

Weidner, E. 1967. *Gestirn-Darstellungen auf babylonischen Tontafeln*. Österreichische *Akademie der Wissenschaften, Philosophisch-historische Klasse, Sitzungsberichte* 254, 2. Vienna: Böhlau.

6

THE INTELLECTUAL BACKGROUND OF THE ANTIKYTHERA MECHANISM

ROBERT HANNAH

IN 1901, A GREEK SPONGE DIVER CHANCED UPON THE WRECK OF an ancient ship off the coast of the small island of Antikythera, to the south of mainland Greece. Divers and archaeologists combined to recover the contents of the wreck, in what constitutes one of the first organized underwater excavations. The resultant collection of Greek sculptures in bronze and marble now resides in the National Archaeological Museum of Athens, where it was put on special display.[1]

Also among the finds was a mass of corroded bronze plates bonded together and encrusted over. The fragments were found to be from a multi-geared instrument, which included plates marked with inscriptions in Greek. At last count eighty-two fragments are now known to belong to it. This instrument has come to be called the Antikythera Mechanism.[2]

In 1972, X-rays were taken of the Mechanism by the Greek radiographer Charalambos Karakalos for the British physicist Derek de Solla Price. These showed that it originally contained more than thirty interlocking, toothed gears and several plates that were interrelated by their capacity to mark time in various ways – an Egyptian calendar, a zodiac dial, and a star calendar, or *parapegma*.[3]

Groundbreaking though this research was, and fundamentally important for our appreciation of the complexity and sophistication of the Antikythera Mechanism, more intensive techniques and ongoing physical reconstructions have allowed investigators to refine and correct much of Price's interpretation and model. From the 1980s, Allan Bromley, from the University of Sydney, and Michael Wright, from the Science Museum in London,

[1] See Kaltsas et al. 2012, de Solla Price 1974. Marchant 2008 provides a popular account of the discovery and more recent developments; see the useful review at Evans 2009.

[2] Freeth 2008.

[3] De Solla Price 1974.

combined as collaborators, working initially from the 1970s' X-rays, but then having their own made. Since Bromley's untimely death, Wright has worked alone, explaining the underlying theory of the Mechanism in a series of articles and manufacturing two of the most detailed physical reconstructions.[4]

Among the most recent investigators of the Mechanism, and by far the largest group, is the Antikythera Mechanism Research Group (AMRG).[5] It is composed of three teams from the UK, Greece, and North America: the academic team; the Hewlett-Packard team; and the museum team. The AMRG is responsible for the most remarkable discoveries to date, making use particularly of X-ray computed tomography and digital optical imaging to clarify the interconnections of the gearing and the inscriptions. As a result of these endeavors, we now know that the Antikythera Mechanism managed to correlate, in an ingenious system of about thirty geared wheels, the motions of the sun and the moon via the nineteen-year Metonic Cycle, and probably of the five planets known to antiquity in epicyclic motion through the zodiac. The instrument could also be used to compute eclipses and had a dial to signal the four-yearly leap day and the two- and four-yearly games festivals at Olympia, Isthmia, Dodona, and Delphi. A *parapegma* coordinated with the dials gave the zodiacal year, the Egyptian calendar, and even a civil calendar.[6] Several reconstructions, both solid and virtual, have been generated as a result of the group's years of study. Further scientific papers have been produced as a result of the group's work by other scholars, situating the Mechanism's workings and theory in the context of Hellenistic Greek astronomy.[7] The methods for tracking planets and predicting eclipses reflect just one of ancient Greece's major developments in science – its capacity to reduce to geometrical models the complex workings of the cosmos as seen by the human eye.

At this point, we can perhaps see better why the Mechanism is so extraordinary in engineering terms: its complex train of gears, moving at different speeds and arranged so as to coordinate otherwise discordant time scales, will not be seen again for a thousand years, when Arabic

4 For references, especially to the now extensive Wright bibliography, see Hannah 2009, to which add Wright 2007.

5 www.antikythera-mechanism.gr (accessed 24 November 2013). Wright is not a member of the AMRG and has continued to publish independently of the group, but he and the AMRG do communicate their findings to each other.

6 Freeth et al. 2006; Freeth et al. 2008.

7 See e.g. Evans, Carman, and Thorndike 2010; Anastasiou et al. 2013; Evans and Carman 2014.

science reintroduces such things.[8] Yet the Arabs learned much of their science from the Byzantine Greeks, who – we can dimly see in the archaeological record – preserved something of the earlier achievements of their Classical ancestors.[9]

The purpose for which the Mechanism was made in the second century BCE, however, remains uncertain.[10] The Mechanism is currently recognized more as a planetarium or orrery, but Price originally identified it as a "calendar computer,"[11] and it still offers abundant means of marking time. Its coordination of the motions of the celestial bodies into one device, as well as its use of different forms of time schedules – two solar (the zodiac and the Egyptian calendar), one luni-solar (the civil calendar), and one sidereal (the *parapegma*) – all suggest that a fundamental function was not telling time, as with a clock; nor measuring time, as with a stopwatch; but marking time, as with a calendar. It was able to do this in terms of the Egyptian calendar and the *parapegma* on the front plate of the dial, and of a Greek civil calendar and the Metonic and Kallippic luni-solar cycles of nineteen and seventy-six years, respectively, on the back plate, where we also find the (Saros and Exeligmos) eclipse cycles. I also note that the Mechanism provides a means for closely regulating the movements of the sun and stars against the wandering calendars of both Greeks and Egyptians.

But this is simply to say what the Mechanism could do, not why it did so. Were there practical purposes to which such data could be put? In the early Hellenistic period, about 300 BCE, in a festival calendar from Hibeh in Egypt (*P. Hibeh* 27), we find a *parapegma* resembling in its details what we know about that of Eudoxus from the fourth century BCE. Here, however, it is structured within a scheme of twelve zodiacal months and then further incorporated into the native Egyptian calendar, as this sample illustrates:

8 For an Arabic representation, see Schoenberg MS LJS478, 73r: http://dla.library.upenn.edu/dla/medren/pageturn.html?id=MEDREN_4921349&doubleside=0&rotation=0&size=3&currentpage=149 (accessed January 18, 2014). I am grateful to Michael Wright for drawing this to my attention.

9 Field and Wright 1985; Field 1990; Wright 1990.

10 On the possible date of the Mechanism c. 200 BCE, see now Freeth 2014. Along with others, I had argued that the forms of the minuscule letters in the Greek inscriptions of the Mechanism argue for a date in the latter half of the second century BCE: Hannah 2009: 31, 160 n. 10. On this issue, see now Freeth 2014: n. S2, where a broader period is recommended "to the end of the third to the beginning of the first century BC, with a preference for the earlier half of this period."

11 De Solla Price 1974.

Tybi [5]	[The sun is] in Aries.
20	Spring equinox, the night is 12 hours and the day is 12 hours, and the feast of Phitorois.
27	Pleiades set in the evening, the night is $11\frac{38}{45}$ hours, the day $12\frac{7}{45}$.

(*P. Hibeh* 27.62–6)[12]

This illustrates better than anything else that has survived the comment by the first-century CE Roman agricultural writer Columella (*On Agriculture* 9.14.12) that the star calendars (*fasti*) of Meton, Eudoxus, and other astronomers were adapted to public (religious) festivals:

> Indeed, in this rural instruction I am now following the calendars (*fastūs*) of Eudoxus and Meton and the old astronomers, which are adapted to the public sacrifices, because that old view, understood by farmers, is better known, and, on the other hand, the subtlety of Hipparchus is not necessary, as they say, for the duller learning of rustics.
>
> (Columella, *On Agriculture* 9.14.12)

What Columella is referring to here as *fasti* ("calendars") is what we call *parapegmata*. In what way they were "adapted to the public sacrifices" for him we cannot tell, but the Hibeh papyrus probably gives us a glimpse of what he means. An extended literary version of adaptation may be seen in the Roman poet Ovid's *Fasti*, which was written a generation before Columella and correlates astronomical data of star rise and star set with Roman festivals in the first half of the year (as that is all that survives of the poem).[13] The term *parapegma* is used at the head of a compilation of star data, which is appended to the astronomer Geminus's *Introduction to Astronomy* of the mid–first century BCE. This compilation presents various "observations" of star rise and star set attributed to a number of Greek astronomers, including Meton, his contemporary and colleague Euctemon, and Eudoxus. Some of the "observations" are accompanied by meteorological data indicating changes in the weather.[14] To what extent these star data represent real naked-eye observations is debated, but my own view is that the earlier ones are more likely to be so, while later ones may be the result of schematic calculation. The accuracy of such data is also a matter of long-standing debate, but current scholarship is

[12] Grenfell and Hunt 1906; Lehoux 2007.

[13] See further Robinson 2007, 2011, 2013.

[14] Hannah 2002; Lehoux 2007.

tending toward the view that they are accurate, within the parameters of ancient observational practice.[15]

The Antikythera Mechanism certainly provides the potential to correlate the *parapegma* with the Egyptian calendar, but it displays no interest, as far as we know to date, in aligning these data with religious festivals. So what other function is possible?

Astrometeorology – weather forecasting by the stars – is a possible function for the Mechanism, and one with cultural and practical significance. Weather forecasting by the stars had a long history throughout antiquity and was noted earlier in relation to the *parapegmata*. Its cultural context in both the Near Eastern and Greek worlds is now being given serious attention.[16] At the start of Greek literature, Hesiod, in his poem *Works and Days*, makes use of the stars as harbingers of changes in the weather, for instance, telling us that when the Pleiades set, gales will begin to blow (Hesiod, *Works and Days* 619–21). The stars were also used to signal via weather changes the safe sailing season in the Classical Mediterranean, as in the same passage from Hesiod (l. 622).[17] In the same period as the Antikythera Mechanism, the second century BCE, is situated the Tower of the Winds in Athens. On its exterior walls is a combination of solar and meteorological data, presented as incised plane sundials and personifications of the Winds sculptured in high relief, while inside the building there appears to have been a sophisticated water clock, whose water channel survives but whose workings still elude us.[18] Yet in comparison, the Antikythera Mechanism provided more functions than the purely astrometeorological.

Astrology remains an option. Through the Mechanism, the rapid calculation of the positions of all the major planets and related phenomena, essential to ancient astrology, could have been performed. Tables from the imperial Roman period survive that record planetary positions with a remarkable degree of accuracy. Until now, it has not been understood how these positions were so accurately recorded, whether it be by observation, or calculation, or a mixture of both.[19] The type of instrument that the Antikythera Mechanism represents arguably provides such a means. Genethialogy, or casting a horoscope for one's birth, must have already developed in the eastern Mediterranean well before the period of the earliest

[15] Fox 2004; Robinson 2009.

[16] Taub 2003. Cf. Lehoux 2007.

[17] See Hannah 1993 for further references in later literature.

[18] Noble and de Solla Price 1968; Gibbs 1976: 342–345, no. 5001; Kienast 1997, 2005, 2014; Schaldach 2006: 60–83.

[19] Neugebauer 1941–1943.

surviving material evidence for it: the earliest surviving horoscopes are (1) the literary horoscope of 72 BCE preserved by the mid–first century CE astrologer Balbillus, who married his daughter into the royal family of Commagene and perhaps thereby acquired access to astrological archives of earlier vintage and (2) the sculpted horoscope of Antiochus I of Commagene at Nemrud Dağ in Turkey from 62 BCE.[20] It seems plausible that this type of astrology would have been in its infancy in the general period assigned to the Mechanism, although a date of manufacture closer to 200 BCE than to 100 BCE would count against this option.

The presence of the civil calendar on the Mechanism, however, then also requires explanation. It indicates that the instrument was intended for a relatively narrow geographical, or more correctly cultural, region of the Doric Greek world. More than this, however, it also suggests that the Mechanism, as its very complexity argues, was probably a bespoke item for a wealthy individual. Where that individual lived is now difficult to trace. The Doric calendar, the scientific ingenuity of the instrument and even the bronze manufacture may recommend places such as Syracuse or Corinth, but the former was captured and plundered by the Romans in 212 BCE. The Romans in 146 BCE razed the latter to the ground, but a date of production for the Mechanism of around 200 BCE could still leave Corinth as a plausible provenance. Iversen (2017) has argued persuasively that the calendar on the Antikythera Mechanism is that of Epiros, which belongs to the Corinthian family of calendars.

The engineering and mechanical tradition within which the Antikythera Mechanism stands includes types of instruments that we may loosely call "planetaria," although we need not locate the prime purpose of the Mechanism in that category of instruments. These devices sought to replicate the motions of the celestial bodies. Best known are Archimedes' sphere and Posidonius' orrery, described admiringly by Cicero.[21] Thus, the Roman on the second globe of Archimedes:

> But this type of globe, on which were set the motions of the sun and moon and of those five stars which are called the planets, or, as it were, the wanderers, could not be represented on that solid globe. And in this the invention of Archimedes was to be admired, because he had thought out how a single revolution should maintain unequal and varied courses in dissimilar motions. When Gallus moved this globe, it happened that the

[20] Beck 2007: 130–131. The editors kindly inform me that the Qumran document 4Q186 attests to horoscopic genethialogy around the mid–first century BCE: Popović 2007.

[21] Cicero, *Rep.* 1.14.21–22, *Tusc.* 1. 63, *Nat. d.* 2. 87–88.

> moon followed after the sun by as many revolutions on the bronze, as it does in so many days in the sky. Therefore also that same eclipse of the sun occurred on the globe, and the moon then came to that turning point which is the shadow of the earth. (Cicero, *Rep.* 1.22)

Undoubtedly, we can now appreciate the engineering skill involved in the manufacture of the globe in light of the inner workings of the Antikythera Mechanism. Indeed, this appreciation increases when one sees the recreation built by Michael Wright, based on aspects of the Antikythera Mechanism.[22] Nonetheless, we should not lose sight of the other cause for admiration that Archimedes' and other spheres drew from their ancient viewers:

> For when Archimedes fastened on a globe the motions of the moon, the sun and the five planets, he effected the same as that god of Plato, who built the world in the Timaeus, so that a single revolution controlled movements dissimilar in slowness and speed. Therefore if in this world things cannot happen without a god, neither could Archimedes have reproduced the same movements upon a globe without divine genius.
> (Cicero, *Tusc.* 1.63)

A proper understanding of technical innovations in antiquity must allow for the metaphysical and religious concerns that informed thought then much more than they do in our more secular and "scientific" culture.[23] Cicero's admiration here is expressed in terms that we would characterize as theological. The explicit reference to Plato's Demiurge creating the world in his dialogue *Timaeus* signals this metaphysical element, as well as reflecting Cicero's personal interest in that work, which he partially translated. The first commentary on the *Timaeus* appeared in the works of Crantor, in the third century BCE, with this particular passage among those that we can be certain were commented on.[24] The fascination with Plato's account of the creation of the world would continue for centuries in antiquity.

For Plato, it was by observing the regular movements of the cosmos that humans could learn to order and regularize their own interior movements. We are granted our sight, he says,

[22] I had the pleasure of seeing this recreation in Michael Wright's workshop in 2013. It is mentioned at www.nytimes.com/2013/06/25/science/archimedes-separating-myth-from-science.html (accessed January 17, 2014).

[23] While discussions of the Large Hadron Collider and its purpose make mention of trying to find the "God particle" through it, this is popular, not scientific, talk.

[24] Dillon 2005: 216–224.

> so that by observing the circuits of intelligence in heaven, we might make use of them for the revolutions of our own thought, which are related to them, though ours are troubled to theirs untroubled; and that, having learned thoroughly and sharing the ability to calculate rightly according to nature, by imitating the completely fixed revolutions of the god, we might settle the wandering revolutions in ourselves.
>
> (Plato, *Tim.* 47b6–c4)

The Antikythera Mechanism represents a far more abstract form of the interrelated movements of the celestial bodies than a regular planetarium or orrery. Roger Beck has discerned a relationship between this abstract form and the Platonic ideal in Greek philosophy. He has argued that the instrument's underlying mathematical formula for the Metonic cycle (19 years = 254 sidereal months = 235 synodic months) "is the intelligible reality behind the relative motions of the visible Sun and the visible Moon . . . But the addition of the little model luminaries is for the purist something of a distraction, a concession to appearances which, even if they can be replicated precisely, are not really worth replicating since in the strictest sense they are unintelligible."[25] We may add that inasmuch as the Mechanism abstracts the workings of the cosmos to number, so time too is thus abstracted. After Plato, Aristotle opined, "This is time, the number of movement with regard to before and after" (*Phys.* 219b1–2).

Let us unpack this, so as to clarify what Beck means. In his *Timaeus* (37c6–38c1), Plato had the creator Demiurge fashion the cosmos in such a way that it resembles as much as possible the eternal, its model:

> Now the nature of this being happened to be eternal, and it was not possible to bestow this character completely on the created. He decided to make a moving image of eternity, and when he organized the heaven, while eternity stays in unity, he made this image eternal but moving according to number; and we call this being time. For there were no days and nights and months and years before the heaven came to be, but together with its construction he made the beginnings of them. These are all parts of time, and the "was" and "will be" are created forms of time, which we unwittingly but wrongly attribute to the eternal being; for we say that it "was," it "is" and it "will be," but the truth is that only "is" belongs to it according to the true argument.
>
> (Plato, *Tim.* 37d3–38a1)

[25] Beck 2007: 125.

From a Platonic point of view, the material world, called the sensible world as being the result of our perception of it through the senses, was regarded as inherently an inaccurate representation of reality. Instead, reality was to be perceived through the intellect, and the resultant intelligible world lay beyond this earthly, sublunary ("under the moon," the nearest celestial body) world in the celestial realm. This view of knowledge we owe initially to Plato, expressed in his *Republic* and *Timaeus*,[26] and then to the later followers of Plato, the so-called Neoplatonists, of the third-to-fifth centuries CE.[27] Their work has been mined for centuries for its theological content, but there is also a strand of what we may call "scientific Platonism," stretching from Plato's own works and those of his followers in the fourth century BCE to the Neoplatonist Proclus in the fifth century CE.

The case for a scientific, as opposed to a theological, strand in Platonism has been made over the years by scholars interested in different scientific endeavors – in mathematics and astronomy, both popular and interrelated projects in natural philosophy in antiquity, but also in the more recondite area of optics.[28] That a case needs to be made at all for a natural science in Plato is a result of the philosopher's own differentiation, noted earlier, between the world of the senses – the sensible world – and hence the natural world, on the one hand, and the world of the intellect – the intelligible world – which was to be regarded as the real world, on the other. The study of the sensible world constituted the philosophy of nature, while the investigation of the intelligible world was the subject of theology. The two realms, of the senses and of the intellect, have usually been regarded in scholarship as distinct to the point of separation, with the world of the intellect providing the only true knowledge, while the world of the senses was too debased to provide knowledge.[29] Marije Martijn, however, has recently argued that this older view, which is underpinned by a rejection of the value of knowledge via the senses, is mistaken. Instead, she proposes, there is in the Neoplatonic philosophy of nature defined by Proclus, the most scientifically oriented of the Neoplatonists,[30] a fundamental continuity between the sensible world (the "world of generation") and the intelligible world. In other words, knowledge can be gained from the sensible world.[31]

[26] Smith 1981: 76–77.

[27] For useful summaries of the Neoplatonists and Proclus in particular, and his historical and philosophical contexts, see Pedersen 2012; Tarrant 2006: 1–9; Siorvanes 1996: 1–47. I wish to thank Dr. Stefan Pedersen for assistance in this area.

[28] See e.g. Smith 1996; Smith 1981; 1982; Lloyd 1978; Sambursky 1965.

[29] See e.g. Sambursky 1965: 6–7.

[30] See Lloyd 1978.

[31] Martijn 2010.

This ability to derive knowledge from the sensible world has been recognized also by A. Mark Smith in the ancient science of optics. He has argued for four principal assumptions inherent in Greek scientific methodology:

(1) "that all change or flux, insofar as it manifests irregularity, is just an 'appearance' or illusion."
(2) "that beneath the appearances there lies a real, intelligible world that is utterly simple, changeless and eternal."
(3) that this intelligible world "is a true Euclidean locus or 'space,' within which things are really what they are by virtue of their spatial attributes and relationships. Within such a locus, moreover, the only real relationships are those most basic ones obtaining between and among points, and they are mathematically expressible in terms solely of rectilinear distances and angles."
(4) that "these relationships are assumed to be immanent in, and thus immediately inferable from, the appearances."

"In short," Smith concludes, "if properly understood, appearances do not really deceive us at all. They betoken a deeper and simpler reality that is rationally accessible to us in terms, finally, of straight lines and rectilinear angles."[32]

Mathematics and geometry provided media through which Plato and his followers could more closely approach the reality of the intelligible Forms. The Antikythera Mechanism, as much as any constructed object could do, brings us that bit closer to the Platonic metaphysical reality.

The astronomer Geminus states:

> It is assumed in all astronomy that the sun, the moon and the five planets move at uniform speed in circular fashion and in a contrary manner to the cosmos. For the Pythagoreans, who were the first to enter into such investigations, assumed the movements of the sun, the moon and the five planets to be circular and uniform. For they would not accept disorder, with regard to divine and eternal things, such as would make them move at one time more swiftly, at another time more slowly and at another time stand still (which indeed they call the stationary points of the five planets). For no-one would accept such irregularity of motion even of a decent and orderly man in his journeys. For the necessities of life are often causes of slowness and swiftness for men. But for the incorruptible nature of the stars it is not possible for any cause of swiftness or slowness to be adduced.

[32] Smith 1982: 224.

> Therefore, they proposed thus: how the phenomena might be accounted for by means of circular and uniform movements.
>
> (Geminus, *Introduction to Astronomy* 1.19–21)

Simplicius, in the mid–sixth century CE, reported that

> Plato ... set ... this problem for all keen about these matters: by what assumed, uniform and ordered movements the phenomena can be saved in relation to the movements of the planets.
>
> (Simplicius, *On Aristotle's On the Heavens* 488.18–24)

With the Antikythera Mechanism, we have a visualization of this assumption of uniform and circular movements.[33] It is important that we do not lose sight of this cosmological perspective of ancient astronomers when trying to understand the Antikythera Mechanism.

From the time of Plato onward, Greek astronomers were not simply empiricists seeking to describe the apparent phenomena, as we arguably find in the *parapegmata*, or to predict eclipses and planetary positions, with however complex an instrument, such as the Antikythera Mechanism. Nor did they seek to derive, inductively, a system from those particular, empirical data of star positions, eclipses or planetary positions. Rather, their purpose was to provide a theoretical basis, an overarching system, into which the observable phenomena, especially regarding the planetary system, could be fitted.[34]

The primary function of the Antikythera Mechanism remains something of a mystery. It may well be that the practical functions that I have rehearsed here prove not to be possible, alone or in combination. It is a curious instrument, bearing witness on the one hand to extraordinary engineering skills, yet also, it seems, to antiquated scientific knowledge even in its own time (if it utilized Babylonian rather than state-of-the-art Greek knowledge). It displays an interest in the major Greek games festivals (Olympia, Delphi, Isthmia), yet also in a minor one (Dodona). It provides the means to correlate a variety of time-marking schedules – solar, luni-solar, and sidereal – but apparently not to the same purposes as the related *parapegmata* known elsewhere in the Greek and Roman worlds. Its use of a distinct civil calendar suggests a very specific end-user. It has been suggested that it is a bespoke

[33] On the question of whether the Mechanism incorporates the details of epicycles, see Evans, Carman, and Thorndike 2010.

[34] Pedersen and Hannah 2002; Pedersen 2012. That the underlying system of the Mechanism might be Babylonian, not Greek, as suggested by Evans, Carman, and Thorndike 2010, does not undermine the idea put forward here that a preconceived worldview governed the form of the Mechanism.

item made as a showpiece.[35] But made for whom and displayed to whom? The functionality of the Mechanism argues, I believe, for a utilitarian purpose, albeit at an élite level of Greek society. Whatever it was made for, it is also a product of a particular worldview of the cosmos, which seems to align well with Platonic cosmology. The Antikythera Mechanism brings us closer than any other surviving instrument from antiquity to a full realization of this philosophical ideal: "as above, so below."

Bibliography

Anastasiou, M. et al. 2013. "The Astronomical Events of the Parapegma of the Antikythera Mechanism." *Journal for the History of Astronomy* 40: 173–186.

Beck, R. 2007. *A Brief History of Ancient Astrology*. Oxford: Blackwell.

De Solla Price, D. J. 1974. "Gears from the Greeks." *Transactions of the American Philosophical Society* 64/7. Philadelphia: American Philosophical Society.

Dillon, J. M. 2005. *The Heirs of Plato: A Study of the Old Academy (347–274 B.C.)*. Oxford: Clarendon Press.

Evans, J. 2009. "The Saga of the Antikythera Mechanism." *Journal for the History of Astronomy* 40: 362–364.

Evans, J., C. C. Carman, and A. S. Thorndike. 2010. "Solar Anomaly and Planetary Displays in the Antikythera Mechanism." *Journal for the History of Astronomy* 41: 1–39.

Evans, J. and C. C. Carman. 2014. "Mechanical Astronomy: A Route to the Ancient Discovery of Epicycles and Eccentrics." In N. Sidoli and G. van Brummelen (eds.), *From Alexandria, Through Baghdad: Surveys and Studies in the Ancient Greek and Medieval Islamic Mathematical Sciences in Honor of J.L. Berggren*. Berlin: Springer, 145–174.

Field, J. V. 1990. "Some Roman and Byzantine Portable Sundials and the London Sundial-Calendar." *History of Technology* 12: 103–135.

Field, J. V. and M. T. Wright. 1985. "Gears from the Byzantines: A Portable Sundial with Calendrical Gearing." *Annals of Science* 42: 87–138.

Fox, M. 2004. "Stars in the *Fasti:* Ideler (1825) and Ovid's Astronomy Revisited." *American Journal of Philology* 125: 91–133.

Freeth, T. 2008. *The Antikythera Mechanism: Decoding an Ancient Greek Mystery*. Cambridge: Whipple Museum of the History of Science.

Freeth, T. 2014. "Eclipse Prediction on the Ancient Greek Astronomical Calculating Machine Known as the Antikythera Mechanism." *PLoS ONE* 9(7): e103275. doi:10.1371/journal.pone.0103275.

Freeth, T. et al. 2006. "Decoding the Ancient Greek Astronomical Calculator Known as the Antikythera Mechanism." *Nature* 444: 587–591.

Freeth, T. et al. 2008. "Calendars with Olympiad Display and Eclipse Prediction on the Antikythera Mechanism." *Nature* 454: 614–617.

Gibbs, S. L. 1976. *Greek and Roman Sundials*. New Haven: Yale University Press.

Grenfell, B. P. and A. S. Hunt (eds.). 1906. *The Hibeh Papyri*, vol. 1. London: Egypt Exploration Fund.

[35] Cf. www.antikythera-mechanism.gr/media/clips/666 (accessed October 5, 2014).

Hannah, R. 1993. "Alcumena's Long Night: Plautus, *Amphitruo* 273–276." *Latomus* 52: 65–74.

Hannah, R. 2002. "Euctemon's Parapegma." In C. J. Tuplin and T. E. Rihll (eds.), *Science and Mathematics in Ancient Greek Culture.* Oxford: Oxford University Press, 112–132.

Hannah, R. 2009. *Time in Antiquity.* London: Routledge.

Iversen, P. A. 2017. "The Calendar on the Antikythera Mechanism and the Corinthian Family of Calendars." *Hesperia* 86: 129–203.

Kaltsas, N. et al. 2012. *The Antikythera Shipwreck: The Ship, the Treasures, the Mechanism.* Athens: National Archaeological Museum.

Kienast, H. J. 1997. "The Tower of the Winds in Athens: Hellenistic or Roman?" In M. C. Hoff and S. I. Rotroff (eds.), *The Romanization of Athens: Proceedings of an International Conference Held at Lincoln, Nebraska (April 1996).* Oxford: Oxbow, 53–65.

Kienast, H. J. 2005. "La Torre dei Venti di Atene." In E. Lo Sardo (ed.), *Eureka! Il genio degli antichi.* Naples: Electa, 245–251.

Kienast, H. J. 2014. *Der Turm der Winde in Athen.* Archäologische Forschungen 30. Wiesbaden: Reichert Verlag.

Lehoux, D. R. 2007. *Astronomy, Weather, and Calendars in the Ancient World: Parapegmata and Related Texts in Classical and Near-Eastern Societies.* Cambridge: Cambridge University Press.

Lloyd, G. E. R. 1978. "Saving the Appearances." *The Classical Quarterly, New Series* 28: 202–222.

Marchant, J. 2008. *Decoding the Heavens: Solving the Mystery of the World's First Computer.* London: Heinemann.

Martijn, M. 2010. *Proclus on Nature: Philosophy of Nature and its Methods in Proclus' Commentary on Plato's Timaeus.* Philosophia antiqua 121. Leiden: Brill

Neugebauer, O. 1941–1943. "Egyptian Planetary Texts." *Transactions of the American Philosophical Society* 32: 209–250.

Noble, J. V. and D. J. de Solla Price. 1968. "The Water Clock in the Tower of the Winds." *American Journal of Archaeology* 72: 345–355.

Pedersen, S. 2012. *Toward a Celestial Physics: Natural Philosophy and Planetary Motion in the Cosmology of Proclus.* PhD Thesis, Dunedin: University of Otago.

Pedersen, S. and R. Hannah. 2002. "Celestial Dynamics at the Crossroads: Proclus' Reassessment of Plato in the Light of Empirical Science." *Antichthon* 36: 65–79.

Popović, M. 2007. *Reading the Human Body: Physiognomics and Astrology in the Dead Sea Scrolls and Hellenistic-Early Roman Period Judaism.* STDJ 67. Leiden: Brill.

Robinson, M. 2007. "Ovid, the Fasti and the Stars." *Bulletin of the Institute of Classical Studies* 50: 129–159.

Robinson, M. 2009. "*Ardua et Astra*: On the Calculation of the Dates of the Rising and Setting of Stars." *Classical Philology* 104: 354–375.

Robinson, M. 2011. *Ovid, Fasti Book 2.* Oxford Classical Monographs. Oxford: Oxford University Press.

Robinson, M. 2013. "Ovid and the *Catasterismi* of Eratosthenes." *American Journal of Philology* 134: 445–480.

Sambursky, S. 1965. "Plato, Proclus, and the Limitations of Science." *Journal of the History of Philosophy* 3: 1–11.

Schaldach, K. 2006. *Die antiken Sonnenuhren Griechenlands: Festland und Peloponnes.* Frankfurt am Main: Deutsch.
Siorvanes, L. 1996. *Proclus: Neo-Platonic Philosophy and Science.* New Haven: Yale University Press.
Smith, A. M. 1981. "Saving the Appearances of the Appearances: The Foundations of Classical Geometrical Optics." *Archive for History of Exact Sciences* 254: 73–99.
Smith, A. M. 1982. "Ptolemy's Search for a Law of Refraction: A Case-Study in the Classical Methodology of 'Saving the Appearances' and Its Limitations." *Archive for History of Exact Sciences* 26: 221–240.
Smith, A. M. 1996. "Ptolemy's Theory of Visual Perception: An English Translation of the 'Optics' with Introduction and Commentary." *Transactions of the American Philosophical Society, New Series* 86/2.
Tarrant, H. 2007. *Proclus: Commentary on Plato's Timaeus. Volume 1. Book 1: Proclus on the Socratic State and Atlantis.* Cambridge: Cambridge University Press.
Taub, L. 2003. *Ancient Meteorology.* London: Routledge.
Wright, M. T. 1990. "Rational and Irrational Reconstruction: The London Sundial-Calendar and the Early History of Geared Mechanisms." *History of Technology* 12: 65–102.
Wright, M. T. 2007. "The Antikythera Mechanism Reconsidered." *Interdisciplinary Science Reviews* 32/1: 27–43.

7

DIVINE FIGURATIONS OF TIME IN ANCIENT EGYPT

Alexandra von Lieven

While time is always present as an existential fact in our universe, it is nevertheless also "constructed," inasmuch as its exact measuring as well as its social and cultural significance is different in various cultures and historical periods. This partly depends on the technical capacities of the respective cultures. While today even fractures of seconds can be measured and are decisive not only in scientific experiments but also, for example, in the world of sports competitions or economics, they hardly mattered even a few hundred years ago in our own civilization, let alone in antiquity.

1 A Moment Is Not a "Second" – the Egyptian Concept of Time

A good indicator for this is the absence of even a word for "second" in many ancient languages, including Egyptian. There certainly is a word *ꜣ.t* in Egyptian meaning "moment, instant," but it is not anything to be measured or counted in the plural like our seconds. Rather, it means the moment something happens, which is defined not abstractly by a clock, but by the observation of the action happening. Thus, a person "in his moment" means somebody in a rage or otherwise engaging in intense activity.[1] It is often used of the god Seth, usually in comparison with the king on the battlefield, who is "like Seth in his moment." As Seth is among other things the Egyptian

This chapter was written during a Heisenberg Fellowship (reference Li 1846/1-2) by the German Research Council (DFG), to which I would like to express my gratitude. For a recent study with a similar subject, see Quack 2013a. I would like to thank the author for having furnished me with a copy of the manuscript before publication. While certain elements had to be considered in both, I have nevertheless tried to put the emphasis on other aspects, so that our studies may complement rather than duplicate each other. I also would like to thank C. Leitz for having furnished me with the beautiful photos reproduced here in the figures.

[1] De Meulenaere 1982: 139–140 with further literature in note i.

weather god, it is very likely that the "moment" here means a thunderstorm. However, in terms of precise time concepts, it is unclear whether this would imply the whole period of the thunderstorm or just the very moment when a flash of lightning or a thunderclap occurs. The comparison with the battle situation seems rather to advocate the former interpretation, as a battle is also a prolonged action, even if it consists of many single strokes with a weapon. If this assumption is true, it is clear that *ꜣ.t* is an expression for time that does have a duration, yet defies any exact measurement.

From this observation it follows that before tackling the Egyptian concepts of time and its diverse divine figurations, we first must take a look at the basics: how did the Egyptian calendar work, and how was time measured in the first place?

2 Calendar and Measuring of Time

The Egyptian year had 365 days, each made up of twenty-four hours.[2] While this sounds familiar enough, in reality these hours were not all of the same length because of the methods of measuring time.[3] There were several such methods, namely, the tracking of the course of the sun during the day via a shadow clock[4] and the observation of certain constellations,[5] especially the so-called decans,[6] during the night. There were different lists of decans, giving the names and numbers of stars constituting them.[7] As *The Fundamentals of the Course of the Stars*,[8] a well-attested manual on the subject, puts it, the important criterion for a decan was that "one of them rises and another one sets within ten days." Thus, decans were used not only to mark hours by their position in the nightly sky but also to define ten-day "weeks"[9] or rather decades. As these time-measuring methods relied on the visibility of the relevant phenomena in the sky, it is not too surprising that the Egyptians also invented another device in case the sky was cloudy, namely, the water clock.[10]

2 Parker 1950; Depuydt 1997.

3 Borchardt 1920; for a reprint, see Wuensch and Sommer 2013, with review by von Lieven 2013.

4 Borchardt 1920: 26–50, to which add for the shadow clocks Frankfort 1933: 76–80, pl. LXXXII–LXXXIII; von Bomhard 2014: 86–111; Leitz 2014: 489–491, pl. 115–117.

5 Leitz 1995.

6 Quack 2002.

7 Neugebauer and Parker 1960, 1969.

8 Von Lieven 2007a, with some important new observations in von Lieven 2012.

9 There were no seven-day weeks as known to us today in Ancient Egypt.

10 Borchardt 1920: 6–26, 60–63; on the text treated there 60–63, see now von Lieven 2016.

Three decades made up a month, of which the Egyptian year had twelve. They were further organized into three seasons, each composed of four months, called *ꜣḫ.t* "Flooding," *pr.t* "Appearance" (i.e., reappearance of the submerged land), and *šmw* "Heat." Thus, the year proper is constituted of 360 days. Another five, known as the epagomenal days, were added to this. They were clearly believed to be a period of time outside the order of time, as their Egyptian designation as "the five days on top of the year" indicates. As such, they were believed to be the birthdays of the five children of the sky goddess Nut – the important gods Osiris, Horus the Elder, Seth, Isis, and Nephthys.[11] At the same time, they were also frightful, as they were the period when the "wandering (*šmꜣ.iw*)" and "slaughtering (*ḫꜣ.tiw*)" demons of the Dangerous Goddess were scouring the earth, looking for victims.[12] Their background in the real world was a variety of diseases that spread along with the Nile flood, probably partly because of the decay of animals killed by the inundation.

Ideally, the New Year should have started with the heliacal rising of the decan Sothis, a triangle-shaped constellation, with Sirius as its most important star. The Nile inundation should have started at the same time as well, a fact that had enormous importance for the Egyptians. However, in reality, the civil calendar and the religiously important facts of the cycle of nature rarely coincided. The reason is the lack of a leap day every four years. Such a leap day was only introduced in 238 BCE under Ptolemy III with the Decree of Canopus.[13] Apparently, for the Greek mind of the Ptolemies, it was unacceptable that the calendar was always out of tune with reality. A leap day had the added benefit that one could install it as a special holiday for the dynastic cult of the Ptolemaic dynasty itself. This is the calendrical model that was subsequently adapted by Julius Caesar and still lives on in our own calendar, despite some further modifications in later centuries.

3 A Necessary Comment on the Sources

A typical feature of Egyptian religion was its belief that all facts of nature were divine.[14] Almost all of the aforementioned time units therefore had their divine personifications and other connections. These figured prominently in temples, tombs, and on objects such as amulets, particularly but by no means exclusively from the Late and Greco-Roman Periods. One of the reasons

[11] Leitz 1993; Spalinger 1995.
[12] Bommas 1999; on the demons, see von Lieven 2000a: 42–55.
[13] Pfeiffer 2004. See also the essay by Ben-Dov, Chapter 2, this volume.
[14] Von Lieven 2004; Fischer-Elfert 2008.

for this prominence is that in Egypt, science was traditionally conducted by priests in temples. This is comparable to the situation in the monasteries of medieval Europe. Unfortunately, most temple libraries are lost as a result of the conditions of preservation. This, in turn, considerably distorts our view of what was generally known and practiced in Egypt, and how this knowledge and practice were distributed over the various periods. The apparent preponderance of "funerary" attestations, particularly during the earlier periods, has to be evaluated from this perspective. In reality, all these concepts were of enormous relevance for the living, and it was only because of this relevance that they were also adapted for funerary purposes. Especially in terms of dates of first and last attestations, it is vital to keep in mind that most sources from the actual sphere of the living are irretrievably lost, and funerary adaptations underlie a conscious selection by the ancient Egyptians in view of their beliefs about the hereafter.

4 The Year

The year itself could be personified as a goddess, as the word *rnp.t* "year" is also feminine in Egyptian. A similar logic governs most of these personifications, although there are rare exceptions to this rule. The word "year" is etymologically connected to *rnp* "to be young, fresh." Thus, a year is an entity that renews itself cyclically. Not surprisingly then, the start of a new year was in a way considered a repetition of creation. The New Year's day itself was thought to be the sun god Re's birthday.

In line with this idea of a life cycle, a famous myth recounts how humanity rebelled against Re when he became old.[15] One version of this myth as found in the temple of Edfu specifies his age at the time as 363 years, which alludes of course to the phase immediately before the end of a year.[16] Interestingly, in terms of a year, day 363 would be the third epagomenal day, reputed to be the birthday of Seth.[17] While Seth is often represented as a helper of the solar god against the latter's snake enemy Apopis, he is nevertheless infamous for the murder of his brother Osiris, whose reign he usurped. Moreover, he is said to have been born in an unnatural and violent way. Therefore, it is clear that this third epagomenal day was significant and the age of 363 years was not chosen arbitrarily. In fact, the respective text clearly presents Seth as the enemy behind this rebellion. In other sources, however, the sun god is

[15] Hornung 1991.
[16] Chassinat 1931: 109–132; translation: Kurth 2014: 190–222.
[17] Derchain 1978.

believed to have reigned for 7,000 years,[18] a figure that does not seem to hold any astronomical significance but is clearly intended to sound like a very long time.

The goddess "The Good Year (*Rnp.t-nfr.t*)"[19] is mainly known from a long series of invocations in the temple of Edfu.[20] Most likely, the name is just an epithet of another, more important goddess: plausible candidates would be Isis, Hathor, or Sakhmet, who can all be linked to the astral entity Sothis, whose rise heralds the New Year.

5 Chronocrats as Rulers of Each Day

Much more popular than a single representation of the whole year were sets of personifications or guardian deities for each day of the year, the so-called chronocrats.[21] Such sets can be found in several temples of the Greco-Roman period. Particularly well known are those from Dendara, where each day has a male and a female protector associated with it. Judged by their iconography, being represented as hawk-headed men, the male chronocrats are perhaps connected to Horus, the partner of the local main goddess Hathor. The snake-shaped female chronocrats are definitely forms of Hathor and the Dangerous Goddess in general. The association of the Goddess with the concept of the chronocrats is even more impressive with respect to the famous standing and sitting Sakhmet statues, originally from the Mut temple in Karnak, which Jean Yoyotte interpreted as early examples of chronocrats.[22] As they have unfortunately been dispersed all over the world for centuries, it is difficult to assess the original concept underlying them. However, the sheer number of preserved statues – in fact, more of them are still unearthed occasionally – support the proposal that at Karnak there were originally 365 seated and 365 standing statues.

6 Deities of the Months

While in the Old Kingdom, depictions of the three seasons are more frequent than representations of the twelve months individually, later, the months themselves become much more popular.[23] They could be represented in

[18] Sauneron 1983: 171–174; Meeks 2006: 183.
[19] Not to be confused with deified Renpetneferet, the younger sister of the 3rd dynasty sage Imhotep, a mortal woman who accidentally held the same name as the goddess. On these deified personages, see von Lieven 2007b.
[20] Germond 1986.
[21] A list of attestations is found in Leitz 2002: 137, n. 3.
[22] Yoyotte 1980.
[23] Quack 2013a: 75–85; Kaper 2014.

Figure 1 Hippopotamus goddesses of the months, temple of Athribis (© Athribis-Projekt Tübingen)

different ways. Apart from depictions of one personification for each, groups of protective deities were also associated with them. But the most important and widespread depiction of the months is certainly the group of twelve hippopotamus goddesses attested in several late temples, sometimes even with detailed descriptions of the materials their statues should consist of (Figure 1).[24] Moreover, their heads are sometimes made different from one another.

[24] Mendel 2005, to which add Leitz, Mendel, and El-Masry 2010: 502, 504, 506, pl. 124, 126.

As with the chronocrats, there probably is a three-dimensional attestation for the month-hippopotamuses too, as early as the New Kingdom. One of the treasures of the Museum August Kestner in Hanover (inv. no. 2616) is the upper part of a tiny glass figure of a standing, lion-headed hippopotamus made of blue glass with the cartouche of king Amenhotep II.[25] Until now, it has been interpreted as a depiction of the goddess Taweret/Thoeris. However, this goddess usually has a normal hippopotamus head. The lion head of this piece, clearly identifiable as such by its whiskers, rather points in the direction of the month goddesses, where this iconography is quite common. While they may have hippopotamus heads in some depictions, in others they have the heads of human women, vultures, and lionesses.[26]

In her publication about the glass figure, Birgit Schlick-Nolte again identifies it as Thoeris and mentions the month goddesses only in a footnote but does not give them further consideration, probably because it would be the first attestation of any of them. However, this is not a valid argument, as many other concepts thought to be only late inventions can by now be positively proven to have existed much earlier than previously thought.[27] Her other argument – that the month goddesses would have been of minor importance not worthy of such a piece of exceptional quality – is equally invalid in view of their prominent position in later temples.

There, according to the material assembled by Daniela Mendel,[28] lion heads are attested in three temples for the following months (as indicated by the hieroglyphic inscriptions next to them):

Philae	*Kom Ombo*	*Deir el-Bahari*
3rd *ꜣḫ.t*		
4th *pr.t*	3rd *pr.t*	3rd *pr.t*
1st *šmw*	1st *šmw*	
	2nd *šmw*	
	1st epagomenal day	

Thus, despite some variation, there are two months to which as many as two sources each assign a lion's head. These months are the 3rd *pr.t* and the 1st *šmw*. Fortunately, some of the lists also indicate the material from which these figures of deities should be fashioned. In fact, it was quite usual to differentiate the personifications of temporal units not only by their

25 Schlick-Nolte, Werthmann, and Loeben 2011: 28–32.
26 Mendel 2005: 124 n. 268.
27 For a clear example, see von Lieven 2001.
28 Mendel 2005.

iconography but also by their association with different materials, usually metals or minerals. The same practice is also known from the decans.

Further figures are attested from the temples of Armant and El Qal'a, although none have lion heads. As to the materials of the figures, for the 3rd *pr.t*, Kom Ombo and Armant both give *bḳs-ʿnḫ* "magnetite" or "hematite" (El Qal'a is unfortunately not preserved). For the 1st *šmw*, Kom Ombo, El Qal'a and Armant indicate *čḥn.t* "faience."[29] However, for the first epagomenal day, both Kom Ombo and El Qal'a require *ỉnr wčḥ*[30], literally "cast stone," most likely signifying glass;[31] as mentioned, this day has a lion's head in Kom Ombo as well. All other months with lion's heads are out of the question because of their materials. Thus, the Hanover glass hippopotamus in all likelihood depicts the protective goddess of the first epagomenal day.

The existence of one such figure alone is of course unlikely. Rather, this figure indicates that the full system already existed in the early New Kingdom, at around 1438–1412 BCE. Coincidentally, for the decans that period apparently also provides the first attestation of the ophiomorphic iconography of the so-called Sethos I B-family: the throne of a divine figure dated by a cartouche to the successive reign, that of Thutmosis IV.[32] Thus, for two sets of time-related divinities, a particular iconographical scheme can be proven to have already existed in the early eighteenth dynasty.

The famous ceiling in the tomb of Senenmut further proves that the months were already important in this period.[33] It shows among other astronomical motives twelve pie chart–like circles with twenty-four compartments each, each circle labeled with a month's name. However, these circles lack any protective deities. Such protective deities of the months, but without circles (which are unique to Senenmut in the preserved material), are for the first time depicted on the astronomical ceiling of the Ramesseum, about 180 years later.[34] There, however, they are not hippopotamus-shaped.

[29] I.e. Egyptian faience, not identical with modern European faience.

[30] Pantalacci 1995: 194 reads instead *ỉnr bḫn* "Greywacke," but Mendel's reading of the child hieroglyph as *wčḥ* is much more plausible, since there is a well-attested homophonous word for "weaned child." An explanation for a reading **bḫn* to the contrary seems hard to come by.

[31] To the single attestation for *ỉnr n wčḥ* in the annals of Thutmoses's III cited by Mendel 2005: 13, n. 29 and 23, n. 47 should be added *mfkꜣ.t wčḥ* "cast turquoise" and *ḫsbč wčḥ* "cast lapis lazuli" as designations for green and blue glass respectively (Harris 1961: 110).

[32] Published with good photos by Schlögl 2013, who unfortunately misinterpreted the iconography. For a correct interpretation of a very similar figure dated under Takelot III, see Brandl and Quack 2013.

[33] Neugebauer and Parker 1969: pl. 1 (again early 18th dynasty).

[34] Neugebauer and Parker 1969: pl. 5 (Ramses II A) (early 19th dynasty).

Moreover, they are not only female, but there are also male gods among them. At least one of the female ones, the protector of the 3rd *šmw*, is called *ꜣIp.t-ḥm.t*, a name similar to that of the later protector of the 1st *šmw*. Of course, this could be mere chance. In the Ramesseum, she is shown as a woman wearing the Red Crown.

Obviously, this goddess figures in the lists of hippopotamus goddesses of the month as a hippopotamus, not a woman. However, the comparison with the decans, with their different "families" (i.e., differing traditions labeled by Otto Neugebauer and Richard Parker after their earliest attestation known at the time) and their different iconographies used side by side, should be a clear warning against taking these variations as arguments for a much later date of the hippopotamuses of the months. Each group could simply have had a functionally different purpose. Since the hippopotamus figures were to be made from different materials, it is likely they had an apotropaic function and were used for protection rituals. The iconographic similarity with Thoeris is probably not coincidental either, as Thoeris is, among other things, a protective deity for pregnancy and childbirth. The month goddesses could thus quite naturally be understood as the protectors of those humans born within their respective month.

At least in the later periods, the month goddesses are called *šps.wt* "noble ladies." The museums in Cambridge, UK, and Brooklyn, New York, own two parts of a doorway of a tomb chapel that once decorated a tomb. The tomb itself dates from the 26th dynasty; the doorjambs, however, belong to the refurbishment of the second owner from the 30th dynasty, who had both himself and the original owner represented on either side of the door.[35] The idea was that he would have done the other man good by restoring the latter's tomb and therefore the two would share the tomb as friends. But not only were the two men united in friendship, their respective *šps.wt* were as well. On each side, a female figure is shown suckling and tending one of the two. Despite their fully anthropomorphic iconography, their names and epithets leave no doubt that the two are actually the goddesses of the third and fourth month of *šmw*, respectively. This is somewhat similar to the medieval European practice of naming a child after the respective Christian saint on whose feast day the child had been born or baptized.[36] Indeed, the goddesses of the months were conceived as tutelary deities of their respective "children."

In Demotic, the word is attested as *špšy.t*, and we find, for example, in a letter to the gods the plea to make the *špšy.t* gracious to the supplicant.[37]

[35] Jansen-Winkeln 1997.
[36] Cf. Vleeming 2010: 104–105.
[37] Hughes 1968.

Similarly, her contentment is promised as a result of a good dream in the dream book Jena 1209.[38] Finally, in the Demotic horoscopes, the term *pꜣ ꜥwỉ špšy.t* "house of the noble lady" is the designation for house V of the *dodekatropos*, a theoretical division of the sky into twelve parts, each of which indicates different aspects of a person's fate such as their children and type of death.[39] House V is the house that is similarly called ἀγαθὴ τύχη in Greek and *bona fortuna* in Latin and is believed to indicate somebody's good fortune.

7 Decades and Decans

The decades of the year are manifest both in the various lists of decans and in the texts on the so-called *Naos of the Decades*, which are in turn closely linked to the decans.[40] This fascinating monument from the reign of Nectanebes (381–363 BCE) gives for each decade of the year five depictions of a *bꜣ*-bird with disk and star in a boat, an armed hawk-headed lion, a crowned ram, a crowned jackal-headed mummy, and finally a mummy on a bier, respectively. Each of these figures is identical for each decade, and each is described in terms of its function in identical labels. The bird is said to be the one on whom water, wind, and crops are dependent in its decade. The armed lion is the form that indicates the decan's responsibility to punish enemies with death in his capacity as a divine messenger. The ram form, to the contrary, is benign and responsible for granting lifetimes. It is explicitly stated that one begs lifetime from him in its respective decade, a point to which we will return later. The standing mummy indicates the decan's cult image in the temple receiving offerings. The reclining mummy finally indicates the decan's own burial in the necropolis,[41] and, consequently, it is the one of whom a good burial is asked. In front of the bird, for each decade there is an indication of the calendrical date within the ideal calendar plus the stock formula "Presenting offerings by the king to this god in Iatnebes (i.e., the place where the *naos* was originally set up), to protect the country from *ꜥb*." The exact meaning of the latter word is unclear. While "impurity" has been proposed, this seems somewhat too harmless in view of the writing with the flame determinative (). Perhaps "inflammation" in the sense of "pestilence of one or the other sort" would be more correct. It would also fit the

38 Zauzich 1980: 96–98, pl. 8.
39 Von Lieven 1999: 123.
40 Leitz 1995: 3–57; von Bomhard 2008; Quack 2010.
41 I.e., when the decan is supposed to be temporarily "dead" during the respective constellation's period of invisibility.

fact that illnesses of all sorts were exactly the most feared kind of dangerous influence ascribed to the decans.

In front of the other four figures is a text, which varies for each decade, indicating what might happen in this decade through the agency of the respective decan god. For example, for the thirtieth decade, that is, for the second month of the heat season (*šmw*), days 21 to 30, we are told, "The great god at the beginning, it is he who creates massacres among the rebels among all the living. It is he who creates fear of the king in the hearts of all the great ones of the foreign lands. It is he who beats all the crooked ones because of the lord of the temple as daily ration."

Neither the figures themselves nor the accompanying text are differentiated, except for the calendrical dates and the predictions for the respective decade. No individual names or iconographic forms of the decans are given, although such forms did exist[42] (Figure 2); as already stated, since the ideal year and the real year were out of tune most of the time, it was impossible to indicate a precise decan in a work intended as an eternal calendar. The precise prediction therefore depends on the decade of the calendar and not on the actual decan, even though there are also manuals attested that clearly connect the predictions with the decans themselves.[43] These however date from a later time, at least in their preserved copies. The *Naos of the Decades* has enthusiastically been claimed as the earliest monument of astrology in Egypt, but as shown, the matter is, strictly speaking, calendrical prognostication, not astrology proper.[44] On the other hand, there are even earlier Egyptian attestations for true divination by stars, particularly from the rising of Sirius.[45] To conclude this part of our discussion, it should be mentioned that the *naos* shows actually thirty-seven decades instead of the thirty-six as expected. The last one is not a decade proper but a representation of the five epagomenal days. The latter were, for obvious reasons of practicality, also furnished with additional decans.

8 Deities of the Hours

The naming of the goddesses of the hours follows different traditions. These goddesses are particularly prominent in the cosmographic treatises that organize the sun god Re's journey through the night or the day. The journey is

[42] Neugebauer and Parker 1969: 105–167; Osing 1998: 187–197; von Lieven 2000b; Quack 2002.

[43] Quack 2002.

[44] Lehoux 2007: 123–126.

[45] Von Lieven 1999: 99–105.

Figure 2 Ophiomorphic decans of the so-called Sethos I B-family, temple of Dendara, ceiling of pronaos (private photo C. Leitz)

divided into twelve hours, separated from one another by a heavily guarded gate. The various treatises indicate the names for the gates and sometimes their guardians. For example, the *Amduat*,[46] a composition detailing the sun's nightly journey through the netherworld, calls the third hour of the night "Who cuts up souls" and the gate of this hour "Robber." The *Book of Gates*,[47] a similar text, calls the third gate "Lady of feeding"; there is no hour goddess here but only several gatekeepers with different names.

The hour goddess in the *Book of the Night*[48] has the same name as the one in the *Amduat*, but the gate name is different. To keep track of these details, the Egyptians at some point started to compose encyclopedic works in their temple libraries, listing all such variant names. Unfortunately, because of the

[46] Hornung 1987, 1992, 1994; translation: Hornung 1989: 57–194.
[47] Hornung 1979–80; translation: Hornung 1989: 195–308.
[48] Roulin 1996.

loss of most temple libraries, we do not know when this process of collecting knowledge in encyclopedic form started. An important example that has been preserved comes from the temple library of Tebtynis in the Fayum; it was preserved because the place was later temporarily abandoned. This library dates to the second century CE, but remnants of papyri from at least the Ptolemaic period have recently been excavated at the same site. The age of the master copies of such works is difficult to pinpoint, as they are mostly lists with not much grammar on which to exercise historic-linguistic dating. The text in question, known as *Onomasticon Tebtynis I*,[49] gives no pictures but rather only several name lists of time-related deities. The first one opens with the title "[Guidance for knowing the na]mes of the hours of the day," followed by a list of twelve hours. For each, its function, its name, and the respective tutelary god are indicated. The names closely follow the liturgical text known as *Ritual of the Hours*.[50] Thus, for the first hour of the day, the following text is given: "Who lets appear the beauty of Re, 1st hour: 'Who appeases for herself.' She stands for Shu, variant: Maat."

The indication of a variant is noteworthy because it shows that Egyptian scholars compared different master copies.[51] In this particular instance, it is to be expected that in the archetype of the text, only an ostrich feather 𓆄 would have been written in place of the divine name. This evaluation makes a hieroglyphic master copy very likely. Later scribes transposing the text into hieratic that uses a fuller orthography interpreted this as either the god Shu or the goddess Maat; both names could be written with the feather but would have been complemented differently. As a matter of fact, for reasons of conventional orthography, "Shu" is likely the original reading. It is therefore no coincidence that this name is indicated first.

Next comes a list, "Knowing the names of the gates of the netherworld." About the first we learn that it is "Lady of Sparkling, 1st gate, the god within it is 'Bull of Light.'" In a similar fashion, the following list contains the names of the hours of the night. A second list with hours of the day was apparently appended as a variant. Unfortunately, the bad preservation of the text prevents us from fully appreciating the subtleties of the distinction. This loss is even sadder for the following section, which also treated the days of the months and their feasts.

In pictorial representations, the hour-goddesses are often shown with either a disk or a star on their heads; sometimes it is also a star within

[49] Osing 1998: 198–204.

[50] Graefe 1995, for a continually updated preliminary edition see Graefe 2014.

[51] This practice is well attested throughout Egyptian textual history. Compare the remarks in von Lieven 2007a: 298–299.

Figure 3 Hours of the day with sun disks, temple of Dendara, ceiling of pronaos (private photo C. Leitz)

a disk. Monuments that differentiate between day and night such as the coffin of Peftjauneith[52] or the ceiling panels in the temple of Dendara (Figures 3, 4), assign the stars to the hours of the night and the disks to those of the day. The symbols are allusions to the means of measuring the hours. In a similar fashion, the illustrations of the *Books of Night and Day* on the ceiling of the sarcophagus hall in the tomb of pharaoh Ramses VI,[53] on the part of the night, paint little stars on the sky goddesses' body; on the part of the day, we find little sun disks, not too surprisingly twelve in number.

9 Solar and Lunar Cycles and Their Relationship with Month and Day

Closely linked to the passing of time are the daily cycle of the sun and the lunar monthly cycle. To be sure, a true lunar month is not precisely thirty days long, as the ideal Egyptian month was. Yet, the names for the thirty days of the

[52] Raven 1992, 60–62.
[53] Piankoff and Rambova 1954: pl. 186.

Figure 4 Hours of the night with stars, temple of Dendara, ceiling of pronaos (private photo C. Leitz)

month clearly refer to the lunar cycle.[54] Interestingly, some festival calendars as well as other evidence suggest that a cycle did not necessarily begin with the New Moon or the first visibility, but with the last visibility of the moon.[55] However, the normal concept is that the New Moon is the first day of the lunar month, its first visibility thus being already the second day.

In principle, each day and thus each lunar phase could also be represented by a personification. In practice, however, the days of the waning moon are rarely represented or spoken of. To the contrary, the fifteen days of the waxing moon from invisibility during new moon until the full moon are very important. They are particularly often represented in the form of fifteen deities who enter the moon to fill it.[56] These deities are the Theban *ennead*. Contrary to what the word "*ennead*" would suggest, it need not consist of

[54] Parker 1950: 11–13; Osing 1998: 205–212.
[55] Von Lieven 2007a: 175–177.
[56] Von Lieven 2000a: 127–132, (folding) pl. 5.

nine gods; rather, the word "nine," that is, three times three, implied a notion of completeness, just like the number three denoted the plural. Even more diverse in its forms was the relationship of the sun and its cycle with the passing of time. As already stated, the twenty-four hours of a day were differentiated by their names, as were the gates that delimit each hour in some of the early cosmographic texts. But other features were believed to change by the hour too. Even divinatory questions to the sun god Re varied within the hours of a day.[57] The *Ritual of the Hours* mentioned earlier addresses the sun god differently in each hour, as expressed also in visual depictions. While various compositions and monuments show certain similarities, huge variety in detail still exists. Some sources make the sun god change shape in four phases from morning as a child or scarab, through noon as an adult or four-headed ram and evening as a ram-headed human or aged man to, finally, night as Atum or Osiris. The latter of course implies that the moon, often identified with Osiris, would not just be the nightly deputy of the sun but rather, actually, the night sun.[58] In view of the mysterious union of Re and Osiris as celebrated in the cosmographies of the *Amduat, Book of Gates*, and the famous depiction of the unified Re-Osiris in the tomb of Nefertari,[59] this does not come as a surprise.

Other sources see the sun god shape-shifting every hour or double-hour via different human, semi-anthropomorphic, and animal shapes. One such concept, in which he transforms into twelve animals over twelve double-hours, found its way in Greek language into the astrological treatises that were disseminated via Alexandria into the whole *oikumene*. This concept is known as *dodekaoros* in the Greek treatises and still lives on today, after a good deal of further metamorphosis of a different kind. After a long travel eastward via the Silk Road, it spawned what we know today as the Chinese or more correctly East Asian Zodiac.[60] While the latter is known in the West mainly as a sort of "zodiac" structuring a year, it was and still is also a system structuring a day divided into twelve double hours. While in the Asian conception these double hours are just allotted an animal each, in the original version still present in some Greek sources on the *dodekaoros*, the concept of twelve animals clearly relates to different forms of the sun within these hours.[61]

57 Roccati 1994.
58 A possible example for the latter would be ceiling panel C in Esna (Sauneron 1969, plate after p. 26; von Lieven 2000a: 81–82, 88).
59 Hornung 1976: 53–54, 60; 1983: 85–87.
60 Von Lieven (in press).
61 The development in all its complexity is discussed in von Lieven (in press).

Figure 5 Solar bark of the fourth hour of the day, Apopis shown as an Asiatic, temple of Dendara, ceiling of pronaos (private photo C. Leitz)

Figure 6 Solar bark of the sixth hour of the day (i.e., noon), the sun god being four-headed, temple of Dendara, ceiling of pronaos (private photo C. Leitz)

Not only the sun god himself shifted shapes and names. The decoration of his bark, its crew, and even his enemy Apopis, who constantly tries to attack him, also change in some of the most elaborate depictions of the cycle on the ceilings of Greco-Roman period temples (Figures 5, 6). In the *pronaos* of Dendara, for example, Apopis in some hours takes the shape of a turtle, an Asiatic foreigner, or a donkey, and, most amazingly, he is also once represented by the mere hieroglyphs of his name. Regardless of his shape, a defending god thrusts a spear into his body, or even just into his name, as the case may be. In four of the twelve hourly pictures, however, Apopis is entirely missing. Strangely, he is never shown as a snake, despite that being his original and most common iconography. The donkey, to the contrary, is normally a form of Seth, the god traditionally defending Re from Apopis, but also notorious as murderer of his own brother Osiris. This is a clear indication that the two original antagonists Seth and Apopis have here been identified because of their shared character as divine enemies.

10 Uses of Knowledge on Time

Egyptian priest-scholars meticulously collected all of the previously described facts about time and its aspects.[62] This knowledge, which was mostly kept in handbooks in the temple libraries but also – thankfully – often transferred to the walls or ceilings of monuments, was important for two reasons. One was knowledge for its own sake as a legitimation to act within this framework. The Egyptians believed that humanity, that is, the Egyptian king in particular, and in his place the priests in all the temples, had to help the sun god maintain the order of the universe through rituals.[63] His defense against Apopis was of special importance. Maintaining cosmic order also automatically implied maintaining political and social order, as some texts make clear. But to play one's role in this universal clockwork, one had to know its workings in minute detail. Texts such as, for example, the cosmographic treatise *Amduat*, which describes the course of the sun during the twelve hours of the night, stress the comprehensive importance of knowledge.

The second reason for the importance of detailed information on the different aspects of time was the wish to control it effectively, so that one might protect oneself and others from its effects. For this, again, knowledge of names and shapes was vital, as knowledge gave the power to manipulate even deities. But how could one manipulate time?

One possibility was to propitiate and pacify deities related to time, so that they grant a good year with prosperity, peace, and health and do not send pestilence, hunger, or rebellion. That such features were time related is clear from the texts concerning the disastrous workings of the messenger demons of the Dangerous Goddess during the epagomenal days as well as from the description of the decades on the *Naos of the Decades*.

[62] On these in general, see Osing 1999.
[63] Compare von Lieven 2010.

One could also have manipulated time to gain additional lifetime. Apparently, the Egyptians believed that one could extend the lifetime allotted to an individual by the gods by means of prayer and correct ritual. A number of sources over the millennia indicate this belief. One of the most interesting is at the same time one of the oldest. It is first attested from the lid of the coffin of the lector priest Sesenebef from the Late Middle Kingdom (labeled *Coffin Text* spell 691).[64] Subsequently, the text was incorporated into the *Book of the Dead* as spell 71 according to the modern numbering.[65] However, if one reads the text without prejudice, it is obvious that it was originally intended to be a ritual for the prolongation of life of a living person, which was to be conducted on New Year's day.[66] Interestingly, the later the copies are, the more the text is adapted to fit its new funerary purpose.

The text starts with seven invocations of deities to let the speaker be whole and to release him. Each time, the god answers in the positive, granting the request. Finally, the speaker says:

> Oh you seven knots that carry the balance in this night of counting the Sound Eye (*wḏꜣ.t*), that cut off heads and sever necks, that grab *ỉb*-hearts and tear out *ḥꜣ.tỉ*-hearts, that commit the massacre in the two flaming islands! I know you, I know your names! You know my name, as I know your names! I address you, you address me, you live through me, I live through you! Command me to life, which is in your arms, to hail, which is in your fists! Allot me to this life of the New Year's day! May he [i.e., either the sun god or the personified New Year's day] let many years descend for me on top of my years of life, may he let many months descend for me on top of my months of life, may he let many days descend for me on top of my days of life, may he let many nights descend for me on top of my nights of life, until I go and shine in view of my picture (*snn*)! Air to my nose! My eyes, they see among these inhabitants of the horizon on this day of counting the spoils.

In one New Kingdom *Book of the Dead* manuscript, the papyrus of Tjenena, the text concludes with an interesting rubric, explaining its usefulness in life and after death: "*Concerning him who regularly performs this spell, it means to be whole on earth with* Re *and a beautiful burial with* Osiris ... *A true remedy, (tried) a million times.*"[67]

This spell exemplifies several of the points made before: the importance of the New Year's day, as well as the importance of the name and of knowledge

[64] De Buck 1956: 322–324; Carrier 2004: 1574–1579.
[65] Lapp 2011: 322–378.
[66] Von Lieven (in preparation).
[67] The part in italics is written in red ink in the original.

about the entities to be entreated. Apparently, the seven invocations are spoken over seven knots. Such knot-amulets are well known from other texts as well as from archaeological finds. The number seven is quite common in this context. Usually, the deities invoked are also to be drawn on the material, which is then tied in a knot. While this is not explicitly mentioned here, the allusion to life (*ꜥnḫ*) and hail (*wꜣs*) in the hands of the addressed beings clearly refers to the usual way of depicting deities with the *wꜣs*-scepter and the *ꜥnḫ*-symbol in their hands. Such drawings were also an important reason to fix the iconography of the gods in texts or pictorial sources. Not just each name had to be correct, but also each god had to be depicted in the correct way to make the ritual effective.

Part, if not all, of the seven invocations seems to address forms or hypostases of the sun god. He is ultimately in charge of the various time units – year, month, day, and night – as they are all defined by his cycle. These units are themselves not personified here, except perhaps the New Year's day. It is not entirely clear from the text whether it is Re himself or the personified New Year's day who dispenses extra time for the supplicant. As stated earlier, the latter is the birthday of the sun god; therefore, the close link between the two is only natural. The mentioning of the New Year's day, as well as the grammatical form in the rubric, shows that this type of long-life ritual was performed cyclically for a living person every first day of a new year, at least during the Middle Kingdom and possibly still later. Because of its supposed usefulness after death as well, it was included in the funerary literature of Ancient Egypt, but it by no means originated there.

Similar rituals must have also existed in later periods, because they are alluded to several times. Moreover, a Late Period literary text preserved in P. Vandier has precisely such an endeavor as its subject.[68] There, however, everything is much more complicated, even outright dangerous for the officiating priest. At the same time, the situation is not a regular prophylactic ritual either but rather a ritual to avert a precise death threat already foretold, and its victim is not just anybody but the king himself. In this situation, it seems that a magician actually has to enter the netherworld to beg the life of the king from Osiris – at the price that he himself is not allowed to return to life again. This Alcestis motif, however, was probably never a real belief held by the Egyptians but merely served the plot for a fantastic tale of magic, adventure, and betrayal. Unfortunately, it is unclear how the story ended,

[68] Posener 1985; improved translation in Hoffmann and Quack 2007: 153–160; further comments in Quack 2013b: 71–72.

because, as so often, clever worms and the tooth of time have robbed us of the parts where it gets really interesting. It is not unlikely, though, that the hero Merire eventually returned to life again, as there are indications that would fit a happy ending.

Be that as it may, long-life rituals and begging the gods for more time are attested in sources indicating their real use. One such ritual was presented earlier. Interestingly, in this ritual the gods are told that they and the supplicant were mutually dependent on each other, clearly alluding to their reception of offerings. Even in P. Vandier, Osiris questions Merire about the state of the temples as well as the social conditions in the country. When he hears that everything is in splendid order, he grants the twenty-five-year-old king another seventy-five years by his grace. A similar conception can already be seen in another, earlier tale, the Report of Wenamun, when Wenamun suggests to a Syrian potentate that he could beg fifty more years of life as surplus to the fate allotted to him from the god Amun by sending the wood he had requested for the god's cult bark.[69] This is in line with a hymn to the same god from a collection attested already a century earlier in P. Leiden I 350.[70] There it is said about Amun, "He prolongs lifetime and he shortens it, he gives a surplus to the fate of the one whom he loves (3,17)."

11 Conclusion

In Ancient Egypt, time and its divisions played an important role in many respects. While there was no personification of the abstract notion of "time" itself, the two terms for "eternity," male *(n)ḥḥ* and female *č̣.t*, could occasionally be associated with gods.[71]

More relevant in the present context are however the individual subdivisions of time, namely the year, the seasons, the months, the decades, and the days. All of them were closely associated with different deities. Some of these were major figures of the pantheon, which were hypostasizing in different shapes according to different periods of time, as, for example, the sun god himself. Others however were in their very essence nothing but personifications of time as, for example, the so-called chronocrats. However, even the chronocrats were regarded as forms of the Dangerous Goddess, that is, Hathor or Tefnut, the daughter of Re, and her partner Horus, throughout

69 Schipper 2005: 89, 109, 208–211, pl. XI.

70 Zandee 1948: 58, pl. III; for similar ideas in other texts, see Morenz 1960: 18–19, 23, 28, 31.

71 Assmann 1991: 90–97; Hornung 1991: 85–87, 102–105; von Lieven 2007a: 192.

a year. Thus, ultimately, all personifications of different units of time seem to have been in general hypostases of major deities in specific functions and during specific periods.

Evidence for such time-related deities can be found in textual as well as pictorial sources. The latter demonstrate that iconographic differentiation was a major factor in organizing the individual systems. Such variation could concern the entire form of a figure or just the head with similar bodies for all members of the same group, for example, the twelve hippopotamus goddesses of the months.

Much of this material is to be found in the context of temples, either as a decoration of the walls or ceilings or in manuscripts from temple libraries, which contain manuals of priestly knowledge. Such knowledge enabled the priests to officiate within rituals linked to cosmological cycles, such as the course of the sun. It also gave them power over time, which could be used to manipulate it to the benefit of individual human beings. Thus, time's importance in the real world, for example, in relation to the agricultural cycle opening with the inundation at the ideal start of the year, directly translated into its religious importance.

If the gods personifying certain time divisions were correctly pacified by ritual activity and offerings to secure their protection and goodwill, one could harness their power for one's own good, even requesting extra years, months, or days, on top of the lifetime that one had been allotted by fate. This is demonstrated by a spell preserved in a funerary adaptation among the so-called *Coffin Texts* and later the *Book of the Dead*. This spell affords a unique glimpse into the possible original types of use of knowledge on divine personifications of time. While many of these constructions originally would have served the needs of the king, the attestation of this spell on private documents demonstrates that other members of the elite also had access to such practices. As the Ancient Egyptian culture and religion continuously lasted for at least four millennia, they must have been quite successful at this.

Bibliography

Assmann, J. 1991. *Ägypten – Theologie und Frömmigkeit einer frühen Hochkultur*. 2nd ed. Stuttgart: Kohlhammer.

Bomhard, A.-S. von 2008. *The* Naos of the Decades: *Underwater Archaeology in the Canopic Region in Egypt*. Oxford: Oxford Centre for Maritime Archaeology.

Bomhard, A.-S. von 2014. "Le début du Livre de Nout." *Égypte Nilotique et Méditerranéenne* 7: 79–123.

Bommas, M. 1999. *Die Mythisierung der Zeit: Die beiden Bücher über die altägyptischen Schalttage des magischen pLeiden I 346*. Göttinger Orient-Forschungen IV 37. Wiesbaden: Otto Harrassowitz.

Borchardt, L. 1920. *Die altägyptische Zeitmessung*. Die Geschichte der Zeitmessung und der Uhren 1 B. Berlin: de Gruyter.

Brandl, H. and J. F. Quack. 2013. "A Bichrome Faience Statuette of Bastet from the Reign of Takeloth III." In F. Hoffmann et al. (eds.), *Kleine Götter – große Götter: Festschrift für Dieter Kessler zum 65. Geburtstag*. Tuna el-Gebel 4. N.p.: Verlag Patrick Brose, 67–89.

Buck, A. de 1956. *The Egyptian Coffin Texts VI*. OIP 81. Chicago: University of Chicago Press.

Carrier, C. 2004. *Textes des Sarcophages du Moyen Empire Égyptien II*. Lonrai: Éditions du Rocher.

Chassinat, É. 1931. *Le temple d'Edfou VI*. Cairo: Institut Français d'Archéologie Orientale.

Depuydt, L. 1997. *Civil Calendar and Lunar Calendar in Ancient Egypt*. Orientalia Lovaniensia Analecta 77. Leuven: Peeters.

Derchain, P. 1978. "En l'an 363 de Sa Majesté le Roi de Haute et Basse Égypte Râ-Harakhty vivant par-delà le Temps et l'Espace." *Chronique d'Égypte* 53: 48–56.

Fischer-Elfert, H.-W. 2008. "Weiteres zur Göttlichkeit der Natur – Fragmente eines spräthieratischen Lexikons (Pap. Hal. Kurth Inv. 33 A-C [Halle/Saale])." *Zeitschrift für ägyptische Sprache und Altertumskunde* 135: 115–130.

Frankfort, H. 1933. *The Cenotaph of Seti I at Abydos*. Egypt Exploration Society Memoir 39. London: The Egypt Exploration Society.

Germond, P. 1986. *Les invocations à la bonne année au temple d'Edfou*. Aegyptiaca Helvetica 11. Geneva: Éditions de Belles-Lettres.

Graefe, E. 1995. "Das Stundenritual in thebanischen Gräbern der Spätzeit (über den Stand der Arbeit an der Edition)." In J. Assmann et al. (eds.), *Thebanische Beamtennekropolen: neue Perspektiven archäologischer Forschung. Internationales Symposion, Heidelberg, 9.–13.6.1993*. Studien zur Archäologie und Geschichte Altägyptens 12. Heidelberg: Heidelberger Orientverlag, 85–93.

Graefe, E. 2014. "Das Stundenritual." 2014 version, www.uni-muenster.de/IAEK/org/WMA/graefe/stunden/index.html (accessed September 22, 2015).

Harris, J. R. 1961. *Lexicographical Studies in Ancient Egyptian Minerals*. Berlin: Akademie-Verlag.

Hoffmann, F. and J. F. Quack. 2007. *Anthologie der demotischen Literatur*. Einführungen und Quellentexte zur Ägyptologie 4. Berlin: LIT Verlag.

Hornung, E. 1976. *Das Buch der Anbetung des Re im Westen II*. Aegyptiaca Helvetica 3. Geneva: Éditions de Belles-Lettres.

Hornung, E. 1979–80. *Das Buch von den Pforten des Jenseits I–II*. Aegyptiaca Helvetica 7–8. Geneva: Éditions de Belles-Lettres (published 1984).

Hornung, E. 1983. *Der Eine und die Vielen: Ägyptische Gottesvorstellungen*. 3rd ed. Darmstadt: Wissenschaftliche Buchgesellschaft.

Hornung, E. 1987, 1992, 1994. *Texte zum Amduat I–III*. Aegyptiaca Helvetica 13–15. Geneva: Éditions de Belles-Lettres.

Hornung, E. 1989. *Ägyptische Unterweltsbücher*. 3rd ed. Zürich: Artemis.

Hornung, E. 1991. *Der ägyptische Mythos von der Himmelskuh: Eine Ätiologie des Unvollkommenen*. Orbis Biblicus et Orientalis 46. 2nd ed. Fribourg: Universitätsverlag – Göttingen: Vandenhoeck & Ruprecht.

Hughes, G. R. 1968. "A Demotic Plea to Thoth in the Library of G. Michaelides." *Journal of Egyptian Archaeology* 54: 176–182, pl. XXVIII.

Jansen-Winkeln, K. 1997. "Eine Grabübernahme in der 30. Dynastie." *Journal of Egyptian Archaeology* 83: 169–178.

Kaper, O. 2014. "The Three Seasons in the Soubassement." In A. Rickert and B. Ventker (eds.), *Altägyptische Enzyklopädien: Die Soubassements in den Tempeln der griechisch-römischen Zeit. Soubassementstudien I.* Studien zur spätägyptischen Religion 7. Wiesbaden: Otto Harrassowitz, 517–525.

Kurth, D. 2014. *Edfou VI.* Die Inschriften des Tempels von Edfu. Abteilung I Übersetzungen, 3. Gladbeck: PeWe-Verlag.

Lapp, G. 2011. *Die prt-m-hrw-Sprüche (Tb 2, 64–72): Synoptische Textausgabe nach Quellen des Neuen Reiches.* Totenbuch-Texte 7. Basle: Orientverlag.

Lehoux, D. 2007. *Astronomy, Weather, and Calendars in the Ancient World: Parapegmata and Related Texts in Classical and Near-Eastern Societies.* Cambridge: Cambridge University Press.

Leitz, C. 1993. "Die Nacht des Kindes in seinem Nest in Dendara." *Zeitschrift für ägyptische Sprache und Altertumskunde* 120: 136–165, 181.

Leitz, C. 1995. *Altägyptische Sternuhren.* Orientalia Lovaniensia Analecta 62. Leuven: Peeters.

Leitz, C. 2002. "Die Chronokratenliste von Edfu – Ein Pantheon aus der zweiten Hälfte der 12. Dynastie." *Revue d'Égyptologie* 53: 138–156.

Leitz, C. 2014. *Die Gaumonographien in Edfu und ihre Papyrusvarianten. Ein überregionaler Kanon kultischen Wissens im spätzeitlichen Ägypten.* Studien zur spätägyptischen Religion 9. Wiesbaden: Otto Harrassowitz.

Leitz, C., D. Mendel, and Y. El-Masry. 2010. *Athribis II.* Cairo: Institut Français d'Archéologie Orientale.

Lieven, A. von 1999. "Divination in Ägypten." *Altorientalische Forschungen* 26: 77–126.

Lieven, A. von 2000a. *Der Himmel über Esna: Eine Fallstudie zur Religiösen Astronomie in Ägypten am Beispiel der kosmologischen Decken- und Architravinschriften im Tempel von Esna.* ÄgAbh 64. Wiesbaden: Otto Harrassowitz.

Lieven, A. von 2000b. "Die dritte Reihe der Dekane *oder* Tradition und Innovation in der spätägyptischen Religion." *Archiv für Religionsgeschichte* 2: 21–36.

Lieven, A. von 2001. "Der Himmel über Esna – Nachtrag zu Esna 451." *Göttinger Miszellen* 184: 111–112.

Lieven, A. von 2004. "Das Göttliche in der Natur erkennen: Tiere, Pflanzen und Phänomene der unbelebten Natur als Manifestationen des Göttlichen (mit einer Edition der Baumliste P. Berlin 29027)." *Zeitschrift für ägyptische Sprache und Altertumskunde* 131: 156–172, pls. XX–XXI.

Lieven, A. von 2007a. *Grundriß des Laufes der Sterne: Das sogenannte Nutbuch.* The Carlsberg Papyri 8, Carsten Niebuhr Institute Publications 31. Copenhagen: Museum Tusculanum Press.

Lieven, A. von 2007b. *Heiligenkult und Vergöttlichung im Alten Ägypten.* Habilitation Thesis, FU Berlin (publication as a monograph in preparation).

Lieven, A. von 2010. "'The Soul of the Sun Permeates the Whole World': Sun Cult and Religious Astronomy in Ancient Egypt." *Pandanus* 10 4/2: 29–60 (= A. Gail and J. Vacek [eds.], *Special Issue Sun Worship in the Civilizations of the World*).

Lieven, A. von 2012. "Schweine, Fische, Insekten und Sterne: Über das bemerkenswerte Leben der Dekane nach dem Grundriß des Laufes der Sterne." In M. Geller and K. Geus (eds.), *Productive Errors: Scientific Concepts in Antiquity*. TOPOI – Dahlem Seminar for the History of Ancient Sciences, Max-Planck-Institut für Wissenschaftsgeschichte Berlin, Preprint 430, Berlin: Max-Planck-Institut für Wissenschaftsgeschichte, 115–141 (also online: pubman.mpiwg-berlin.mpg.de/pubman/item/escidoc:643510:4/component/escidoc:644382/P430.pdf [accessed September 22, 2015]).

Lieven, A. von 2013. Review of D. Wuensch and Sommer, K. P. (eds.), Ludwig Borchardt, Die altägyptische Zeitmessung. Reprint mit einer Einleitung von Daniela Wuensch: Was die alten Ägypter über Uhren und Zeitmessung wussten, Göttingen 2013. *British Journal for the History of Science* 46: 716–718.

Lieven, A. von 2016. "The Movement of Time: News from the 'Clockmaker' Amenemhet." In: R. Landgráfová and J. Mynářová (eds.), *Rich and Great: Studies in Honour of A. J. Spalinger on the Occasion of his 70th Feast of Thot*. Prague: Charles University, Faculty of Arts, 207–231.

Lieven, A. von (in press). "From Crocodile to Dragon: History and Transformations of the Dodekaoros." In: D. Brown and H. Falk (eds.), *The Transmission and Appropriation of Astral Sciences in Pre-Islamic Times*.

Lieven, A. von (in preparation). *Die nichtfunerären Ursprünge ausgewählter Sprüche der sogenannten Sargtexte*.

Meeks, D. 2006. *Mythes et légendes du Delta d'après le papyrus Brooklyn 47.218.84*. Mémoires de l'Institut Français d'Archéologie Orientale 125. Cairo: Institut Français d'Archéologie Orientale.

Mendel, D. 2005. *Die Monatsgöttinnen in Tempeln und im privaten Kult*. Rites Égyptiens XI. Turnhout: Brepols.

Meulenaere, H. de 1982. "La statue d'un vizir thébain: Philadelphia, University Museum E. 16025." *Journal of Egyptian Archaeology* 68: 139-144.

Morenz, S. 1960. *Untersuchungen zur Rolle des Schicksals in der ägyptischen Religion*. Abhandlungen der Sächsischen Akademie der Wissenschaften 52,1. Berlin: Akademie-Verlag.

Neugebauer, O. and R. A. Parker. 1960. *Egyptian Astronomical Texts I: The Early Decans*. London: Brown University Press.

Neugebauer, O. and R. A. Parker. 1969. *Egyptian Astronomical Texts III: Decans, Planets, Constellations and Zodiacs*. London: Brown University Press.

Osing, J. 1998. *Hieratische Papyri aus Tebtunis I*. The Carlsberg Papyri 2, Carsten Niebuhr Institute Publications 17. Copenhagen: Museum Tusculanum Press.

Osing, J. 1999. "La science sacerdotale." In D. Valbelle and J. Leclant (eds.), *Le Décret de Memphis*. Paris: De Boccard, 127–140.

Pantalacci, L. 1995. "Compagnies de gardiens au temple d'el-Qal'a." In D. Kurth (ed.), *3. Ägyptologische Tempeltagung: Systeme und Programme der ägyptischen Tempeldekoration*. Ägypten und Altes Testament 33,1. Wiesbaden: Otto Harrassowitz, 187–198.

Parker, R. A. 1950. *The Calendars of Ancient Egypt*. Studies in Ancient Oriental Civilization 26. Chicago: University of Chicago Press.

Pfeiffer, S. 2004. *Das Dekret von Kanopos (238 v. Chr.)*. Archiv für Papyrusforschung Beiheft 18. München: K. G. Saur.

Piankoff, A. and N. Rambova. 1954. *The Tomb of Ramesses VI*. Egyptian Religious Texts and Representations, Bollingen Series XL,1. New York: Bollingen Foundation/ Pantheon Books.

Posener, G. 1985. *Le Papyrus Vandier*. Cairo: Institut Français d'Archéologie Orientale.

Quack, J. F. 2002. *Beiträge zu den ägyptischen Dekanen und ihrer Rezeption in der griechisch-römischen Welt*. Habilitation Thesis, FU Berlin (publication in preparation).

Quack, J. F. 2010. "The Naos of the Decades and Its Place in Egyptian Astrology." In D. Robinson and A. Wilson (eds.), *Alexandria and the North-West Delta. Joint Conference Proceedings of Alexandria: City and Harbour (Oxford 2004) and The Trade, Topography and Material Culture of Egypt's North-West Delta, 8th Century BC to 8th Century AD (Berlin 2006)*. Monograph 5. Oxford: Centre for Maritime Archaeology, 175–181.

Quack, J. F. 2013a. "Zeit, Krise und Bewältigung: Ägyptische Zeiteinheiten, ihre Schutzgötter und deren bildliche Umsetzung." In T. Greub (ed.), *Das Bild der Jahreszeiten im Wandel der Kulturen und Zeiten*. Paderborn: Wilhelm Fink, 73–98.

Quack, J. F. 2013b. "Apports récents des études démotiques à la compréhension du livre deux d'Hérodote." In L. Coulon, P. Giovannelli-Jouanna, and F. Kimmel-Clauzet (eds.), *Hérodote et l'Égypte: Regards croisés sur le livre II de l'Enquête d'Hérodote*, Lyon: Maison de l'Orient et de la Méditerranée, 63–88.

Raven, M. 1992. *De dodencultus van het Oude Egypte*. Amsterdam: De Bataafsche Leeuw.

Roccati, A. 1994. "." In C. Berger, G. Clerc, and N. Grimal (eds.), *Hommages à Jean Leclant I*. Bibliothèque d'Étude 106/1. Cairo: Institut Français d'Archéologie Orientale, 493–497.

Roulin, G. 1996. *Le livre de la Nuit: Une composition égyptienne de l'au-delà*. Orbis Biblicus et Orientalis 147. Fribourg: Universitätsverlag – Göttingen: Vandenhoeck & Ruprecht.

Sauneron, S. 1969. *Le temple d'Esna IV*. Cairo: Institut Français d'Archéologie Orientale.

Sauneron, S. 1983. *Villes et Légendes d'Égypte*. Bibliothèque d'Étude 90. Cairo: Institut Français d'Archéologie Orientale.

Schipper, B. U. 2005. *Die Erzählung des Wenamun: Ein Literaturwerk im Spannungsfeld von Politik, Geschichte und Religion*. Orbis Biblicus et Orientalis 209. Fribourg: Universitätsverlag – Göttingen: Vandenhoeck & Ruprecht.

Schlick-Nolte, B., R. Werthmann, and C. E. Loeben. 2011. "An Outstanding Glass Statuette Owned by Pharaoh Amenhotep II and Other Early Egyptian Glass Inscribed with Royal Names." *Journal of Glass Studies* 53: 11–44.

Schlögl, H. 2013. "Gebet zur Genesung des kranken Königs Thutmosis IV." *Sokar* 27: 52–57.

Spalinger, A. 1995. "Some Remarks on the Epagomenal Days in Ancient Egypt." *Journal of Near Eastern Studies* 54: 33–47.

Vleeming, S. P. 2010. "Some Notes on P. Hawara OI 1." *Göttinger Miszellen* 225: 99–108.

Wuensch, D. and Sommer, K. P. (eds.). 2013. *Ludwig Borchardt, Die altägyptische Zeitmessung. Reprint mit einer Einleitung von Daniela Wuensch: Was die alten Ägypter über Uhren und Zeitmessung wussten*, Göttingen: Termessos Verlag.

Yoyotte, J. 1980. "Une monumentale litanie de granit: Les Sekhmet d'Amenophis III et la conjuration permanente de la Déesse dangereuse." *Bulletin de la Société Française d'Égyptologie* 87/88: 46–75.

Zandee, J. 1948. *De Hymnen aan Amon van Papyrus Leiden I 350*. Oudheidkundige Mededelingen uit het Rijksmuseum van Oudheden te Leiden 28. Leiden: Brill.

Zauzich, K.-T. 1980. "Aus zwei demotischen Traumbüchern." *Archiv für Papyrusforschung* 27: 91–98.

8

THE MOON AND THE POWER OF TIME RECKONING IN ANCIENT MESOPOTAMIA

Lorenzo Verderame

The ancient Mesopotamian calendrical system was based on a year of twelve months, to which a thirteenth month was occasionally added to readjust the lunar cycle with the seasons, and in particular with the heliacal risings of stars.[1] The year began in mid-March after the equinox, coinciding with the barley harvest, and was celebrated in the New Year festival. The beginning of each month was based on direct observation of the rising of the New Moon.[2] Similarly, the intercalation of the thirteenth month did not follow a fixed scheme[3] but was rather established by the king, who seems to have had control over the calendar.[4] This feature, namely royal control over the calendar, finds both its meaning and expression in the association of the moon with kingship, both subject to periodical renewal.

In this chapter, I discuss the Mesopotamian concept of time, from its mythological foundation to its ritual marking by means of festivals, taking the New Year celebrations and the role played in them by the Babylonian *Poem of Creation* (*Enūma eliš*)[5] as the focus of the analysis. This poem is the most important, extensive, and completely preserved mythological and theological work of the Babylonian civilization. Its recitation during the

Many thanks are owed to Jonathan Ben-Dov and Lutz Doering for inviting me to the Durham conference, *The Construction of Time in Antiquity*, and for their valuable comments on the submitted manuscript. Transliteration follows the conventions in Assyriology (Roman characters for Sumerian, italicized for Akkadian). Translations from Akkadian and Sumerian sources not otherwise referenced are mine.

[1] For an overview, see Cohen 1993; Rochberg 1995; Britton 2007; Verderame 2008; Steele 2011.

[2] See in general Steele 2007, 2012.

[3] Britton 2007; Steele 2011: 475–478.

[4] On the connection between royal power and the calendar in Mesopotamia, see Steele 2012.

[5] Lambert 2013.

New Year festival directly links myth and ritual. The oldest copies of the work are dated to the beginning of the first millennium BCE, but its composition goes back several centuries earlier, to the same period when the so-called standard Mesopotamian calendar was adopted or, perhaps, imposed. The fact that *Enūma eliš* was composed around the same time that the standard calendar was adopted allows us to treat the pertinent sources within the same temporal and cultural frame.

Furthermore, I explore the relation between the moon and the king in the light of the evidence provided by the *Poem of Creation*. Academic discourse on ancient Mesopotamian royal institutions has been focused mainly on the various types of rulership existing in the formative period (Early Dynastic). With few exceptions, the relation of specific gods with kingship has been only vaguely sketched. What is more, when it comes to the identification of the king with celestial bodies, studies have only paid attention to solar imagery.[6] The association of the moon with kingship has never been properly examined and will be analyzed in the second part of the chapter.

1 The Creation of Time in the Babylonian *Poem of Creation*

The composition known as the *Poem of Creation* or, from its incipit, *Enūma eliš* ("When above") celebrates the victory of Marduk the god of Babylon over the primeval chaotic forces, as well as the successive creation of the world. The enemy is Tiamat (the personified salty waters or the sea), who together with Abzu (or *apsû*, the sweet waters) constitute one of the primordial couples or opposites who generated the successive genealogy of gods. It is the killing of Abzu by his own offspring that unleashes the rage of Tiamat, who then creates an army of monsters led by Qingu. Marduk defeats the opponents and fashions the world from the body of Tiamat. Having organized the different parts of the world, he makes the stars appear in order to set the calendar, assigning three stars or constellations to each month. Nēberu, possibly the Pole star, an emanation and hypostasis of the same Marduk, is fixed as a guide for the other stars. Nēberu stands at the central path of heaven, governing the other two paths named after the gods Enlil and Ea:[7]

> He fashioned heavenly stations for the great gods,
> And set up constellations, the patterns of the stars.
> He appointed the year, marked off divisions,
> And set up three stars each for the twelve months.

[6] See e.g. Charpin 2013 and Frahm 2013.
[7] The third path, that of An, is not mentioned here; see Horowitz 1998: 252–258, 2007.

After he had organized the year,
He established the heavenly station of Nēberu to fix the stars' intervals.
That none should transgress or be slothful
He fixed the heavenly stations of Enlil and Ea with it.
(*Poem of Creation* V 1–8)[8]

The first time unit to be reckoned is the year, which is bound to and marked by the movements of the stars. This system was central and well known in Mesopotamia, particularly for the determination of festival and ritual activities. The specific mention in the *Poem of Creation* of three stars per month recalls the so-called *Zwölfmaldrei* system and in general the compositions improperly called "astrolabes," which in Akkadian receive the name of "three stars per each (month)," that is, almanacs providing ritual and mythical explanations of the calendar.[9] However, heliacal rising and setting of heavenly bodies must have been known long before the first written star catalogs began to be compiled in cuneiform writing. Their observation and significance for calendrical as well as for occasional ritual performances are documented in scanty but prominent references from the third millennium BCE.[10] According to the Sumerian composition *The Farmer's Instructions*, the sowing season was announced by the rising of certain constellations, marking the beginning of the agricultural year:

When the constellations in the sky are right,
Do not be reluctant to take the oxen force to the field many times.
The hoe should work everything.
(*The Farmer's Instructions* 38–40)

In the twenty-first century BCE, the ruler of Lagaš, Gudea, waits for the rising of a constellation to begin the construction of the Eninnu, the temple of the city god Ninĝirsu. After receiving instructions for the temple building in a dream, Gudea goes to the goddess Nanše to have them interpreted. One of the figures appearing in Gudea's dream is a woman with sheaves in her hair, who holds a silver stylus in her hand and a tablet with the heavenly stars on her lap. The deity Nanše reveals to Gudea that the woman is Nisaba,[11] the goddess of cereals and accounting, who is announcing to him the rising of the star that will mark the beginning of the rituals for the temple building:

8 Lambert 2013: 98–99.
9 Horowitz 2007; see later discussion of the *Astrolabe B*.
10 Reculeau 2002; Verderame 2003, 2008.
11 For the relation of Nisaba with the stars, see Verderame 2008: 124–125.

> Then there was a woman – whoever she was. She . . . sheaves. She held a stylus of refined silver in her hand, and placed it on a tablet with propitious stars, and was consulting it.
>
> . . .
>
> (The young woman . . .) was in fact my sister Nisaba. She announced to you the holy stars auguring the building of the house.
>
> (*Gudea Cylinder A*: IV 19–23 and V 19–26)

These early references may suggest that the marking of large periods of time based on the heliacal rising of stars was the oldest form of time reckoning in Mesopotamia, a conclusion that is corroborated by evidence from other cultures.[12]

While the reckoning of the year is marked by the movement of stars, that of its subunits – the month, the week, and the day – is related to the cyclical movements of the moon. In the *Poem of Creation*, Marduk, having established the course of the stars and the year, entrusts the mark of the daily time to the moon god. Marduk first opens two gates at the opposite sides of the sky vault,[13] which are the points from which the moon and the sun enter and exit when running their daily and nightly course:[14]

> Gates he opened on both sides
> And put strong bolts at the left and the right.
> He placed the heights (of heaven) in her (Tiamat's) belly.
>
> (*Poem of Creation* V 9–11)[15]

The moon god Sîn, dubbed Nannar in the poem, is then created,[16] and the fixing of the day and the month is entrusted to him. Moving in the sky like a crown, the moon will mark the four periods of seven days each, concluding its cycle and the month, disappearing and meeting with the sun to establish the destinies of the country for the next month. The beginning of each of the four quarter-months, particularly the first two quarters during the waxing of

12 Nilsson 1920; Brelich 1954.

13 Horowitz 1998: 266–267.

14 The sun may rise from an opening in the cosmic mountain, which is called Mašû and guarded by the scorpion-men according to the *Epic of Gilgameš* IX 38–45; see Horowitz 1998: 98–100, 331–332.

15 Lambert 2013: 98–99.

16 The *Poem of Creation* was composed to exalt the figure of Marduk and expresses the point of view of his theologians. However, the moon god Nanna(r)/Su'en in third-millennium sources was an old and important deity, one of the seven great gods (Anunna) and leader of the astral triad. The Babylonian writers adopted here the Sumerian name Nanna(r), which is derived from a Semitic root **nwr* "to be splendent," meaning it to be a learned reference to the older Sumerian tradition.

the moon, served to fix the basic frame of the regular religious celebrations and offerings throughout the month:

> He created Nannar, entrusting to him the night.
> He appointed him as the jewel of the night to fix the days,
> And month by month without ceasing he elevated him with a crown,
> (Saying,) "Shine over the land at the beginning of the month,
> Resplendent with horns to fix six days.
> On the seventh day the crown will be half size,
> On the fifteenth day, halfway through each month, stand in opposition.
> When Šamaš [sees] you on the horizon,
> Diminish in the proper stages and shine backwards.
> On the 29th day, draw near to the path of Šamaš,
> . . . the 30th day, stand in conjunction and rival Šamaš."
> (*Poem of Creation* V 12–22)[17]

The comparison of the moon to a crown directly associates its monthly renewal and the regeneration of kingship.[18] This monthly renewal of both the heavenly body and kingship reaches its apex during the celebration of the New Year festival in the first month of the year, Nisan. No wonder then that Nisan is defined as the month of the god Sîn and of the lifting and reassessing of kingship (see later discussion in § 3).

In contrast to the moon, the role of the sun in time reckoning is limited. A section in the *Poem of Creation*, unfortunately partly destroyed, contained the creation and the functions of the sun. In general, however, the preeminent role of the moon in time reckoning relegates the sun and the sun god to a subordinate position, expressed by a filiation relationship (the sun god Utu/Šamaš is the son of the moon god Nanna(r)/Sîn).[19] The calendrical attributes of the sun god are rather limited, as is the attention of the Mesopotamian scribes to the motions of the sun. This divine figure excels in other divine aspects, particularly as god of justice, mercy, war, and divination. These aspects are expressed through the symbolism of the light that can reach everything.[20]

> Your beams are ever-mastering secrets,
> At the brightness of your light, humankind's footprints become vis[ible].
> Your dazzle is always seeking out . . .,
> The four world regions [you set alight] like fire.

[17] Lambert 2013: 98–99.
[18] For the crown (*agû*/aga) of the moon, see Verderame 2002: 60–62; for the relation of the lunar crown to that of the king, see later and n. 48. The shape of the moon is compared also with such other objects as a wagon, a boat, a wheel, a fruit, or a kidney (Stol 1992).
[19] Verderame 2003: 26–29.
[20] Frahm 2013: 99–101.

You open wide the gate of all [sanctuaries],
You . . . the food offerings of the Igigi-gods.
O Šamaš, humankind kneels to your rising,
All countries . . .
Illuminator of darkness, opener of heaven's bosom,
Hastener of the morning breeze (for?) the grain field, life of the land,
Your splendour envelops the distant mountains,
Your glare has filled all the lands.
Leaning over the mountains, you inspect the earth,
You balance the disk of the world in the midst of heaven (for) the circle of the lands.

(*Šamaš Hymn* 9–22)[21]

The relationship between the sun and the moon exceeds the ties of filiation, leading to a sort of syncretism, whereby the two divine figures share a common series of epithets.[22] In particular, the moon's brightness is compared to the sun's light or described with expressions typical of the latter; on the other hand, Šamaš assumes the calendrical powers and duties of his father, the moon god:

O Sîn and Šamaš, gods both,
Sîn of the night, Šamaš of all the day,
You <pronounce> the verdicts of heaven and earth.
You look each day upon the dimensions of day, month, and year.

(*Hymn to Sîn and Šamaš* 1–4)[23]

The excess of this syncretism can be found in the bilingual introduction to the astrological series *Enūma Anu Enlil*, where an "archaizing" pseudo-Sumerian version that attributes to the moon god Su'en its traditional role as a marker of time precedes an Akkadian version where the sun god Šamaš fulfils that role:

(Sumerian version) When An, Enlil, and Enki, the great gods, in their immutable counsel, had established the great *powers* (ME) of heaven and earth and the boat of Su'en and fixed the constant shining of the (Moon) crescent, the birth of the month and the signs in heaven and earth, (then) the heavenly boat (the moon), after having been made to appear, splendidly came out in the sky.

(Akkadian version) Variant: When An, Enlil, and Ea, the great gods, in their immutable counsel, established the great *designs* (GIŠ.HUR) of heaven

[21] Foster 2005: 628.
[22] Verderame 2003: 27.
[23] Foster 2005: 762.

and earth and assigned them to the great gods, then the creation of the day, the renewal of the month and the things related to the observation – *the humanity saw them at the rising of the sun from its door* – these became immutably visible in the midst of sky and earth. (*Enūma Anu Enlil* I §1)[24]

The cosmogony described in the *Poem of Creation* offers an overview of the mythical foundation of time and its reckoning. This idea is confirmed by other cosmogonies and literary parallels,[25] proving that time, or time reckoning, is established in the beginning as one, or the first, of the creative acts in the cosmogonies. While created before humanity, time and its markers are created for humanity.[26] Time and the heavenly bodies are firmly related to each other, the latter being created mainly with the aim of allowing men to reckon time.

The time created in the cosmogonies is thus a human dimension, since the gods precede the creation of (human) time. This last consideration, that is, a different temporality for divine and human events, develops the distinction between mythical and historical time. Mythical time refers to the moment when the divine events leading to the creation and foundation of reality take place. Cosmogony is the point when mythical time ends and historical time begins. Mythical time is therefore necessarily separated from historical time.

Furthermore, the previous and the sequence of heavenly bodies in the *Poem of Creation* (stars, moon, sun) show a hierarchic perspective, suggesting that the observation of rising stars was the main and oldest system of marking the larger time cycle (year).

2 The New Year Festival (*akītu*)

Periodicity and repetition stand at the base of the qualitative discourse on time, in which the main distinction is between everyday, routine normal activity and sacred time.[27] Natural as well as socioeconomic cycles are marked by periods of more intensive religious activity – the festivals, which are the basic, original,

[24] Verderame 2002: 2, 9, 13.

[25] Verderame 2003, 2008.

[26] See in particular the previously mentioned passage of the *Enūma Anu Enlil* introduction; see also Nilsson 1920: 151; Brelich 1954: 36.

[27] In scholarly discussions, historical time or time as perceived by the human community has always been described by means of dichotomies such as sacred/profane, qualitative/quantitative, repetitive/growing, circular/linear. Compare Eliade 1949, on the one hand, and Hubert 1905, on the other, to mention only two of the most influential studies on time in the history of religions; for a general discussion, see Brelich 1954: 43–48; and compare Gingrich 1994; Pirenne-Delforge and Tunca 2003. For a critical overview of the debate on time in social sciences, see Munn 1992.

and crucial moments around which the calendar has been structured.[28] The role of festivals is, on the one hand, to mark the sacred space-time and separate it from ordinary time and, on the other, to regenerate reality by re-enacting the first founding act narrated in the myth.[29]

In Mesopotamia, as in other polytheistic cultures, religious activity covers the entire calendar year. While each month features a major festival, minor feasts were celebrated together with the regular celebrations determined by the lunar cycle. The differentiation of time reaches its most fundamental level with the smallest units, the day and its parts.[30] Days are qualified as good or bad and consecrated to specific gods with their own prohibitions.[31]

The main festival of the Mesopotamian cycle, which became the pivot of the calendar, was the celebration of the *akītu*. This celebration took place twice a year, marking a period of six months that coincides with the two equinoxes and, therefore, with the moments of cereal harvesting and sowing.[32] The oldest reference to these festivals dates back to the mid–third millennium BCE.[33] While older traditions support the preeminence of the *akītu* of the moon god Nanna(r)/Su'en, the *akītu* of other gods is also known, within expressions of local festivals.[34] The long process of political unification of southern Mesopotamia led to the adoption or imposition of an official "standard" calendar, which was a combination of the main local festivals. Thus the first-millennium *akītu*s appear as complex and stratified celebrations, which merged centuries of local as well as other traditions. While *akītu*s of different gods and towns continued to be celebrated, the most important one was that of Marduk in Babylon, which coincided with the beginning of the year.

28 According to Brelich 1954: 53–54, "periodical festivals are regularly connected with a 'crisis' that in substance always means the disintegration – or the risk of disintegration – of an order and implies the need to restore the compromised order . . . or to establish a new one . . . the 'new' order must rely on pre-existing and permanent models (archetypes) which are fixed in the stable forms of the myth and ritual. The need to link up to a permanent order explains the recourse to festival celebrations, that is to say, to the contact with the 'sacred time' which is external to the tottering human world" (translation from the Italian is mine).

29 Here is not the appropriate place to discuss the relation of myth and ritual; see Kirk 1973 and Jacobsen 1975.

30 Verderame 2008: 128–130.

31 Most of the hemerologies that have reached us must, however, be considered as calendars or almanac of a specific group of persons performing religious duties, see in general Brelich 1954: 43–48; Gingrich 1994; and, for ancient Mesopotamia, Livingstone 2007.

32 For the *akītu*, see in general Bidmead 2002 and Zgoll 2006 with previous bibliography; for the second *akītu*, see Ambos 2013.

33 The earliest reference is in an administrative text from Fara (*TSŠ* 881: ii 1'–2'); see Cohen 1993: 401.

34 Cohen 1993: 400–453; Bidmead 2002.

During the *akītu*, on the fourth day, the *Poem of Creation* was recited. The strict correlation between the mythical composition and the celebration emerges in the ritual acts, which recall the mythological narrative at different levels. The celebration is the enactment of Marduk's fight against chaos and of his subsequent fashioning of the world. The mythical composition and the rituals of the New Year festival are intertwined, since the former founds and instructs the latter, and the latter repeats and recalls the former.

Theological and cultic explanatory works of the first millennium BCE further substantiate the relation between myth and ritual. The Assyrian and Babylonian commentators offer the equation of each ritual act and object with episodes from the narrative of the *Poem of Creation*. The meaning of most of these explanations escapes us, having little or apparently random relation with the *Poem of Creation*:

> It is said in *Enūma eliš*: When heaven and earth were not created, Aššur came i[nto being.][35]
>
> [The *brazie*]*r* which is lighted in front of Mullissu, and the sheep which they throw on the brazier and which the fire burns, is Qingu, when he burns in the fire.[36]

Within the stratified layers of cultic ceremonies, it is, however, possible to identify the focal point of the festival. The core of the celebration is the "absence" of Marduk, which recalls different religious motives such as the periodical disappearance of the "dying gods" or celestial bodies, especially the moon. The Assyrian and Babylonian commentators, in their explanatory texts, explain this disappearance of the celebrated god by suggesting that he was taken prisoner or even killed:[37]

> [The Akītu House where] he goes, is the house at the edge of (the place of) the ordeal; they question him there.
>
> [Nabû, who] comes [from] Borsippa, comes to greet his father, who has been taken prisoner. [Bēlet-il]ī, who roams the streets, is looking for Marduk: "Where is he kept prisoner?"
>
> [Zarpanītu], whose hands are stretched out, prays to Sîn and Šamaš: "Let Bēl live!"
>
> [Bēlet-ilī] who goes away, is going to the graveyard and looking for him.
>
> [The ath]letes who stand at the gate of Esaggil are his guards; they are appointed over him, and guard [him].
>
> ...

35 Livingstone 1989: 85 n. 34 ll. 54ff.
36 Livingstone 1989: 93 n. 37 ll. 9ff.
37 von Soden 1955; Ambos 2010; see n. 41.

Enūma eliš, which is recited and chanted in front of Bēl in Nisan, concerns his imprisonment.[38]

Marduk's return, which is depicted as liberation from captivity or as resurrection from death, reestablishes the order and thus renews the yearly cycle recalling the mythical foundation of the world through the recitation of the *Poem of Creation*.

From a ritual point of view, the expulsion of Marduk is marked by a spatial separation of the god from his usual seat, the temple in the city. During the festival, the god's statue was led in procession to a temple outside the city wall, known as the house or temple of the *akītu* (*bīt akīti*). In Mesopotamian ideology, it was the presence of the god in his house, the temple, which ensured the cosmic order within the civilized space enclosed by the city wall. Outside the wall lay the chaotic space, the lair of demons and populations perceived as uncivilized (nomads).[39] This space, the steppe, is where the *akītu* temple was situated and whither the god retired, an event that caused a momentary lapse of order.[40]

The festival contains ritual ceremonies that close a cycle (the old year), lead through a momentary period of crisis and lapse of order, and finally mark the beginning of a new cycle/year. These three stages are marked by the events the main god undergoes during the festival (see accompanying table). The phase of separation is represented by the disappearance of Marduk and by his successive spatial and social liminality, which is expressed through his remaining outside of the city wall in the *akītu*, his status as a criminal, or even his death.[41] The festival ends with the reintegration of the god into his

[38] Livingstone 1989: 82 n. 34 ll. 7–12, 84 l. 34. The mention of the graveyard and of the athletes links this passage to a funerary context.

[39] Verderame 2011.

[40] The mythological composition known as the *Poem of Erra* describes how Marduk, at the moment of retiring from his throne to get his armor restored, leaves the world to chaos. Spatial separation marks the momentary lapse of order caused by both periodical and occasional removal of the ruler from kingship. In the substitute king ritual (*šar pūḫi*), a ritual performed on the occasion of particular lunar eclipses, the "real" king is sent away from the court and dubbed as the "peasant," thus ideally locating him in the open field (Verderame 2013); see n. 43 and 54. In the "house of ablution" (*bīt rimki*) ritual, the king goes through seven structures (lit. "houses") built in the steppe. The "house of ablution" was performed not only during the substitute king ritual but also each month before the disappearance of the moon (Laessøe 1955: 95–98, 101–102), thus confirming the association of the kingship with the moon and their mutual monthly renewal here discussed.

[41] Van Gennep 1909: 129–131; Brelich 1954: 91–92; Turner 1969: 94–130. The practice of robbery and the state of outlaw are part of initiation rites during the seclusion and license phase; van Gennep 1909: 130–131. For the idea of imprisonment in ancient Mesopotamian rituals, see Ambos 2010.

former status and powers. It is in this final phase that the renewed order is marked by ritual acts such as temple lustrations or the pronouncement of oracles for the incoming year.

End of the old cycle	Separation	Marduk's absence
Passage	Seclusion / liminality	M.'s captivity / death
Beginning of a new cycle	Post-liminality / reintegration	M.'s return

While the festival celebrates Marduk's "absence" and return through the recitation of the *Poem of Creation* on a mythological level, and the procession of his statue to the *akītu* temple on a ritual level, his human representative, the king, undergoes the same fate:

> The king, who wears his jewellery and roasts young virgin goats, is Marduk, who wearing his armour bur[ned] the sons of Enlil and Ea in fire.
>
> The king, who stands on the podium with a [heart] in his hand, while the singer chants *To the Western Goddess*, is Marduk, [who], with his bow in his hand, casts down Ea, while Venus was ascendant in front of him.
>
> The king, who opens the vat in the race, is Marduk, who [defeat]ed Tiamat with his penis.
>
> [The ki]ng, who with the high priest tosses the cake, is Marduk (with) Nabû, [who ...] vanquished and crushed Anu.[42]

The momentary inversion of order that turns the god into a criminal, judged and beaten, results in a dismissal of the king, who is stripped of the royal *insignia*, slapped on the cheek, and obliged to confess his sins before being reestablished in his role.[43]

3 The King Is the Moon

The festival shows how the complexity and stratification of elements have blurred the original agricultural and calendrical meaning of the ritual. The springtime equinox and the New Moon, when this festival took place, are associated with the beginning of a cycle (the year), the crossing of a main critical moment (harvest), and the suspension and renewal of divine order through its representative, the king. The original nature of the festival remains scantly attested. Lunar and stellar motifs may be identified "underneath" the rituals in indirect references or symbolical acts, such as the

[42] Livingstone 1989: 94 n. 37 ll. 16–19.

[43] A momentary dismissal of the king, followed by the successive re-enthronement, also took place during the substitute king ritual (Verderame, 2013: 317–321); see n. 40 and 54.

painting of stars on the temple ceiling on the fifth day after the ritual cleansing[44] or the explanatory text that mentions the flight of the god up to the *ziqqurat* and his successive removal from there,[45] which may recall the myths of segregation of heavenly bodies.

These associations also emerge from other compositions of the same period. The so-called *Astrolabe B* is an almanac, whose Akkadian name recalls the star creation passage of the *Poem of Creation*, "three stars per each (month)."[46] This composition is composed of two parts, the second of which is a star catalog. The first part is a bilingual (Sumerian-Akkadian) compilation of cultic explanations arranged by month. First, the name of the month and the constellation, which rises heliacally during that month, are provided; then follows a cultic and theological commentary; the association of the month with a specific god closes the section. The composition begins with the first month, Nisan (March-April).

Month "of setting the podium off the border," Field constellation, podium of An,	Month "first" (or: "first fruit"), Field constellation, throne (var. seat) of An,
Lifting of the podium, (re)establishment of the podium,	In Nisan the king is lifted, the king is (re)established,
Propitious beginning of the year of An and Enlil,	(in Nisan) the star of An, propitious beginning of the year of An and Enlil,
Month of Nanna(r), the first born of Enlil.	Month of Sîn, the first born of Enlil (var. An).
Astrolabe B §1–5 (Sumerian)	*Astrolabe B* §6–9 (Akkadian)

For Nisan, the first month, when the New Year was celebrated, the constellation is that of the Field (α, β, γ *Pegasi* and α *Andromedae*), which recalls the centrality of agricultural work during this month. The Sumerian version refers to a podium ($bara_2$-an-na "podium of the sky / of [the sky god] An"), a term contained in the very Sumerian name of the month ($^{iti}bara_2$-zag-ĝar);[47] this term may refer to both the divine and royal seats. The main point is the action that the podium undergoes: it is first lifted and then reestablished. In the

44 This may refer to the mythical creation of stars in the *Poem of Creation* as well.
45 Livingstone 1989: 82 n. 34 ll. 13–14.
46 Casaburi 2003; for further discussion on the "three stars per each (month)" system see above.
47 The meaning of the Sumerian month name is still debated. While the first element ($bara_2$) has the basic meaning of "elevated place, podium, altar," the second part (zag-ĝar) has been interpreted as a unique term for "sanctuary," based on the first-millennium lexical list correspondence between the Sumerian zag-ĝar-ra and the Akkadian *ašírtu* and therefore possibly a later learned association, or as "to place off to the side," for zag "border" and ĝar "to place"; see Cohen 1993: 80–81. The latter interpretation fits the idea of separation proposed in our analysis of the New Year festival, and it is grounded on the reference of a podium removal and reestablishment mentioned in *Astrolabe B* §2.

Akkadian version, this sequence is said to apply to the king rather than a podium. This ritual action marks the interruption and successive reintegration of the stability and order associated with the podium or the king as symbols of order. It recalls the interruption of the order that precedes the beginning of the New Year cycle, symbolized in the *akītu* festival through the absence and return of the god and, in the human world, through the removal and re-enthronement of the king. The propitious moment is then marked by the reference to the main gods An and Enlil. Finally, the month is associated with the moon god Nanna(r)/Sîn, confirming the connection of the heavenly body with cyclical renewal and kingship. Here the connection with kingship is strengthened by the reference to primogeniture, which is at the base of the inheritance and monarchic succession right.

Despite the baroque complexity of the New Year festival, this association remains a vivid, sound, and fundamental concept providing legitimization for both the power of the king and the calendrical system. This idea is clearly revived and synthesized by the direct words sent to the Assyrian king Esarhaddon by his main counselor (*ummânu*), Issar-šumu-ēreš (seventh century BCE). While restoring the temple of the moon god in Harran, the king discusses with his experts the new arrangement of his statue, as well as of those of the two crown princes (Assurbanipal and Šamaš-šumu-ukīn), in relation to the statue of the moon god:

> If it [is acceptable] to the king, my lord, the large royal statues should be erected on the right and the left side of the [Moon] god. The statuettes of the king's sons should be s[et up behind] and in front of the Moon god. The Moon, lord of the cr[own],[48] will (then) every month without fa[il], in rising and [setting], unceasingly send h[appy] signs of long-lasting days, steady reign and increase in power to the king, my lord.[49]

The identification of the king with the moon, particularly through the symbol of the crown and the idea of radiance, is clearly stated in a passage from an apotropaic ritual (first millennium BCE). While in the previously quoted letter, it was the moon that strengthens the king through its light, here it is the king who supports the people and brings them light through his crown:

[48] "Lord of the crown" is a common epithet of the moon god, shared seldom with other gods, particularly Anu; compare n. 18.

[49] Parpola 1993: 12f. n. 13 ll. r. 2'–17'.

(It is) the king, son of his personal god, who, like the moonlight, supports the population,
Bearing radiance on his head like the new moon.
(*Utukkū lemnūtu* XVI 83–84)[50]

The association of the moon with the king, specifically through the deployment of light terminology, serves the purpose of the text in terms of narrative and metaphor. The passage, in fact, is contained in a ritual for the king against the demons that is introduced by a *historiola* describing an assault on the moon god's rule of heaven. The offender is Ištar, who, under her astral aspect (Venus), shares the duty of controlling heaven with both the moon god Sîn and the sun god Šamaš. Ištar plots against Sîn's leadership and unleashes the Seven demons, who attack and obscure the moon, causing an eclipse:

Sîn, beloved of mankind's descendants (and) of the inhabitants of the land,
Became dimmed [in] (his) brightness and became still,
Darkened both [night and] day, no longer residing in the seat of his rule.
They are the evil gods, messenger of their Lord An.
As an accessory to evil, they are always agitated at night,
Always looking for trouble.
They have risen like winds from the midst of heaven against the land.
(*Utukkū lemnūtu* XVI 38–44)[51]

The disappearance of Sîn, that is, the eclipse of the moon, results in an absence of rule and order, which throws the world into chaos:

An and Enlil called to them:
"They have darkened Sîn in the midst of heaven,
They have torn off his corona
Stripped off his adornments,
They have darkened his beloved face."
The gods fell to the ground,
The wild animals have become disturbed,
The whole of the people is confused.
(*Utukkū lemnūtu* XVI 109–116)[52]

The moon god is humiliated, confined, deposed, and deprived of his *insignia* and the crown,[53] quite similarly to the way Marduk and the king

[50] Geller 2007: 181, 253.
[51] Geller 2007: 179, 252.
[52] Ibid., 182–183, 254.
[53] See previous discussion and n. 18 and 48.

are treated during the New Year festival. The *historiola* lays the foundations of an apotropaic rite unrelated to the calendar. The main point, therefore, is not the cyclical renewal of both the moon and kingship, but the identification of the moon with the king as well as their loss of power caused by the demons' attack. The leadership of the moon is then reestablished by the intervention of Ea and Marduk, whose rituals and incantations are suitable for protecting and restoring the leadership of the king as well. In this case, the cyclical renewal of the moon god, who finally prevails over the demons, is a further element that strengthens the effectiveness of the procedure.[54]

4 Conclusion

In the ancient Mesopotamian calendar, various systems of time reckoning coexist and overlap. The moon marks the beginning of the day and, with its phase cycle, the month and its subunits. The succession of light and darkness divides the day into two parts, which in turn are divided into three subunits (watches) according to motion of the sun and the moon. The "yearly" cycle of the sun and the heliacal rising of stars mark longer periods such as the seasons and allow resetting the lunar year.

The motion of the heavenly bodies not only determines time but also marks the intersection of crucial moments in the cyclical activities, which are commemorated by festivals in the calendar. The relevance of the calendar for cultic and economic activities is crucial, as the political and religious leadership shows a strong control over time reckoning. In the Mesopotamian cultures, this results in the direct control the king exerts over intercalation. On a ritual and mythological level, it manifests in the association of the renewal of kingship with the passage from the old to the new cycle (year), that is, in the New Year festival. In the mid–second millennium BCE, the evolution of the political and religious systems led to the rise of "national" gods, that is, Assur in Assyria and Marduk in Babylonia. These underwent a process making them the rulers over the pantheon and assimilating the functions and features of other gods (syncretism). In the same period and through the same process of unification, a "standard" calendar was created and adopted. One of the main results of these processes is that the national god becomes the protagonist of the New Year festival, during which the king identifies himself with that god. The crucial features of this celebration,

54 The association of the dimming light of the moon with the power of the king is further developed in the Assyrian ritual of the substitute king (*šar pūḫi*) performed after a lunar eclipse; see Verderame 2013: 317–321, with previous bibliography; see also n. 40 and 43.

however, continue to show their pristine relation with the calendar and its driving forces, that is, the heavenly bodies, and in particular with the moon god Nanna(r)/Sîn, whose periodical cycle marks the Mesopotamian calendar. The idea of renewal, strictly associated with the moon phases, led to the direct identification of the moon god with the kingship since the beginning of Mesopotamian civilization, an association that clearly survived until the first millennium BCE.

Bibliography

Ambos, C. 2010. "Ritual Healing and the Investiture of the Babylonian King." In W. Sax, J. Quack, and J. Weinhold (eds.), *The Problem of Ritual Efficacy*. New York: Oxford University Press, 17–44.

Ambos, C. 2013. *Der König im Gefängnis und das Neujahrsfest im Herbst*. Dresden: ISLET.

Bidmead, J. 2002. *The Akitu Festival: Religious Continuity and Royal Legitimation in Mesopotamia*. Piscataway: Gorgias.

Brelich, A. 1954. *Introduzione allo studio dei calendari festivi*. Rome: Edizioni dell'Ateneo.

Britton, J. P. 2007. "Calendars, Intercalations and Year-Lengths in Mesopotamian Astronomy." In J. M. Steele (ed.), *Calendars and Years: Astronomy and Time in the Ancient Near East*. Oxford: Oxbow, 115–132.

Casaburi, M. C. 2003. *Tre-stelle-per-ciascun(-mese). L'astrolabio B: edizione filologica*. Naples: Università degli Studi di Napoli "L'Orientale."

Charpin, D. 2013. "'I Am the Sun of Babylon': Solar Aspects of Royal Power in Old Babylonian Mesopotamia." In J. A. Hill, P. Jones, and A. J. Morales (eds.), *Experiencing Power, Generating Authority: Cosmos, Politics, and the Ideology of Kingship in Ancient Egypt and Mesopotamia*. Philadelphia: University of Pennsylvania Press, 65–96.

Cohen, M. E. 1993. *The Cultic Calendars of the Ancient Near East*. Bethesda: CDL.

Eliade, M. 1949. *Le mythe de l'éternel retour: archétypes et répétition*. Paris: Gallimard.

Foster, B. R. 2005. *Before the Muses: An Anthology of Akkadian Literature*. 3rd ed. Bethesda: CDL.

Frahm, E. 2013. "Rising Suns and Falling Stars: Assyrian Kings and the Cosmos." In J. A. Hill, P. Jones, and A. J. Morales (eds.), *Experiencing Power, Generating Authority: Cosmos, Politics, and the Ideology of Kingship in Ancient Egypt and Mesopotamia*. Philadelphia: University of Pennsylvania Press, 97–120.

Geller, M. J. 2007. *Evil Demons: Canonical Utukkū Lemnūtu Incantations*. Helsinki: Neo-Assyrian Text Corpus Project.

Gingrich, A. 1994. "Time, Ritual and Social Experience." In K. Hastrup and P. Hervik (eds.), *Social Experience and Anthropological Knowledge*. London: Routledge, 125–134.

Horowitz, W. 1998. *Mesopotamian Cosmic Geography*. Winona Lake: Eisenbrauns.

Horowitz, W. 2007. "The Astrolabes: Astronomy, Theology, and Chronology." In J. M. Steele (ed.), *Calendars and Years: Astronomy and Time in the Ancient Near East*. Oxford: Oxbow, 101–113.

Hubert, H. 1905. *Étude sommaire de la représentation du temps dans la religion et la magie: Avec un rapport sommaire sur les conférences de l'exercice 1904–1905* . . . École Pratique des Hautes Études, Section des Sciences Religieuses. Paris: Imprimerie nationale.

Jacobsen, T. 1975. "Religious Drama in Ancient Mesopotamia." In H. Goedicke and J. J. M. Roberts (eds.), *Unity and Diversity: Essays in the History, Literature, and Religion of the Ancient Near East*. Baltimore: Johns Hopkins University Press, 65–97.

Kirk, G. S. 1973. *Myth: Its Meaning and Functions in Ancient and Other Cultures*. Cambridge: Cambridge University Press; Berkeley: University of California Press [paperback; first published 1970].

Lambert, W. G. 2013. *Babylonian Creation Myths*. Winona Lake: Eisenbrauns.

Laessøe, J. 1955. *Studies on the Assyrian Ritual and Series bît rimki*. Copenhagen: Munksgaard.

Livingstone, A. 1989. *Court Poetry and Literary Miscellanea*. Helsinki: Helsinki University Press.

Livingstone, A. 2007. "The Babylonian Almanach: A Text for Specialists?" In B. Groneberg and H. Spieckermann (eds.), *Die Welt der Götterbilder*. Berlin: De Gruyter, 85–101.

Munn, N. D. 1992. "The Cultural Anthropology of Time: A Critical Essay." *Annual Review of Anthropology* 21: 93–123.

Nilsson, M. P. 1920. *Primitive Time-Reckoning: A Study in the Origins and First Development of the Art of Counting Time among the Primitive and Early Culture Peoples*. Lund: Gleerup.

Parpola, S. 1993. *Letters from Assyrian and Babylonian Scholars*. Helsinki: Helsinki University Press.

Pirenne-Delforge, V. and T. Tunca (eds.). 2003. *Représentations du temps dans les religions*. Geneva: Droz.

Reculeau, H. 2002. "Lever d'astres et calendrier agricole à Mari." In J. M. Durand and D. Charpin (eds.), *Recueil d'études à la mémoire d'André Parrot*. Paris: SEPOA, 517–538.

Rochberg, F. 1995. "Astronomy and Calendars in Ancient Mesopotamia." In J. M. Sasson (ed.), *Civilizations of the Ancient Near East*, vol. III. New York: Scribner, 1925–1940.

Soden, W. von 1955. "Gibt es ein Zeugnis dafür, daß die Babylonier an die Wiederauferstehung Marduks geglaubt haben?" *Zeitschrift für Assyriologie und vorderasiatische Archäologie* 51: 130–166.

Steele, J. M. 2007. "The Length of the Month in Mesopotamian Calendars of the First Millennium BC." In J. M. Steele (ed.), *Calendars and Years: Astronomy and Time in the Ancient Near East*. Oxford: Oxbow, 133–148.

Steele, J. M. 2011. "Making Sense of Time: Observational and Theoretical Calendars." In K. Radner and E. Robson (eds.), *The Oxford Handbook of Cuneiform Culture*. Oxford: Oxford University Press, 470–485.

Steele, J. M. 2012. "Living with a Lunar Calendar in Mesopotamia and China." In J. Ben-Dov, W. Horowitz, and J. M. Steele (eds.), *Living the Lunar Calendar*. Oxford: Oxbow, 373–387.

Stol, M. 1992. "The Moon as Seen by the Babylonians." In D. J. W. Meijer (ed.), *Natural Phenomena: Their Meaning, Depiction and Description in the Ancient Near East*. Amsterdam: Noord-Hollandse Uitgevers Maatschappij, 245–276.

Turner, V. W. 1969. *The Ritual Process: Structure and Anti-Structure*. Chicago: Aldine.

van Gennep, A. 1909. *Les rites de passage*. Paris: E. Nourry.

Verderame, L. 2002. *Le Tavole I-VI della serie astrologica Enūma Anu Enlil*. Messina: Di.Sc. A.M.

Verderame, L. 2003. "Riferimenti astrali nella mitologia sumero-accadica." In S. Colafrancesco and G. Giobbi (eds.), *Cosmology Through Time: Ancient and Modern Cosmologies in the Mediterranean Area Conference Proceedings*. Milan: Mimesis, 25–32.

Verderame, L. 2008. "Le calendrier et le compte du temps dans la pensée mythique suméro-akkadienne." *De Kêmi à Birīt Nāri* 3: 121–134.

Verderame, L. 2011. "L'immagine della città nella letteratura sumerica." In R. Dolce and A. Pellitteri (eds.), *Città nel Vicino Oriente e nel Mediterraneo: Linee di storie e di simboli dall'antichità ad oggi*. Palermo: Flaccovio, 99–126.

Verderame, L. 2013. "Means of Substitution: The Use of Figurines, Animals, and Human Beings as Substitutes in Assyrian Rituals." In C. Ambos and L. Verderame (eds.), *Approaching Rituals in Ancient Cultures*. Pisa: Fabrizio Serra, 301–323.

Zgoll, A. 2006. "Königslauf und Götterrat: Struktur und Deutung des babylonischen Neujahrsfestes." In E. Blum and R. Lux (eds.), *Festtraditionen in Israel und im Alten Orient*. Gütersloh: Gütersloher Verlagshaus, 11–80.

9

TOWARD A PHENOMENOLOGY OF TIME IN ANCIENT GREEK ART

SeungJung Kim

The time which we have at our disposal every day is elastic; the passions that we feel expand it, those that we inspire contract it; and habit fills up what remains. Marcel Proust, *Within a Budding Grove* (1919)

Every impression is double and the one half which is sheathed in the object is prolonged in ourselves by another half which we alone can know.
Marcel Proust, *Time Regained* (1927)

1 Introduction: The Tale of Two Eras

Walking down the Boulevard des Capucines on a wintry day in 1896 Paris, one would have encountered a line forming outside the Salon Indien du Grand Café. Every thirty minutes, people would be jostled into the salon to witness the latest technological curiosity, the Cinématographe Lumière and, in particular, the (in)famous moving image of a locomotive: *L'Arrivé du train à la Ciotat*. This seemingly banal, fifty-second film by Louis Lumière that shows a train pulling into the station was at once a staggering sensation, and over the past century has solidified its iconic status as the founding myth of cinema's birth.[1] The legend, of course, is a familiar one: the moving image of the train dashing toward the crowded audience caused such terror and panic amid them, who, feeling "physically threatened," were said to have "jumped up

This article derives from parts of my PhD dissertation (Kim 2014). I am indebted to my former dissertation advisor Prof. Ioannis Mylonopoulos for his generous guidance; I also thank Profs. Francesco De Angelis, Helene Foley, Mark Stansbury-O'Donnell, and Francis Dunn for sharing their keen expertise on reading versions of this article, although any faults or omissions are solely my own. Finally, I owe much to the late Natalie B. Kampen for her unfaltering support from the very inception of this project.

[1] Loiperdinger (2004), who dubs the episode "cinema's founding myth," reexamines evidence for the anecdote.

from their chairs in shock" or "threw themselves back in their seats in fright," "as they feared getting run over."[2]

The Athenian citizens attending the City Dionysia on a balmy spring evening in 492 BCE would have witnessed the production of *The Fall of Miletus* by the tragedian Phrynichus. Most likely the very first factual drama of the Western tradition, *The Fall of Miletus* was put on stage only two years after the Persians had actually captured the Ionian city. The ever so fresh horrors at the dawn of the Persian Wars must have made the contemporary audience particularly sensitive to the dramatized event, as if it were presently unfolding before their very own eyes, judging from their visceral reaction as it is recounted by Herodotus (6.21.10): "the whole audience at the theatre burst into tears and fined Phrynichus a thousand drachmas for reminding them of a calamity that was their very own; they also forbade any future production of the play."[3]

While the degree of truthfulness of these various accounts may not be immediately verifiable, the "grain of truth" that lies behind these legendary *succès des scandales* is clearly one that has to do with the notion of embodied viewing: the phenomenological, *bodily* response to the sensational, overpowering effect of perceiving either the moving image of a train or a narrative display of a painful event fresh in one's personal or collective memory.[4] As a cornerstone of Merleau-Ponty's phenomenology, the notion of embodiment – that the body and its intersensory dynamics inform consciousness – has become a contemporary topic of active scientific research.[5] A fuller understanding of the anatomy of embodied perception, and especially its effect on time perception, will no doubt require detangling the interdisciplinary web of neurobiological, cognitive, psychological, and philosophical methodologies. For the purpose of this chapter, however, it will be sufficient to note that (1) it is this physicality of the interaction between object (artwork) and subject (viewer) that informs the latter's cognition at the moment of perception, (2) this interaction necessarily happens in space *and* in time, and (3) it is the mechanism of embodiment informed by the space-time matrix of this interaction that instinctively conveys the intended temporal message. In short, the chapter will focus on

[2] Loiperdinger 2004: 90–91, quoting in turn Karasek 1994: 154; Burch 1990: 39; Eisner 1975: 104; Sadoul 1956: 27.

[3] Trans. Grene 1987: 416–417.

[4] See Favorini 2003, for a discourse on the "historical plays," especially those of Aeschylus, in the context of the interplay between memory and history.

[5] Merleau-Ponty 1962. For a review of past and recent neuroscientific research on embodiment, see Price, Peterson, and Harmon-Jones 2012, and references therein.

how time is *felt instinctively* in the works of art, rather than how it is *read* through an act of interpretation. Methodologically speaking, this is to displace the top-down model, which would read into the artworks a certain *re-presentation* of the metaphysics of time that was expounded by one Greek philosopher or another; rather it seeks to construct, bottom up, the *phenomenology* of time, the temporal relationships of the lived experience in ancient Greece that are *presented* in the works of art.[6]

Let us come back to the two anecdotes in question. Despite being separated by almost two and a half millennia, they intersect at two notable points: (1) the portrayal of *actuality*, which harbors an indexical relationship to reality, and (2) the temporal immediacy and the quality of *presentness* in both the content of the work and the response of the viewer. Just as Lumière's film portrays a real, physical train that was in existence at the time, Phrynichus's play treats a historical subject matter, an event that took place only two years prior to its production, which was a staggering novelty in itself.[7] The temporal immediacy in the contemporary subject matter also resonates in the urgency of the viewer response, whether the audience members had burst into tears on that very spot or jumped up from their seats at the sight of the train approaching. An important premise common to both circumstances is the novelty of the respective genres. Film and tragedy were both startlingly new media in their respective times, which accounts for the naiveté of the spectators, untrained in their viewing habits, and which consequently allowed the unexpected embodied responses to be particularly heightened.

The parallel drawn between these anecdotes situated in late nineteenth-century Europe and early fifth-century BCE Greece, respectively, extends well beyond their local symptomatic resemblances to a more global societal

[6] The opposition between the terminologies *re-presentation* and *presentation* with reference to works of art is to evoke the discourse of the so-called Iconic Turn (see Moxey 2008); concomitantly, this endows the corollary notion of object agency to the artworks themselves, as they are no longer limited to a certain intermediary status as a mere window through which one can access the societal, cultural, or philosophical ideals of the time of their production. Rather, the artworks become active agents that inform the social actors' relationship with these larger currents (cf. Neer 2002).

[7] It has become a platitude to allude to the Greeks' hostility toward actuality as a viable subject for the arts, be it visual or dramatic, and Phrynicus's anecdote is often taken to corroborate this trend. The distinction between "historical tragedy" and "mythical tragedy" is thus often downplayed on the ground that the Greeks perceived their mythical past and epic tradition as history (e.g., Snell 1928); however, as Hall (1996: 7–9) rightly notes, the Greeks did seem to distinguish between the immediate past of contemporary history and their distant mythical past, and that Phrynicus's initial failure followed by a number of successes is a testimony to the genre itself as experimental and flexible enough to assimilate both subject matters (see also Debnar 2005: 4–5).

attitude toward time. "In the *spleen*, time becomes palpable," wrote Walter Benjamin, "the minutes cover a man like snowflakes."[8] Although referring to Baudelaire's poetry specifically, Benjamin was, as pointed out by Mary Ann Doane, offering an elegant appraisal of the general turn-of-the-century attitude toward an experience of time and its representations in relation to early modernity.[9] With the rapid pace of industrialization and mechanization of the work environment as well as the growing urgency of urban lifestyle and mobility came something of a revolution in how time was perceived, structured, and reified. It is hardly an overstatement to say that modernity itself was perceived as a temporal demand. The last decade of the nineteenth century saw a universal diffusion of personal wristwatches, and the world standardization of time took place with an impetus from railway travel and telegraphy. Time was indeed palpable and was being felt everywhere, enabled by new technological innovations that visibly represented it. At the center of this visual landscape that partook in a sea change in the perception of temporality, of course, stands the invention of cinema, which made possible a new visual access to time with its "perfect" representability and archival ability.

At the dawn of democracy in the late sixth- and early fifth-century BCE Greece, we see a similar, if not more fundamental, change in the way time concepts and temporality were perceived, understood, and integrated into the society at large. History was written for the first time and tragedy was also born: new genres of literature, which focused, to an unprecedented degree, on the present and recent past.[10] Compared to the epic or mythopoetic traditions of Homer and Hesiod, they unmistakably initiated a new way of thinking about time.[11] Francis Dunn paints a compelling picture of an even more prominent change in the later fifth-century attitude toward time, which he terms the "present shock": a societal attitude toward focusing on the rapidly changing uncertainties of the present, while rejecting the authority of the past.[12] This shift in the general attitude toward the present and its

8 Benjamin 1968: 184.

9 Doane 2002: 4.

10 Regarding Greek notions of the past, see in particular Grethlein 2010, for contextualizing them in the framework of memory studies. Marincola et al. 2012 also contains an interesting collection of the most recent treatments on the subject.

11 Tragedy as a developed genre treats, with few exceptions, mythological subject matters. Unlike the case of the epic bard, however, whose very act of reciting the epic narrative implies the absence of the protagonists, the tragic stage *presentifies* both the heroes and their stories "as trials [taking] place before [their] very eyes, adopting the form of real existence in the immediacy of the performance" (Vernant 1988: 243).

12 Dunn (2007: 2) evokes the term "future shock," coined by Alvin Toffler in 1970, referring to the cataclysmic rate of cultural change in the 1960s that left people inadequately prepared for confronting the future. Similarly, Dunn argues for intellectual and political upheavals

uncertainties was, according to Dunn, accompanied by an increasing complexity in articulating temporality. This is shown not only in new literary narrative techniques but also in major breakthroughs achieved in the medical sciences and astronomy, as well as civic, religious, and political time-keeping systems.[13] Such a phenomenon parallels, in many ways, the previously mentioned temporal revolution witnessed with the development of modernity in the late nineteenth century. And the domain of visual culture, as I argue later, is certainly no exception.

The current study is but an exploratory step toward constructing a more comprehensive, macroscopic relationship between time and the image in the Late Archaic and Classical Greece. The elemental challenge in constructing a plausible interconnection between the perceived temporality in the works of art and the lived temporality of the society that produced them hinges on the notion of causality. Eric Csapo and Margaret Miller's commendable essay remains thus far the sole attempt that seeks a broader temporal discourse in the context of the visual arts of fifth-century Greece; they argue that the shifting notions of political time largely fueled by the change from oligarchy to democracy at the turn of the fifth century BCE were responsible for the finer articulations of, and the growing complexities in, visual narrative structures.[14] Csapo and Miller are explicit in striving to avoid the developmental grand narrative, most notably typified by Hermann Fränkel.[15] Their binary model of time, however – the past-oriented archetypal, "aristocratic temporality" giving way to a new present-oriented phenotypal, "democratic temporality" – also remains something of an essential dualism.[16] Moreover, viewing these cultural modes as all controlling in the Foucauldian sense has the notable disadvantage of effacing human agency and disproportionately subscribing to a top-down model of cultural production.[17]

resulting from the series of events in the last two decades of the fifth century BCE that left the Athenians immersed in a "disorienting present."

[13] Dunn (2007) examines a synchronic cross section of literary outputs of the later fifth century BCE, concentrating on Thucydides, Euripides, the sophist Antiphon, and the Hippocratic Treatise; his first chapter entitled "Civic Time" treats the supporting evidence regarding the time-keeping mechanisms or methods.

[14] Csapo and Miller 1998.

[15] Fränkel 1960 (first published 1933) stands at the dawn of the discourse on ancient Greek temporality; he maintains that "temporal awareness" only appears after Homer and does not fully develop until Aeschylus and tragedy. See also Romilly 1968, for a treatment of time in tragedy, whose progression from Aeschylus to Euripides is seen as a kind of modernization.

[16] Csapo and Miller's (1998) binary model is built on the dichotomy long recognized by van Groningen (1953: 93–108), who posits coexisting, conflicting notions of time, which were "mythical" and "historical"; this resonates also with Vidal-Naquet (1986), who distinguishes between "divine" and "human" time in Greek thought.

[17] Dunn 2007: 8.

The current chapter's titular call to phenomenology with respect to the study of time registers on at least three levels, spanning both methodological and historical concerns. First, it indicates an effort to minimize the structural rift created by the causal model of cultural influence. In other words, it aims to be critical of the hierarchical understanding that systematic cultural values and ideas are higher forces imprinting themselves on the specific cultural products, such as artworks. This comes closest to the original Husserlian project of articulating the conscious experience of lived-time as a prerequisite for the Newtonian, scientific notion of time's reality;[18] analogously, the viewer's experience of time vis-à-vis the specific product (be it visual, material, or literary) supersedes a higher order of a theoretical notion of time that may prevail in the society at large.[19] Second, it is also a reference to the recent development in the philosophy of history and art history that adopts a phenomenological framework centered on the notion of *presence*.[20] This is a notion of time registering purely on the level of disciplinary practice: by prioritizing the viewer's experience with the physical presence of the artwork, and by ascribing an inherent *anachrony* to the art object's ontological status (assuming a multiplicity of temporalities that coexist); these newer methodologies tend to collapse the alterities of the historical horizons, allowing us direct epistemological access to their perceptual significance in the ancient Greek world.[21] And finally, on a more literal level, that is, historically speaking, I argue that the mode of visual perception itself that informs the viewer's temporality undergoes a radical change in the late sixth and fifth centuries BCE to become more instinctively "phenomenological." Here, the term is specifically referring to the Merleau-Pontian notion of embodiment; we not only see new modes of representations that explicitly engage viewer participation, but the perceptual process itself also becomes integrally informed by a physical, bodily understanding of what is depicted. The temporality of the artwork thus becomes more immediate, sensorial, and intimately connected to an embodied, phenomenological understanding at the very moment of perception.

18 Husserl 1991 (originally 1893–1917).

19 The principle of phenomenological reduction, or bracketing epistemological priors of the external world, provides the methodological basis for Husserl's philosophy and is structurally relevant for our study (see Boehm 1965 and Orth 1984, for an overview on Husserlian reduction).

20 See Moxey 2013, for the most recent art historical discourse centered on the notion of *presence*, the concept of which is championed by historians such as Eelco Runia (Runia 2006a, 2006b).

21 See, in particular, Ankersmit 2005, for the notion of "sublime historical experience."

2 The Lysippan Kairos: A New Reconstruction

We return to the modernist analogy once more, this time with the philosophy of time. It was thinkers of the Continental tradition at the turn of the twentieth century, notably Henry Bergson, who pioneered the non-Cartesian view of time with his *durée*, inseparable from his notion of intuition, and thus paving the way for later phenomenological inquiries of the early and mid-twentieth century.[22] Heidegger is no exception: in criticizing what he calls "vulgar" time, which is precisely this Cartesian, uniform, linear time, he proposes *Dasein*'s original temporality as "authentic" time, whose starting point is the limitation posed by one's own death.[23] This dual notion of time – having a qualitative flipside to the quantitative Cartesian characteristics – took on many forms throughout the twentieth century and continues to this day in many new areas of scholarly inquiries.[24] These tensions between the chronological, homogenous time and the value-laden, phenomenological aspect of time are, in many ways, a reformulation of the prevalent dual notion of time that already existed in Classical Greece, as it is noted by John Smith:

> Classical literature reveals two Greek words for "time" – *chronos* and *kairos*. These words embrace the uniform time of the cosmic system, the time which, according to Newton, *aequabiliter fluit,* and the time of opportunity or "occasion" come and gone which marks the significant moments of historical action.[25]

If *chronos* is time as measure, as it is defined by Aristotle as the "number of motion with respect to the before and after" (*Phys.* 4.11, 219b1–2), *kairos* is the irreproducible moment of the "now," the moment that demands decision and human action with utmost urgency. The Hippocratic treatise *Precepts* begins with the line: "Every *kairos* is a *chronos*, but not every *chronos* is a *kairos*."[26] Far from being antithetical, the two concepts are thus thought to be intimately interlocked: *kairos* presupposes *chronos*, while *chronos* without *kairos* fails to explain the critical points of human experience. Indeed Plato, in his fourth book of *Laws*, declared, "*Tyche* (chance/fortune) and

[22] Bergson 1910 (first published in 1889); see also Husserl 1991 (originally 1893–1917); Merleau-Ponty 1962; cf. Deleuze 1994.

[23] Heidegger 1927.

[24] Most notably in cognitive science and psychology, the science of internal time or time perception is a categorical subject of inquiry (e.g., Block 2014).

[25] Smith 1969: 1.

[26] Jones 1923: 1.313–315.

kairos (opportunity/propriety) cooperate with God in the control of all human affairs" (*Leg.* 4, 709b).[27]

That being said, it is surprising how rarely time itself was personified and represented in works of art in antiquity. And it was *kairos,* unlike its quantitative counterpart *chronos,* that was endowed with a personified visual form. As the primary notion of time that governed the particularities of human affairs, *kairos* showcases at once a deeply phenomenological engagement with the world and a present-oriented temporal immediacy that is pregnant with human agency. Declared unteachable through set principles, this multilayered, empirical concept was epitomized as a one-of-a-kind statue by Lysippos in the third quarter of the fourth century BCE (c. 334).[28]

The original freestanding Lysippan bronze statue, whose existence is made abundantly clear in both visual and textual sources, may have been first created at Pella in Macedonia for Alexander the Great.[29] The two most often cited textual references – a Hellenistic epigram by Poseidippos (142.12 A–B), written merely fifty years after the creation of the statue, and the much later fourth-century CE passage of *Descriptiones* (6) by Kallistratos – bear witness to Lysippos's genius behind the wonder (θαῦμα) of his creation, striking the viewer with speechlessness (ἀφασία) at the spectacle. Himerios even credits Lysippos for "enrolling Kairos among the gods" and "expounding his nature through the image" that he fashioned.[30] There is indeed suggestive evidence that Kairos became a divine figure by the mid–fifth century BCE; Pausanias (5.14.9) mentions the existence of his altar at Olympia and invokes the Hymn to Kairos by the fifth-century BCE poet Ion of Chios, who is the first to identify him as the youngest son of Zeus.[31]

The complex iconography of the Lysippan Kairos presented a special challenge to ancient authors, whose *ekphraseis* went beyond mere vivid descriptions of its appearance and offered various interpretative meanings

[27] Smith (1969: 8) discusses Plato's dialectic, contrasting *kairos* and *tyche* as opposites.

[28] See Poulakos 2002: 89–90, for the issue of teachability of *kairos* in rhetoric.

[29] Moreno 1995: 190; literary evidence variously suggests Pella, Sikyon, or Olympia as its original location, as well as the Lauseum at Constantinople, where at least one version was seen in post-antiquity. Tzetzes, *Chiliades* 10.266–272 (Byzantine) is the only source that mentions Alexander the Great explicitly in conjunction with the Lysippan Kairos. It is suggestive, however, that the author of the near-contemporary epigram on the statue, Poseidippos, was also from Pella.

[30] Himerios, *Eclogae* 14.1.

[31] Hereafter when referring to the concept, the italicized, lowercase (*kairos*) is used, and the personification or divinity or the statue itself is denoted by the non-italicized uppercase version (Kairos).

Figure 7 (a) Kairos by Lysippos. Roman relief copy. Second century CE. Turin. Museo di Antichità di Torino, Inv. n. 610 (Photograph © Archivio del Polo Reale di Torino). (b) The Lysippan Kairos: 3D digital reconstruction, profile view (© SeungJung Kim and Dave Cortes)

behind its attributes.[32] The handful of visual remains we have resonate relatively well with the mental reconstruction of the Lysippan Kairos from the various sources: a winged male figure, in full bloom of his youth, stands precariously as if on tiptoe, on a kind of sphere, with wings sprouting from his ankles, as he holds a blade in his hand, on which he balances a scale. The peculiar arrangement of his hair is of particular interest – the abundant locks flow in tresses over his forehead, while the back of his head is bald, so that one might grab him easily at the encounter, but not once he has passed.

Unlike other celebrated Greek statues, however, whose original appearances in bronze are preserved in full-scale, three-dimensional Roman marble copies, the visual evidence for the Lysippan Kairos consists only of reduced-sized, two-dimensional renderings: three marble reliefs, of which only one, the Roman relief copy from Turin (Figure 7a), is preserved in full, and a handful of gems.[33] The two other relief fragments from Trogir in Dalmatia and the Athenian

[32] Johnson (1927: Appendix 1, nos. 33–39) lists seven literary sources, two of which are in Latin (Ausonius, *Epigrammata* 33 and Phaedrus, *Fabulae* 5.8), naming the statue Occasio and Temporis, respectively. The other five are Kallistratos, *Descriptiones* 6; Kedrenos, *Historiarum Compendium* 322; Himerios, *Eclogae* 14.1; Poseidippos 142.12 A–B; Tzetzes, *Chiliades* 10.266–272.

[33] *Reliefs*: Turin Relief, Turin, Museo di Antichità n. 610; Trogir fragment, Göttingen, Archäologisches Institut A 1642 (Abramic 1930: pl. 1, figs. 2, 4; Moreno 1987: 125, pl. 66); Acropolis fragment, Athens, AkrM 2799 (Walter 1923: 124; Abramic: 1930: pl. 1, fig. 3); *Gems*: London, Robinson (Furtwängler 1900: pl. 43, no. 49); London, BM 1200 (Furtwängler 1900: pl. 43, no. 50); London, BM 1199 (Furtwängler 1900: pl. 43, no. 51).

Acropolis are both dated to the first century BCE; the former preserves most of the body above the knees, and the latter shows only the left leg above the ankle with a set of wings attached. All three reliefs show generally good consistency in style and iconography. The notable absence of a full three-dimensional replica, however, is often regarded with suspicion, which led some scholars to fall back on the textual sources, which they consider as rhetorical displays rather than accurate descriptions.[34] Even if the material existence of the statue is fully acknowledged, scholarly focus has predominantly been on the psychological dimension of the "reader" of these *ekphraseis*; the viewing of the statue itself, if mentioned at all, is taken note of as a subsidiary act to support a sophisticated text-based interpretation – here, the process of reading and decoding the statue's meaning is but "supported" by the visual image.[35]

Motivated by the lack of attention given to the phenomenological dimension of the sensory act of viewing, this chapter presents the result of a new visual, digital reconstruction in its full three-dimensionality that has never been attempted before. This calls for an equally sophisticated understanding of the physically present embodied viewer, whose interaction with the actual three-dimensional statue unfolds both in space and in time. Only then, it is argued, the form, nature, and meaning of the statue reveal their true colors, understood instinctively by the viewer at the moment of encounter.

The 3D digital reconstruction of the Lysippan Kairos (Figures 7b and 8; frontispiece) was a collaborative project carried out with the graphic artist and sculptor Dave Cortes.[36] While the details of the reconstruction are being prepared in a separate publication, a few notable points may be raised here.[37] The digital Kairos was modeled closely on the Turin relief, while approximating the appearance of the Trogir and Athens fragment when possible.[38] The addition of the globe under Kairos's left foot is the only major departure from the Turin relief, in which he is shown as if "walking" on tiptoe along the ledge.[39] But careful inspection of the Turin relief reveals that his right foot is in fact slightly lifted off the ground, unlike

[34] Johnson 1927: 164; Ridgway 1997: 304; see also Prauscello 2006.

[35] See Prauscello 2006 and Goldhill 1994; Zanker (2004) exemplifies the scholarly explorations on ancient "modes of viewing" that give primacy to extant textual evidence.

[36] Cortes Studio (cortesstudio.com); main software tool used was *Zbrush* (Pixologic, Inc.), widely used in current 3D animation and the movie industry.

[37] A full description of the procedure is also available in Kim 2014: 22–40.

[38] Especially, the overall appearance of the body has been slimmed down to approximate the Hellenistic fragment from Trogir, following the likelihood that the slightly stubby appearance of the Turin Kairos may be a product of the later Neo-Attic style. See Carinci 1985–1986, for attributing the Turin relief to an end-piece of a sarcophagus, possibly of Attic origins.

[39] The size of the globe in the reconstruction is arbitrary.

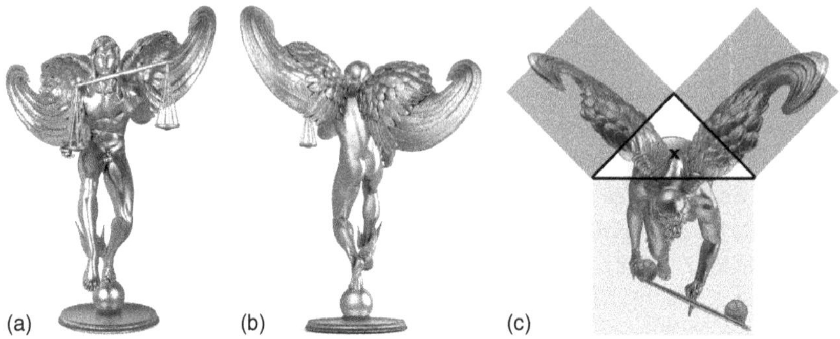

Figure 8 The Lysippan Kairos, 3D reconstruction: (a) Frontal view. (b) Rear view. (c) Overhead view, isosceles triangle around the gravitational center (© SeungJung Kim and Dave Cortes)

the ball of his left foot that presses firmly on the ledge, attesting to the fully weight-bearing character of the latter. Kallistratos explicitly states that Kairos was "standing on tiptoe on a kind a sphere," which would resonate with his right leg hovering in the air, completely weight free. This is indeed corroborated by an onyx gem in London (BM 1772).[40] Moreover, works from postclassical antiquity invariably turn Kairos into a figure "on wheels," attesting to his fleeting nature and mobility, which have likely developed out of the sphere mentioned by Kallistratos.

The absence of any reference to the sphere in the epigram by Poseidippos and its omission on the Turin relief – the two most widely used pieces of evidence for the Lysippan Kairos – have encouraged some to overlook its existence.[41] But this is to ignore the finer details seen on the relief itself; moreover, 3D sculpting and animation tools used to test the weight distribution of the Turin Kairos found it to be entirely compatible with the final reconstruction balanced on a single point of contact under his left foot. It is also argued here that Poseidippos, although failing to mention the sphere explicitly, may give us a vital clue not only for its existence but also for its function.

The twelve-line epigram of Poseidippos expounds the nature of the Lysippan Kairos by engaging the "talking" statue in a question-and-answer session with a passerby. The third line of the epigram presents a question to Kairos, followed by the corresponding answer:

– τίπτε δ᾽ἐπ᾽ ἄκρα βέβηκας (Why do you stand on tiptoe?)
– ἀεὶ τροχάω (I am always running.)

[40] Moreno (1995: 190) also maintains that the globe was part of the original sculpture.
[41] Poseidippos simply relates that Kairos stands "on tiptoe."

It is noteworthy that the verb translated as "running" is not τρέχω, the common verb form "to run," but instead τροχάω, which is a rare form that also harbors the meaning of "revolving" or "turning."[42] The only other conspicuous usage is found in the well-known contemporaneous work *Phaenomena* by Aratos, a comprehensive poetic description of the celestial phenomena. Aratos uses the same verb τροχάω three times to describe certain constellations moving or, rather, "revolving" around a celestial circuit. Another cognate τροχάζω, which harbors the meaning "to run swiftly like a wheel (τρόχος)," may indeed explain the post-antique transformation of Kairos's iconography on wheels.

The possibility of the Lysippan Kairos being a rotating statue, which has been hitherto completely neglected, albeit slim, is certainly worth entertaining. A rotating statue would not only make Poseidippos's unusual choice of the verb explicit, it would also readily explain the novel configuration of Kairos balancing his entire weight on a single point atop a sphere. The Lysippan Kairos, moreover, is situated at the dawn of the Hellenistic Period, during which mechanical artworks proliferated, which would have included self-rotating *automata* often reconstructed on a globe.[43] The grand procession of Ptolemaios Philadelphos, which showcased the famous colossal *automaton* of Nysa, who would stand up and pour a libation of milk and sit down again, is dated to around 280 BCE, only fifty years after the Lysippan Kairos. Although evidence for early *mechanische Kunstwerke* is scanty at best, it is certainly not out of the question that the Lysippan Kairos could have been one of the earliest rotating statues in the history of mechanical statuary. Furthermore, Kallistratos comments on the wings on his feet as denoting his swiftness, and as "causing the revolution of ages," as "he rides on the seasons" (*Descriptiones* 6.4) – if not hinting at his physical rotational capacity, it certainly suggests that cyclical motion had at least a metaphysical import with regard to the statue.

The digital reconstruction of the Lysippan Kairos, which explicitly expands the two-dimensional renderings into the third dimension, includes a number of extrapolations that are a combination of informed decisions and artistic intuition based on anatomical knowledge of the sculptor. This makes the reconstruction more than a simple re-creation; rather, it is a new kind of modern *visual ekphrasis* that literally bestows vividness to the original and adds a completely new dimension to our understanding of it.

42 Rolley (1999: 339) comments on this rare verb form: "qui évoque l'idée de tourner autant que celle de courir: ce n'est pas une erreur."

43 The puppet theater of Dionysos by Heron of Alexander moved automatically on wheels, which were connected to internal pulleys that rotated the Nike figure at the summit and maenads dancing in circles around Dionysos, whose *kantharos* issued wine and thyrsos sprinkled milk (Athenaios, *Deipn.* 5.198–199; see also von Hesberg 1987: 67–68).

Let us first examine how Lysippos might have achieved the impossible feat of gravitationally balancing the immensely top-heavy figure on a single tiptoe. First, Kairos's semi-crouching posture reflects the intuitive impulse to lower the center of gravity for stability. This posture with the outstretched arms allows the entire profile to zigzag from the highest tip of his wings, all the way down to the ball of his left foot, dispersing the weight laterally, creating an even larger margin for balance.[44] A corollary feature of his concave abdominals, which are exceptionally contracted, implies that he is concentrating all his energy into his center of gravity for further stability. Interestingly, this feature has also been characterized as physiologically resembling the turning point of a respiration, indicating the moment of transition from exhale to inhale.[45]

Second, the uneven levels of Kairos's outstretched arms mirror those of his legs. In other words, to keep one's right leg lifted up high, it is natural to counter this movement by lowering the upper parts of the body on the same side, so as not to generate torque that will cause one to keel over. This stance naturally causes the back muscles to contract unevenly, creating an S-curve in Kairos's spine (Figure 8b). This, in turn, causes the wings – thought to be a physiological extension of the back – to skew accordingly, making his right wing dip down below the left. It is worth noting that the resulting position of the wings seen from profile (Figure 7b) is remarkably consistent with what is shown on the Turin relief (Figure 7a). The reconstructed Kairos, seen especially from the front (Figure 8a), thus clearly visualizes the break in lateral symmetry that had not been hitherto visible from extant profile views, breathing life and rhythmic movement into the figure. Moreover, the level of the wings parallels the tilting rod of the scale as well as that of the shoulders, in contrast to the opposite tilt in the level of the knees: an overall triangular shape thus inscribed – a playful elaboration (and clearly an exaggeration) of the Polykleitan *contrapposto*.

Third, the reconstructed profile view (Figure 7b) shows that the central axis (dotted line) clearly demarcates the bulk of Kairos's body from the magnificent set of wings on his back. Structurally speaking, the key function of these wings is thus to achieve the overall balance of the entire piece, and at

[44] For bronze-casting methods in the Classical period of Greece, see Mattusch 1990, 1994, especially for the evidence regarding the lost-wax technique as the primary method used in this period. The indirect method of lost-wax casting using piece molds would have been appropriate for laterally dispersed extremities of the Lysippan Kairos. The growing use of indirect casting in the Hellenistic period has been associated with the increasing popularity of reproductions of Classical Greek originals (Treister 1996: 330–331).

[45] Moreno 1995: 190.

the same time allowing the body itself to have an autonomous impression of moving forward. The overhead view (Figure 8c) distinctly reveals that the entire composition is structured around an isosceles triangle: the two wings at right angle compensate for the larger bulk that is Kairos's body. Related to the Pythagorean triangle, it is simply the most beautiful and minimalistic way of achieving harmonious balance. The relevance of Pythagorean numerology to certain aesthetic principles, especially that of the Canon of Polykleitos, is sufficiently known in existing scholarship.[46] And *kairos* itself was, in fact, not an alien concept to the Pythagoreans, as it embodied the "virginal" prime number 7, resonating with the fidgety, elusive, shy figure of the youth, caught in the transition between two stages of life.[47]

The exact point of equilibrium for the 3D reconstruction was found by calculating the mass distribution proportionate to the surface area of the sculpture, consistent with the bronze-casting technique. Without a doubt, such finely tuned balance for hollow bronze could not have been effectively transferred to the heavy and solid medium of stone, which would have been unsustainably front heavy, since the body occupies the bulk of the volume. Turning this around, let us imagine the viewer confronting the original Lysippan sculpture frontally (Figure 8a).[48] It would have immediately struck the viewer as being caught in the act of rushing toward him, as Kallistratos aptly puts (6.3): "though standing still it showed that it had the possibility of starting off, and *deceiving one's eye*, conveying the impression that it possessed the power of motion forward." Kallistratos's qualification of the deceptive quality of Kairos now becomes understandable as the physical reality of the front-heavy statue at odds with its seeming equilibrium. Kairos is thus once again caught at the boundary, this time between *stasis* (reality) and movement (appearance).

With another glance at the details, the viewer would notice the razor-blade thrust on him, a sign that he is "sharper than any sharp edge," as Poseidippos explains, or as Tzetzes puts, "a warning not to disregard time." The blade has been read as a metaphor for the ephemeral nature of the moment and compared to the Aristotelian notion of the *now*, which has no duration but is an abstract point that delineates the past and the future.[49] Just as the edge of the blade divides the two arms of the scale, which tips over to

46 See in particular, Raven 1951; Pollitt 1974: 14–22; Stewart 1978a.

47 Aristotle, *Metaph.* 1, 985b30; 1, 990a23; 13, 1078b22; see also De Vogel 1966: 113–122; Burkert 1972: 467.

48 For an animated turntable showing the reconstructed Lysippan Kairos, see the website www.kairotopia.com.

49 Schädler 2003.

one side as Kairos himself gently nudges one of the plates toward him, the ephemeral *kairos*-moment, the *now*, slips away from the viewer. At this point, Kairos has turned a little and reveals the tumbling locks on his forehead.[50] The opportunity literally presents itself for the viewer to reach out and grab the tresses of his hair. But a moment later, as the sculpture furthers its rotation and reveals the balding back of his head, the viewer is invariably left with a sudden anxiety, an understanding that there is nothing to grasp, that the opportune moment is already gone, while the menacing wings create additional distance. When asked about his baldness, Kairos answers, "Because none whom I have raced by on my winged feet will now, though he wishes it sore, take hold of me from behind."[51] The 180-degree rotation of the statue can thus be understood as Kairos "racing by"; the short time that it takes for the viewer to see the back of the statue is then equated to the ephemerality of what Kairos stands for.

The phenomenological exercise presented here, of an embodied viewer whose perception is informed by the intersensory dynamics of his or her bodily experience vis-à-vis the rotating statue, makes it clear that the topochronic relationship between the viewer and the sculpture is integral to the message conveyed. Himerios singles out the wings on Kairos's feet as an explicit device for "concealing the fact that his weight does not rest on earth, though seeming to touch the ground."[52] The emphasis on this deceptive weightlessness, just as Kallistratos offers with motion, underscores precisely the tension between what the viewer is supposed to feel and what the physical reality of the statue is – a monumental top-heavy sculpted material precariously perched on a single point. Even the scale, conventionally interpreted as the iconography of "time as justice," is wobbling on the edge of a blade: there simply cannot be a clearer message that this particular representation of *kairos* is as much about physical balance as it is about a moment in time. The *tour de force* of the Lysippan Kairos lies in how effectively it conveys its message to the embodied viewer's cognitive reality in the temporal sense, through the precarious balance achieved in the physical reality in the spatial sense.

3 Kairos: Medicine and the Visual Arts

The temporal definition of *kairos*, as the right time for action that will lead to success, permeated fifth- and fourth-century BCE Greek thought, whether it

50 See the figure on the frontispiece.
51 Poseidippos (142.12 A–B).
52 *Eclogae* 14.1.

was rhetoric, philosophy, ethics, athletics, or medicine. It is well known, however, that the term might have originated from a spatial rather than a temporal sense of opportunity: Homeric usage as the "right place to stab for killing," and Hesiodic usage as "due measure."[53] A rich body of examples attests to this atemporal sense of *kairos* that sustains throughout to Roman times, variously translated as "apposite," "propriety," "decorum," "right proportion," and especially that of "balance." Combining both the notion of "proper measure" and the "right time," the two key components of *kairos*, Plato constructs the doctrine of virtue as the mean between two extremes.[54] Aristotle, whose ethics is based on the particularity and the contingent, further develops this notion. For Aristotle, real justice and its effectiveness come not from rigid application of theoretically defined laws, but from the spontaneous and fair adaptation of these laws to the variability of the situation; this, following the rhetorical principles of Gorgias, and of the sophistic rhetoric in general, is entirely the domain of *kairos*.[55]

The atemporal notion of *kairos* as the "right measure" thus preserved itself into an ethical dimension and by extension to aesthetics, as governed by principles of harmony and symmetry, lending relevance to the practice of visual arts. The principle of proportion that governs both the good and the beautiful is epitomized in Plato's *Timaeus*: "Everything that is good is beautiful and the beautiful always proportionate; according to a living creature that is to possess these qualities must be well-proportioned" (87c). It is also notable that a thread of this atemporal aspect of *kairos* can be seen in a more practical consideration in relation to Hippocratic medicine. Recent attention to the notion of *kairos* in the Hippocratic corpus – as the root *kair-* appears 264 times according to a database count – reveals that both senses of *kairos* as *timing* and *propriety* constitutes the core of the empirical practice of Hippocratic medicine.[56] Mastering *kairos*, or knowing when to treat an illness with the appropriate measure for each particular case, is key to the situational determinism based on observation of facts and the wide body of collected medical knowledge.

[53] For early spatial uses of *kairos*, see Trédé 1992: 25–31.

[54] Levi 1923: 277–279; see also von Leyden 1964: 40–42, who associates *kairos* with the notion of the eternal in Plato, in contrast to the "moving likeness of eternity" that is *chronos* (Plato, *Timaeus* 37D).

[55] Guillamaud 1988: 367; James Kinneavy's seminal essay (Kinneavy 1986) triggered an active inquiry into the concept of *kairos* in rhetoric as it culminated in the edited volume, Sipiora and Baumlin 2002.

[56] Eskin (2002: 99) uses the Ibycus system (including Thesaurus Lingua Graecae) to arrive at this number; see also Roth 2008: 86–91.

A particular usage of *kairos*-as-balance in the Hippocratic corpus is that of the opposing elements in the body, such as humidity and dryness, required for a healthy state. Such balance of opposites is particularly germane to the Lysippan Kairos: the balance between left and right, up and down, front and back, wings and body, altogether creating a *tour de force* of gravitational equilibrium. To this notion also belong the "boundary" qualities of Kairos, between motion and stasis, exhale and inhale, young and old, as well as past and future.[57]

The Hippocratic treatises show explicit connections between *kairos* and two other concepts, μέτρον (measure) and ἀκρίβεια (detail or precision). Evoked often in relation to dietary or pharmacological concerns, seizing the *kairos* in medicine means to find the right measure (μέτρον) of components (food/medicine) that will yield the precision (ἀκρίβεια) of treatment required for a person to become healthy.[58] The connection between *kairos*, measure and precision, or the concept of συμμετρία (proportion), is thus already well established, as noted by both Monique Trédé and Andrew Stewart and as explicated in the key passage from Plutarch's *Moralia* (45C): "Now in every piece of work, beauty is achieved through many numbers (πολλοὶ ἀριθμοί) coming to a congruence (καιρός) under some system of proportion (συμμετρία) and harmony (ἁρμονία)."[59] And a Galenic treatise in the second century CE reiterates this connection of medical insight explicitly with the principles of the Canon of Polykleitos.

It is, however, to the latter term of exactitude (ἀκρίβεια) to which I would like to draw attention. The term's well-attested importance in the medical arts can be expanded to the late fifth-century obsession for it in the *techne* (arts) in general.[60] To render as much ἀκρίβεια as possible in the pursuit of any *techne*, one must take recourse to three things: measure (μέτρον), number (ἀριθμός), and weight (σταθμός).[61] Of this triad of concepts, both measure and number have been well associated with the concept of *kairos* for the visual arts through the previously mentioned passage of *Moralia*. The last of these, weight, however, has seldom been considered in connection with artistic practice, which is somewhat surprising given its pivotal function in the

[57] The Hippocratic corpus is contemporary with the Lysippan Kairos, which allows us to sidestep some of the unnecessary assumptions that Polykleitan scholars would have to make to explore similar relationships between the practice of the arts and the state of medical knowledge (cf. Métraux 1995; Leftwich 1995).

[58] *On Regimen in Acute Diseases* (*Acut.*) 1.2; see Eskin 2002: 112 n. 8, for pharmacological uses of *kairos* as right measure.

[59] Cf. Stewart 1978b; Trédé 1992: 171.

[60] Trédé 1992: 164–165; see also Kurz 1970: 62–87.

[61] On this triad of concepts, see Genzmer 1952 and Heinimann 1975.

physicality of the sculpture. As we know, the Polykleitan *contrapposto* fundamentally relies on the internal distribution of weight in the body and the external, corporeal articulation as a result of that distribution. This is pronounced even further with the Lysippan Kairos since, as amply argued earlier, the distribution of weight itself and the resulting precise, physical equilibrium are explicitly at the core of its sculptural expression.

Pliny's account of Lysippos, most likely taken from Lysippos's follower Xenokrates of Athens (third century BCE), lists ἀκρίβεια along with συμμετρία and ῥυθμός (movement or composition) as the main contribution of the Lysippan canon.[62] Here ἀκρίβεια is usually translated as "attention to detail," which is extrapolated then to "realism."[63] Instead, I argue that ἀκρίβεια as precision or exactitude, intimately linked with *kairos* in the Hippocratic corpus, is realized in the Lysippan Kairos through the precise distribution of weight. And of course, weight is in turn, as mentioned earlier, one of the triad of concepts that are necessary conditions for ἀκρίβεια in the pursuit of *techne*.

In one of the principal Hippocratic treatises entitled *Ancient Medicine*, references abound to the impossibility of proper μέτρον to attain the level of perfect ἀκρίβεια, because of the diversity of factors that come into play. The "grande originalité" of this treatise, as Trédé puts it, is to have defined the notion of μέτρον as αἴσθησις τοῦ σώματος.[64] The exact meaning of the latter phrase has been in much debate among scholars, ranging from feeling or sensitiveness of the body of the patient, to the sense perception of the doctor when examining or interpreting the patient's body.[65] Regardless of the exact translation, it is clear that the meaning brings a qualitative sense to the process of μέτρον, which in fact becomes an intuitive, phenomenological method to arrive at "une exactitude presque parfait."[66] No quantitative, mechanically repeatable procedure can cater to "every individual's nature":[67] "La seule précision qu'on puisse viser est celle du *kairos*," claims Trédé, which is to say that *kairos* is the *only* qualitative, intuitive principle by which a precise, correct outcome can be achieved in each individual case.

The phenomenological dimension is thus quite explicit in the phrase, "αἴσθησις (sense-perception, sensation) τοῦ σώματος (of the body)," understood as a direct methodology of arriving at a *kairos*. The prime importance

62 Pliny the Elder, *Nat.* 34.65.
63 Stewart 1978b: 168; Stewart 1993: 32.
64 Trédé 1992: 165–169.
65 See Trédé 1992: 166, and references therein.
66 *Vet. Med.* chap. 12 as quoted by Trédé (1992: 170).
67 πρὸς ἑκάστου φύσιν (*Acut.* 1.2).

for *kairos* in medical treatments, therefore, of the bodily sensation and the intuition based on that bodily experience positively corroborates the need for a phenomenological understanding of the Lysippan statue as an embodiment of this concept. In other words, since the notion of *kairos-as-balance* subsumes a bodily experience, the visual manifestation of this notion could not be fully understood without a bodily understanding of its physicality. And the genius of Lysippos lies in fully exploiting that phenomenological dimension to deliver the intended message.[68] The sculptor's trademark of ἀκρίβεια is seen in his Kairos, not as "realism" as conventionally understood but as the perfection achieved in the gravitational balance on a single point of contact, which is at once physically experienced and cognitively translated as the fleeting instance of the "now." We turn once again to the Hippocratic treatise attesting to the narrowness of both spatial and temporal qualities of *kairos*, and the need for phenomenological intuition to pin-point it in both space and time:

> Time (*chronos*) is that wherein there is opportunity (*kairos*), *and opportunity (kairos) is that wherein there is no great time* (*chronos*) . . . knowing this, one must attend in medical practice not primarily to plausible theories, but to experience combined with reason. For a theory is a composite memory of things apprehended with the sense-perception. (*Precepts*, Jones 1923: 1.313–15)

> *In medicine the correct measure* (*kairos*) *is narrow* . . . Correct measure (*kairos*) is the following: to administer as much food as, being administered will be mastered by the body . . . [and] this is the correct measure (*kairos*) the physician must recognize. (*Places in Man*, Potter 1995: 8.89)

4 The Pioneer Group: Beginnings of the Phenomenological Mode of Viewing

The last two decades of the sixth century BCE mark a radical shift in the style and content of Greek vase painting. With the advent of the red-figure technique, the well-defined artistic coterie of vase painters called the "Pioneer Group" visibly changed the landscape of pictorial representation – they were not only extremely popular and prolific, proudly signing their names, in some cases providing self-portraits and referring competitively to one another's works through meta-commentaries, they also conducted

[68] Regardless of whether the Lysippan Kairos was a rotating statue, the specific iconography of the balding back of the head makes it quite explicit that the viewer was at least expected to walk around the statue, with similar phenomenological effect.

audacious experimentations with representations of the body, the spatial relationships and the morphology of the nude male athlete. The Pioneers' keen interest in anatomy encouraged traditional scholarship to place their achievement in the context of the developmental grand narrative of pictorial naturalism of the human figure, driven primarily by the medium of sculpture. This narrative is a familiar one to the student of ancient Greek art: the Archaic style sees a rupture with the so-called Greek Revolution identified around the year 480 BCE, heralding the Classical period, which then culminates with the Polykleitan Canon another generation later.[69] While the limited scope of this chapter precludes extensive commentary on the theory of periodization, its uses and its misuses, suffice it to say that I argue for an independent trajectory for the medium of vase painting – one that is not merely auxiliary to the preestablished, formalist grand narrative. The so-called revolution, if it existed for vase painting, is precisely for that which the Pioneers seem to have achieved in a systematic manner: an explicit awareness of the spatio-temporal structure of their medium and a keen exploration of the ensuing phenomenological relationship with the space-time of the viewer. This is particularly evident in their unprecedented sensitivity toward capturing a veritable moment that was not just cognitively understood but also fundamentally *felt* – tapping into the sensory realm of embodied viewing.

We open our investigation with the well-known "Swallow vase" at the Hermitage (Figure 9).[70] Attributed to the circle of Euphronios and dated to c. 510 BCE, the *pelike*, characterized by heavy pot-bellied contours, shows one of the most engaging scenes in the history of vase painting. From left to right, we have a youth, a bearded man, both seated and half-draped, and a standing, nude *ephebe* (adolescent). A small bird is caught above them in mid-flight, positioned slightly off center as if in a diagonally upward movement. All three characters dramatically point upward with their gazes unmistakably turned toward the bird. The seeming synchronicity of all the pointed fingers, demanding our attention to the bird in flight, focuses not only our *spatial* attention to a single point, but also our *temporal* understanding of the transient nature of the moment.

The difficulty in pinpointing a bird in flight, especially as we shall come to see, a swallow, whose flight pattern is notoriously irregular, puts the transitory temporal focus of the vase in relief. Much fruitful discussion regarding

[69] See e.g. Robertson (1992: 20–42), whose extensive treatment on the Pioneers is entitled "A time of ferment," which encapsulates this idea of positioning the Pioneer Group vis-à-vis the established narrative.

[70] Saint Petersburg, Hermitage 615 (*ARV*² 1594.48).

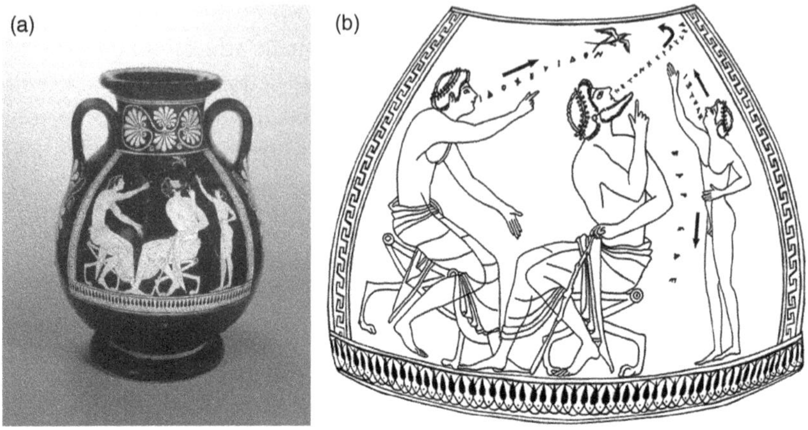

Figure 9 The so-called Swallow vase. (a) Attic red-figure *pelike* attributed to the circle of Euphronios, c. 510 BCE. H. 37.5 cm. The State Hermitage Museum, St. Petersburg, Inv. no. GR-8057 (B-2352) (photograph © The State Hermitage Museum; photo by Yuri Molodkovets). (b) Direction of the inscriptions (drawing by H. Anh Thu Nguyen)

the vase has ensued in recent times, but mostly without consideration for its multilayered temporality. Instead these discussions revolve around either the polysemic nature of the bird as an omen, a "sign" (*semeion*), or the complex intersection between orality, literacy, and visuality and its sociopolitical meaning in the context of the symposium.[71] But as we further peel the layers of time encoded behind this visual staccato of the instant, it becomes clear that the image is all about a newfound fascination with time.

The accompanying inscriptions, in fact, form an integral part of the visual composition, constituting lines in a dialogue that stream out of each character's mouth, whose order seems evident (Figure 9b): the leftmost seated youth, first spotting the bird, demands to his fellow company: "Look! A swallow!" We, as readers, trace the letters in sequence to the final letter *N*, grazing the tip of the swallow's tail. Thus the act of reading, taking a finite amount of time, brings us into the scene as an active participant of witnessing the swallow. We then read, taking the words out of our own mouth, the confirmation offered by the older man, "Yes indeed, by Herakles!" The final tip of this second inscription curls around as if to point back at and frame the swallow in question, stalling its movement forward and allowing us to linger our gaze on the bird once again. A third time, the standing *ephebe*, reaching upward as if to compensate for his short height, energetically extends his

[71] See, in particular, Neer 2002: 63–64; Immerwahr 2010; Steiner 2013.

hand and utters his boyish, pithy remark, "There she is!" The excitement seems to be coupled with the relief that the boy too, at a short instant later, was fortunate enough to spot the bird. The ephemeral *kairos*-moment is all too palpable in this seemingly naive, yet poignant *mise en scène*, as the viewer will instantly evoke the all-too-familiar occurrence of missing a transitory event at the beckoning of one's company to look. The three cognitively separable instants that can only be made sense of by the act of reading the inscriptions, and thus participating repeatedly in the act of witnessing, are in dynamic tension with the visual momentariness of the entire event. In other words, these three statements would have been fired in rapid succession, but the bird caught in midair represents a true instant. The analogous modern philosophical notion of the "specious present" – which results from a cognitive disjuncture between the perceptive experience of the present (always occupying a duration) and the Aristotelian definition of the "now" (an extensionless, abstract point of time) – and resolving its paradox were, in fact, the constitutive drive for Husserl's project of the phenomenology of time-consciousness.

Unlike the dialogues pointing up toward the bird, the meta-discourse, most likely from the vase painter himself, points down toward the ground, thus formally distinguishing itself from the rest: ἔαρ ἤδη "[It is] already spring."[72] This critical information, standing outside the narrative of the action, provides the overall temporal backdrop, and by doing so the vase painter subtly inserts himself into the discourse on time. The significance of the swallow is now obvious as heralding the particular season.[73] Moreover, the temporal qualifier ἤδη, explicitly aware of our act of viewing, refers to the time of viewing both internal and external to the image, merging the viewer's experience with the event depicted. It is as if to say, now that *we* have also spotted the swallow, spring has presently, or already, descended on us. And while this temporal specificity of the particular season is brought to the fore, it contrasts against the implied cyclic flow of the seasons, as the *ephebe* becomes a youth and, eventually, an older man.[74] Three ages of man are depicted here coexisting, and resisting the passage of time, as they participate together in a dramatic witnessing of the onset of spring, which

72 The phrase has been variously attributed to the *ephebe* or the seated man, but I agree with Immerwahr (2010: 576), who follows Guarducci (1987: 3.468) and Richter (1958: 15) in calling it a "title," or Lissarrague (1992: 201) who calls it a "comment" by the vase painter.

73 Neer (2002: 63–64) interprets the vase as a kind of riddle, or a visual pun at tension with the popular Attic proverb "One swallow does not a springtime make."

74 Steiner 2013: 64; see Davidson 2006, for discussion on age-class in Athens and its consequences for a marked change in the temporality of Classical Greece.

Figure 10 The so-called Jumpers vase. Attic red-figure *pelike* attributed to Euthymides (circle of Euphronios). c. 520–510 BCE. H. 31.1 cm. (a) Obverse. (b) Reverse. Museum of Fine Arts, Boston. Juliana Cheney Edwards Collection, 1973.88 (photograph © 2015 Museum of Fine Arts, Boston)

is the season of Eros. The vase's pederastic undertones, of which there is scholarly consensus, resonate with the viewer's own context of the symposium where the vase would have been used, further enhancing the shared temporality between the characters and the viewer.[75]

Explicit awareness of the viewer's participation in the interplay of temporal layers can be seen in another contemporaneous *pelike* in Boston attributed to Euthymides, well known for its depiction of two jumping youths accompanied by a double-*aulos* player (Figure 10).[76] Unlike the amphora whose lower part tapers down into a slim and elegant foot, the bottom-heavy *pelike* was a new shape invented around this time, and the Pioneers were clearly interested in exploiting this form to their advantage.[77] The Hermitage vase, for example, echoes the weight distribution of its form by giving visual weight to the lower half pictorially, and thereby it reinforces the pyramidal focus

[75] Working within the pederastic framework, Immerwahr (2010) and especially Steiner (2013) explore the vase's connection to literary and sympotic culture.

[76] Boston, Museum of Fine Arts 1973.88.

[77] Shapiro (1997) reflects on the shape of the *pelike* and its suggestive relationship to "banausic" subject matter.

given to the swallow above. The Boston *pelike*, on the other hand, inverts this composition to emphasize the particular moment captured: by dominating the visual field in the upper half and leaving the expansive lower part a black negative space, the youths seem to be jumping even higher, defying the gravitational pull encoded in the shape of the vessel that resembles a paunchy wineskin, heavy with liquid, squatting against the floor. Moreover, the rhythmic repetition of their anatomy on the obverse (Figure 10a) echoes that of the vase itself: their arms impersonate the handles; the angle of their bent knees resonates with that of the corner of the frame. Imagine thus lifting the vessel by its handles as if one is lifting the youths by their arms, boosting their jump to even greater heights.

It is when the viewer turns to encounter the other side of the vase that the temporal discourse becomes evident.[78] Immediate recognition of the dominant repetitive elements creates tension between the viewer's short-term memory and present perception, which unfailingly encourages comparison. On the reverse (Figure 10b), it seems that the same two youths are in mid-jump again, seen from the back this time, accompanied by most likely the same double-*aulos* player. The inscription labels two names for the jumpers on the obverse, Aineas and Kallipides, and gives only one name for the piper on the reverse, Smikythion, consistent with the notion that the same three figures are indeed shown twice. The only variations are the position of the jumpers' arms and the level at which the *aulos* is held, suggesting that they are slightly different moments of the same general setting and seen from different angles. It is almost as if the jumpers inhabit the space within the transparent walls – the concept of the Renaissance window taken quite literally.[79] The meta-discourse on this vase is also quite notable: the obverse harbors the common *kalos* inscription "Leagros is handsome," while the reverse presents an uncanny literal response, and that in retrograde, as if to echo the fact that the jumpers are seen from the back: "Leagros is handsome, yes indeed!"

Once it is clear that the two sides of the vase are to be read in close connection with an explicit temporal sequence as utterance and response, the slight difference in the poses on each side explicitly translates into two instants. Perhaps we are seeing two consecutive jumps, performed as part of the so-called *bibasis*, a Spartan jumping contest, whose winner would have

[78] One must recall that the two sides of the vase were never intended to be seen simultaneously. For an animated turntable of the digitally reconstructed vase, see again www.kairotopia.com.

[79] Steiner (2007: 4) calls it a "playful, mind game," since "of course the pelike is not transparent."

performed the most number of successful jumps.[80] Jumping figures are rarely found before the Boston *pelike*, but isolated cases did exist, such as the mid–sixth-century Corinthian *aryballos*.[81] This small Archaic vessel shows a certain Pyrrhias in mid-jump, as a leader of a chorus and the winner of the *bibasis* contest. But given the nature of the *bibasis*, whose defining feature was the exhaustive repetition of jumping, ambiguity is inherent in the temporality of the jumping figure. It is as if Pyrrhias is shown in the constant state of jumping rather than caught in the moment of a single jump. The Boston *pelike*, however, breaks this degeneracy once and for all by fixing the specificity of two separate moments, portrayed from two different angles. Moreover, their synchronized postures with minute but detectable variations reinforce the idea of the specific moment that is "coordinated." The revolutionary aspect of the Boston *pelike*, therefore, is the conscious employment of a temporal delay as a topochronic function of the viewer vis-à-vis the object. Knowing that the viewer or reader will require a finite duration from one sighting or reading to the next, the vase painter encoded two temporal specifications, as well as two viewpoints in accordance with the time and position of the viewer. This kind of participatory awareness had simply never been seen before in Greek visual culture.

5 Conclusion

The few examples investigated here are but a prelude to a copious body of works of art – not only initiated by the Pioneer Group in vase painting that largely belong to the private context but also including public sculpture as well as monumental painting – that showcase a radically different temporal engagement with the contemporary viewer. In the last decade of the sixth century BCE, the Pioneers, and most conspicuously Euphronios, continued to explore with unparalleled sensitivity the types of embodied viewing that informed the exact moment captured, whether with the critical moments of an athletic maneuver or common bodily tasks.[82] We have seen that the Lysippan Kairos, at the end of the Classical period, offered the precise solution to gravitational equilibrium in spatial terms as the epitome of the elusive temporal concept that can only be understood with somatic intuition. It is suggested here that the dawn of this mode of *kairotic* practice of art, epitomized in the later personification of the concept, occurs as early as the end of the sixth century BCE, starting with the Pioneer Group. And it does

80 Pollux (4.102) relays that a Laconian girl once set a record by jumping 1,000 times.

81 Corinth Museum C-54-1; see also Roebuck and Roebuck 1955: pls. 63–64; Amyx 1988: 560 n. 17.

82 Kim 2014: 122–169.

so largely by exploiting the bodily struggle against gravity as a lever to engage the viewer phenomenologically. Whether it is the intra-elite competitions in the context of the symposium or the ultimate aristocratic ideal of athletic competency that the majority of the intended viewership would have possessed, there is little doubt that the *kairos*-moment was felt quite instinctively to the embodied viewer, engaging and using such imagery in their proper contexts.

Finally, we come full circle as we conclude with another fundamental aspect of *kairotic* temporality, which not only underlies its phenomenological premise but also harks back to the anecdotes related at the beginning of this chapter – that is, its deep-seated focus on the present, and presence, by extension. The structure of embodied viewing necessarily presupposes the temporal present and spatial presence of the agents of perception – both the perceiver and the perceived – as a prerequisite for generating meaning. Such a focus on the present, as a basic principle of *kairos*, is seen also quite literally in the fifth-century phenomenon of *Historienbilder*, or official representations of actuality, commemorated on large-scale media, such as the Tyrannicide statue group or the Marathon Painting in the Stoa Poikile.[83] A convincing analysis – in the larger context of shifting societal notions of time – has yet to be offered as to why and how this distinct interest in the present contemporary or recent past became manifest in the visual realm, which was otherwise notoriously reluctant to break with its mythographic tradition. Just as André Bazin elucidates the impact of the photographic, and by extension, the cinematic image, by contextualizing their novel and irrevocable indexical relationship to actuality in the nineteenth century, the emergence of *Historienbilder* at the turn of the fifth century BCE should be held as an equally important milestone, marking an ontological shift in the history of Greek imagery.[84] These imageries of actuality introduced a radically new mode of active viewing from a phenomenological standpoint, not only in relation to embodiment, and even physical emulation, but also with regard to the activation of memory, both individual and collective.[85]

The profoundly novel engagement with temporality that *kairos* embodies, pregnant with human agency, empirical and sensory in nature and with a present-oriented ontology, found expression in the visual arts in the Late Archaic and Classical periods, not as an intellectual engagement of interpretative understanding but as an intuitive engagement of sensorial

[83] Hölscher 1973, 1998; Csapo and Miller 1998.
[84] Bazin 1960.
[85] Kim 2014: 53–121.

experience. The literal idea of *kairos* as a fixed ephemeral moment in time is effectively conveyed by the intimate dialogue between the topology of the medium and the awareness of the viewer's temporality, giving full agency to the artworks as a tangible expression informed by human experience and subjectivity.

Bibliography

Abramic, M. 1930. "Ein neues Kairos-Relief." *Jahreshefte des Österreichischen Archäologischen Instituts in Wien* 26: 1–8.

Amyx, D. A. 1988. *Corinthian Vase-Painting of the Archaic Period*. Berkeley and Los Angeles: University of California Press.

Ankersmit, F. R. 2005. *Sublime Historical Experience*. Stanford: Stanford University Press.

Bazin, A. 1960. "The Ontology of the Photographic Image." *Film Quarterly* 13/4: 4–9.

Benjamin, W. 1968. *Illuminations*. Trans. H. Zohn. Ed. H. Arendt. New York: Schocken.

Bergson, H. 1910. *Time and Free Will: An Essay on the Immediate Data of Consciousness*. London: Allen & Unwin.

Block, R. A. (ed.). 2014. *Cognitive Models of Psychological Time*. New York: Psychology Press.

Boehm, R. 1965. "Basic Reflections on Husserl's Phenomenological Reduction." *International Philosophical Quarterly* 5: 183–202.

Burch, N. 1990. *Life to Those Shadows*. London: British Film Institute.

Burkert, W. 1972. *Lore and Science in Ancient Pythagoreanism*. Cambridge, MA: Harvard University Press.

Carinci, F. 1985–1986. "Eros e Anteros: Alcune osservazioni a proposito di un rilievo della Galleria Colonna." *Rivista dell'istituto nazionale d'archeologia e storia dell'arte*, S. III: VIII-IX, 63-109.

Csapo, E. and M. C. Miller 1998. "Democracy, Empire, and Art: Toward a Politics of Time and Narrative." In D. Boedecker and K. A. Raaflaub (eds.), *Democracy, Empire and the Arts in Fifth-Century Athens*. Cambridge, MA: Harvard University Press, 87–125.

Davidson, J. 2006. "Revolutions in Human Time: Age-Class in Athens and the Greekness of Greek Revolutions." In S. Goldhill and R. Osborne (eds.), *Rethinking Revolutions Through Ancient Greece*. Cambridge: Cambridge University Press, 29–67.

De Vogel, C. J. 1966. *Pythagoreans and Early Pythagoreanism: An Interpretation of Neglected Evidence on the Philosopher Pythagoras*. Assen: Van Gorcum.

Debnar, P. 2005. "Fifth-Century Athenian History and Tragedy." In J. Gregory (ed.), *A Companion to Greek Tragedy*. Oxford: Blackwell, 3–23.

Deleuze, G. 1994. *Difference and Repetition*. Trans. P. Patton, New York: Columbia University Press.

Doane, M. A. 2002. *The Emergence of Cinematic Time: Modernity, Contingency, the Archive*. Cambridge, MA: Harvard University Press.

Dunn, F. 2007. *Present Shock in Late Fifth-Century Greece*. Ann Arbor: University of Michigan Press.

Eisner, L. 1975. *Die dämonische Leinwand*. Frankfurt am Main: Fischer.
Eskin, C. 2002 "Hippokrates, *Kairos*, and Writing in the Sciences." In P. Sipiora and J. S. Baumlin (eds.), *Rhetoric and Kairos: Essays in History, Theory, and Praxis*. Albany: State University of New York Press, 97–113.
Favorini, A. 2003. "History, Collective Memory, and Aeschylus' Persians." *Theatre Journal* 55/1: 99–111.
Fränkel, H. 1960. *Wege und Formen frühgriechischen Denkens*. 2nd ed. Munich: Beck.
Furtwängler, A. 1900. *Die antiken Gemmen: Geschichte der Steinschneidekunst im klassischen Altertum*. Leipzig: Giesecke und Devrient.
Genzmer, E. 1952. "Pondere, numero, mensura." *Revue Internationale des droits de l'Antiquité* 1: 469–494.
Goldhill, S. 1994. "The Naive and Knowing Eye: Ecphrasis and the *Culture* of Viewing in the Hellenistic World." In S. Goldhill and R. Osborne (eds.), *Art and Text in Ancient Greek Culture*. Cambridge: Cambridge University Press, 197–223.
Grene, D. 1987. *Herodotus: The History*. Chicago: University of Chicago Press.
Grethlein, J. 2010. *The Greeks and Their Past: Poetry, Oratory and History in the Fifth Century BCE*. Cambridge: Cambridge University Press.
Guarducci, M. 1987. *Epigrafia greca i–iv*. Rome: Istituto poligrafico e zecca dello Stato, Libreria dello Stato.
Guillamaud, P. 1988. "L'essence du *kairos*. " *Revue des études anciennes* 90: 359–371.
Hall, E. 1996. *Aeschylus, Persians*. Warminster, UK: Aris and Phillips.
Heidegger, M. 1927. *Sein und Zeit*. Translated as *Being and Time* by J. Macquarrie and E. Robinson (1962). Oxford: Basil Blackwell.
Heinimann, F. 1975. "Mass, Gewicht, Zahl." *Museum Helveticum* 33: 183–195.
Hesberg, H. von 1987. "Mechanische Kunstwerke und ihre Bedeutung für die höfische Kunst des frühen Hellenismus." *Marburger Winckelmann-Programm*: 47–72.
Hölscher, T. 1973. *Griechische Historienbilder des 5. und 4. Jahrhunderts v. Chr.* Würzburg: K. Triltsch.
Hölscher, T. 1998. "Images and Political Identity: The Case of Athens." In D. Boedecker and K. A. Raaflaub (eds.), *Democracy, Empire and the Arts in Fifth-Century Athens*. Cambridge, MA: Harvard University Press, 153–184.
Husserl, E. 1991. *On the Phenomenology of the Consciousness of Internal Time (1893–1917)*. Trans. J. B. Brough. Dordrecht: Kluwer.
Immerwahr, H. R. 2010. "Hipponax and the Swallow Vase." *American Journal of Philology* 131/4: 573-587.
Johnson, F. P. 1927. *Lysippos*. Durham, NC: Greenwood Press.
Jones, W. H. S., trans. 1923. *Hippocrates, Volume I: Ancient Medicine. LCL 147*. Cambridge, MA: Harvard University Press.
Karasek, H. 1994. "Lokomotive der Gefühle." *Der Spiegel* 52: 154.
Kim, S. 2014. *Concepts of Time and Temporality in the Visual Culture of Late Archaic and Classical Greece*. PhD Dissertation, Columbia University, New York.
Kinneavy, J. 1986. "Kairos: A Neglected Concept in Classical Rhetoric." In J. D. Moss (ed.), *Rhetoric and Praxis: The Contribution of Classical Rhetoric to Practical Reasoning*. Washington, DC: Catholic University of America Press, 79–105.
Kurz, D. 1970. *Ἀκρίβεια: Das Ideal der Exaktheit bei den Griechen bis Aristoteles*. Göppingen: Kümmerle.

Leftwich, G. V. 1995. "Polykleitos and Hippokratic Medicine." In W. Moon (ed.), *Polykleitos, The Doryphoros, and Tradition*. Madison: University of Wisconsin Press, 38–51.

Levi, D. 1923. "*Kairos* in Greek Literature." *Rendiconti della Reale Accademia Nazionale dei Lincei, Classe di scienze morali, storiche e filologiche* 32: 260-281.

Leyden, W. von 1964. "Time, Number, and Eternity in Plato and Aristotle." *The Philosophical Quarterly* 14/54: 35–52.

Lissarrague, F. 1992. "*Graphein*: écrire et dessiner." In C. Bron and E. Kassapoglou (eds.), *L'image en jeu: De l'antiquité à Paul Klee*. Yens-Sur-Morges: Université de Lausanne, 189–203.

Loiperdinger, M. 2004. "Lumière's Arrival of the Train: Cinema's Founding Myth." *The Moving Image* 4/1: 89–119.

Marincola, J., L. Llewellin-Jones, and C. Maciver (eds.). 2012. *Greek Notions of the Past in Archaic and Classical Eras*. Edinburgh: Edinburgh University Press.

Mattusch, C. C. 1990. "The Casting of Greek Bronzes: Variation and Repetition." In M. True and J. Podany (eds.), *Small Bronze Sculpture from the Ancient World*. Malibu: J. Paul Getty Museum, 125–144.

Mattusch, C. C. 1994. "The Production of Bronze Statuary in the Greek World." In G. Hellenkemper Salies et al. (eds.), *Das Wrack: Der antike Schiffsfund von Mahdia*, vol. 2. Cologne: Rheinland-Verlag, 789–800.

Merleau-Ponty, M. 1962. *Phenomenology of Perception*. Trans. C. Smith, London: Routledge & Kegan Paul.

Métraux, G. P. R. 1995. *Sculptors and Physicians in Fifth-Century Greece*. Montreal: McGill-Queen's University Press.

Moreno, P. 1987. *Vita e arte di Lisippo*. Milano: Il Saggiatore.

Moreno, P. 1995. *Lisippo: l'arte e la fortuna*. Milano: Fabbri.

Moxey, K. 2008. "Visual Studies and the Iconic Turn." *Journal of Visual Culture* 7/2: 131–146.

Moxey, K. 2013. *Visual Time: The Image in History*. Durham, NC: Duke University Press.

Neer, R. 2002. *Style and Politics in Athenian Vase-Painting: The Craft of Democracy*. Cambridge: Cambridge University Press.

Orth, E. W. 1984. "Phenomenological Reduction and the Sciences." In K. K. Cho (ed.), *Philosophy and Sciences in Phenomenological Perspectives*. Phaenomenologica 95, Dordrecht: Martinus Nijhoff, 152–170.

Pollitt, J. J. 1974. *The Ancient View of Greek Art*. New Haven and London: Yale University Press.

Potter, P., trans. 1995. *Hippocrates Vol. VIII: Places in Man [. . .]*. Loeb Classical Library. Cambridge, MA: Harvard University Press.

Poulakos, J. 2002. "*Kairos* in Gorgias' Rhetorical Compositions." In P. Sipiora and J. S. Baumlin (eds.), *Rhetoric and Kairos: Essays in History, Theory, and Praxis*. Albany: State University of New York Press, 89–96.

Prauscello, 2006. "Sculpted Meanings, Talking Statues: Some Observations on Posidippus 142.12 A–B (= XIX G–P) ΚΑΙ ΕΝ ΠΡΟΘΥΡΟΙϹ ΘΗΚΕ ΔΙΔΑϹΚΑΛΙΗΝ." *American Journal of Philology* 127/4: 511–523.

Price, T. F., C. K. Peterson, and E. Harmon-Jones. 2012. "The Emotive Neuroscience of Embodiment." *Motivation and Emotion* 36/1: 27–37.

Raven, J. E. 1951. "Polyclitus and Pythagoreanism." *Classical Quarterly* 45: 147–152.
Richter, G. M. A. 1958. *Attic Red-figured Vases*. 2nd ed. New Haven: Yale University Press.
Ridgway, B. 1997. *Hellenistic Sculpture: The Styles of ca. 331-200 B.C.* Madison: University of Wisconsin Press.
Robertson, M. 1992. *The Art of Vase-Painting in Classical Athens*. Cambridge: Cambridge University Press.
Roebuck, M. C. and C. A. Roebuck. 1955. "A Prize Aryballos." *Hesperia* 24/2: 158–163.
Rolley, C. 1999. *La sculpture grecque*. Paris: Picard.
Romilly, J. de. 1968. *Time in Greek Tragedy*. Ithaca: Cornell University Press.
Roth, A. D. 2008. *Reciprocal Influences Between Rhetoric and Medicine in Ancient Greece*. PhD Dissertation, University of Iowa.
Runia, E. 2006a. "Presence." *History and Theory* 45/1: 1–20.
Runia, E. 2006b. "Forum: On Presence 1. Spots of Time." *History and Theory* 34: 305–316.
Sadoul, G. 1956. *Geschichte der Filmkunst*. Frankfurt am Main: Fischer.
Schädler, U. 2003. "Kairos – der unfruchtbare Moment." In P. C. Bol (ed.), *Zum Verhältnis von Raum und Zeit in der griechischen Kunst*. Möhnesee: Bibliopolis, 171–182.
Shapiro, A. 1997. "Correlating Shape and Subject: the Case of the Archaic Pelike." In J. H. Oakley, W. D. E. Coulson, and O. Palagia (eds.), *Athenian Potters and Painters*. Oxford: Oxbow Books, 63–70.
Sipiora, P. and J. Baumlin (eds.). 2002. *Rhetoric and Kairos: History, Theory and Praxis*. Albany: State University of New York Press.
Smith, J. E. 1969. "Time, Times, and the 'Right Time': *Chronos* and *Kairos*." *Monist* 53/1: 1–13.
Snell, B. 1928. *Aischylos und das Handeln im Drama. Philologus Supplementband 20.1*. Leipzig: Dieterich.
Steiner, A. 2007. *Reading Greek Vases*. Cambridge: Cambridge University Press.
Steiner, D. 2013. "Swallow This: A Pelike Within Late Archaic Song and Visual Culture." *Helios* 40/1–2:41–70.
Stewart, A. F. 1978a. "The Canon of Polykleitos: A Question of Evidence." *Journal of Hellenic Studies* 98: 122–131.
Stewart, A. F. 1978b. "Lysippan Studies 1: The Only Creator of Beauty." *American Journal of Archaeology* 82/2: 163–171.
Stewart, A. F. 1993. *Faces of Power: Alexander's Image and Hellenistic Politics*. Berkeley and Los Angeles: University of California Press.
Trédé, M. 1992. *Kairos: L'à-propos et l'occasion. Le mot et la notion, d'Homère à la fin du IVe siècle avant J.-C.* Paris: Editions Klincksieck.
Treister, M. Y. 1996. *The Role of Metals in Ancient Greek History*. Leiden: Brill.
van Groningen, B. A. 1953. *In the Grip of the Past: Essay on an Aspect of Greek Thought*. Leiden: Brill Archive.
Vernant, J.-P. 1988. "The Tragic Subject: Historicity and Transhistoricity." In J.-P. Vernant and P. Vidal-Naquet (eds.), *Myth and Tragedy in Ancient Greece*. Trans. J. Lloyd. New York: Zone Books, 237–247.

Vidal-Naquet, P. 1986. *The Black Hunter: Forms of Thought and Forms of Society in the Greek World*. Trans. A. Szegedy-Maszak. Baltimore: Johns Hopkins University Press.

Walter, O. 1923. *Beschreibung der Reliefs im kleinen Akropolismuseum in Athen*. Vienna: Hölzel.

Zanker, G. 2004. *Modes of Viewing in Hellenistic Poetry and Art*. Madison: University of Wisconsin Press.

10

WOMEN'S BODIES AS METAPHORS FOR TIME IN BIBLICAL, SECOND TEMPLE, AND RABBINIC LITERATURE

Sarit Kattan Gribetz

> "Pregnant existence entails, finally, a unique temporality of process and growth in which the woman can experience herself as split between past and future."[1]

> "The last days of pregnancy – sometimes stretching to agonizing weeks – are a distinct place, time, event, stage. It is a time of in between. Neither here nor there. Your old self and your new self, balanced on the edge of a pregnancy. One foot in your old world, one foot in the new world. Shouldn't there be a word for this state of being, describing the time and place where mothers linger, waiting to be called forward?"[2]

Mourning the destruction of the Second Temple, the author of 4 Ezra attempts to reconcile present disappointment with hope for an anticipated redemption through a series of revelatory visions and dialogues between Ezra and the angel Uriel. The text, which is set in Babylon after the destruction of the First Temple, describes Ezra's fear that God has postponed the end of days from its original timing because of the people's improper behavior, and that as a result redemption will not occur at its

I presented parts of this chapter at the 2013 Society of Biblical Literature Annual Meeting in Baltimore and at a conference hosted by the Department of Theology at Fordham University in April 2014. Many thanks to Mika Ahuvia, Orit Avishai, AJ Berkovitz, Fannie Bialek, Kathryn Pfisterer Darr, George Demacopoulos, Molly Farneth, David Grossberg, Martha Himmelfarb, Karina Martin Hogan, Eduard Iricinschi, Lance Jenott, Lynn Kaye, AnneMarie Luijendijk, Naphtali Meshel, and Adele Reinhartz for questions, sources, and ideas, and especially to Jonathan Ben-Dov and Lutz Doering for their thoughtful and meticulous editing.

[1] Young 1990: 160.

[2] Studelska 2012.

preordained time.[3] The archangel Uriel answers Ezra: "Go and ask a woman who is with child if, when her nine months have been completed, her womb can keep the child within her any longer."[4] Here, the angel evokes the metaphor of pregnancy and birth to capture the inevitability of redemption at the end of days – at the proper time.[5] When Ezra answers that a woman cannot delay birth, the angel continues: "In Hades the chambers of the souls are like the womb. For just as a woman who is in travail makes haste to escape the pangs of birth, so also do these places hasten to give back those things that were committed to them from the beginning."[6] Considering himself as living during the "labor pains" of the end-times, and dealing with the anxiety that the promised messianic end might *not* come, the author of 4 Ezra tries to assure his skeptical reader, through the angel's words, that salvation will occur at the appropriate time. In fact, redemption will transpire as soon as possible, just as a woman tries her best to birth her baby in haste to avoid a prolonged, and increasingly painful, labor.[7] The souls are being kept only temporarily in the depths of the earth, as a fetus is housed only for a short time in its mother's womb.[8] The metaphor that links eschatological time with birth thus captures a number of aspects: first, the specific time (the day and hour) of a child's birth is, like redemption, unknown and unexpected, but birth, like redemption, is inevitable; second, the time of birth cannot be hastened or postponed; third, the labor that precedes birth is painful and thus would

[3] 4 Ezra 4:39; trans. Metzger 1983: 531. On related themes, see Hogan 2011; Flannery 2012; de Long 2012; and Iricinschi 2013. Iricinschi observes that the author of 4 Ezra "starkly emphasizes the value of womb metaphors for a proper understanding of the timing of divine decisions, which – while wholly natural – remain fundamentally mysterious" (758).

[4] 4 Ezra 4:40.

[5] To be sure, metaphors of childbirth were used to other ends in antiquity as well. Buell 1999: 54–60 discusses the use of metaphorical procreation to describe the process of thinking in Plato's *Theaetetus* (210c), in which Socrates serves as the midwife who assists the learner (Theaetetus) and teacher (Theodoros) through pregnancy and labor. She explains, "when Theaetetus expresses distress over such questions as 'What is knowledge?' Sokrates reassures him that 'those are the pains of labor, dear Theaetetus. It is because you are not barren but pregnant' (148e)" (55). Ephrem's meditations on paradise also include vivid imagery of wombs, childbirth, and nursing in erotic and maternal contexts, which often allude to the end of days; McVey 2001, 2003.

[6] 4 Ezra 4:42; cf. 4 Ezra 5:46–49. Hogan 2011 and Tromp 1969 point out that the analogy between the underworld and the womb is a common trope in biblical texts, including Gen. 3:19; Ps. 139:13–15; Sir. 40:1; and Tg. Neof. on Gen. 3:19.

[7] The birthing metaphor works in conjunction with the idea of reaping the harvest in 4 Ezra 4:26–32. Hogan 2011: 78 emphasizes that the use of birth and agricultural metaphors are invoked here to stress that, like these natural processes, the end-times are predetermined by divine providence.

[8] Hogan 2014.

not be drawn out unnecessarily, but it also signals the imminence of the better days of redemption that will soon – ideally – follow.

Metaphors of women's bodies were applied to time and temporality in ancient Jewish sources in a number of ways. In this chapter, I explore the use and development of such metaphors of female bodies – and specifically women's bodies at different stages of maternity and motherhood (menstruation, pregnancy, labor, and birth) – to describe various temporal processes, and to tease out the ways in which these processes gained metaphorical significance in descriptions of the eschatological end-times and historical time as well as in discussions of calendars and calendrical time:

(1) In the example presented earlier, the metaphor of labor and birth is applied to eschatological time. Themes linking the suffering of labor to the chaos of exile, and the catharsis of birth to an expected redemption, are already found in the Hebrew Bible and are further developed with regard to the approaching end of time and the anticipation of a new age in Second Temple literature, New Testament texts, and rabbinic sources.[9]
(2) By extension, in some of these same sources women's bodies are also employed metaphorically to articulate ideas about historical time, the very beginning of time, and the progression of history.
(3) In rabbinic texts, which are generally less focused on redemption and messianic expectations than the authors of earlier sources,[10] metaphors of women's bodies are applied more prominently to the calculation and regularization of daily and monthly time, that is, to the cycle of the moon, the sanctification of new months, and the luni-solar calendar.

In all of these sources, the temporality of women's (motherly) bodies becomes an apt metaphor for capturing the abstract and often fleeting idea of time.[11] What is particularly fascinating is the application of such physical metaphors concerning women's bodies to describe inherently *intangible* ideas about time, such as waiting, anticipating, delaying, accelerating, progressing, and eventually fulfilling. Such metaphors proved to be an evocative – and effective – rhetorical strategy for expressing ideas about time in ancient Jewish contexts.

[9] I use the term "eschatological" loosely here, as the concept of the "end-times" evolved significantly during antiquity (Collins 2009). Meyers, Craven, and Kraemer 2000: 298–299, 318, 326, 352, 375, 400–402, 456, 486 survey the motif of a woman in labor in biblical sources.

[10] Schiffman 2006.

[11] Lakoff and Turner 1989 discuss the use of grounded metaphors, which are often derived from physical experiences, to "conceptualize the nonphysical" (59); Aaron 2001 elaborates on specifically biblical metaphors.

Metaphors, however, are not merely rhetorical. In their work on metaphors, George Lakoff and Mark Johnson argue the following:

> The concepts that govern our thought are not just matters of the intellect. They also govern our everyday functioning, down to the most mundane details. Our concepts structure what we perceive, how we get around in the world, and how we relate to other people. Our conceptual system thus plays a central role in defining our everyday realities. If we are right in suggesting that our conceptual system is largely metaphorical, then the way we think, what we experience, and what we do every day is very much a matter of metaphor.[12]

In their discussion of the metaphor "time is money," Lakoff and Johnson demonstrate how the metaphor illuminates Western culture's view of time as a commodity and a limited resource as well as the interconnectedness between time and money.[13] They write:

> Corresponding to the fact that we *act* as if time is a valuable commodity – a limited resource, even money – we *conceive* of time that way. Thus we understand and experience time as the kind of thing that can be spent, wasted, budgeted, invested wisely or poorly, saved, or squandered . . . This isn't a necessary way for human beings to conceptualize time; it is tied to our culture. There are cultures where time is none of these things.[14]

In ancient Jewish culture, one of the dominant metaphors used for time is that of a woman's body and its cycles (menstruation, pregnancy, labor and birth, as well as fertility).[15] If we follow Lakoff and Johnson's metaphor theory, such metaphors function as a way of ordering how ancient Jewish culture thought about time, and in turn how ancient Jews regarded their relationship with time. So while it is possible to understand the use of women's bodies in metaphors of time as yet another case of women being "good to think with," to invoke Claude Lévi-Strauss, applying Lakoff and Johnson's work here compels us to grant that metaphors do more than simply

[12] Lakoff and Johnson 1980: 3; in some cases, metaphors even "structure the actions we perform" (4).

[13] Lakoff and Johnson 1980: 7–9.

[14] Lakoff and Johnson 1980: 8–9. The example of the metaphor "time is money" is one of several the authors discuss to further their argument that metaphors have real-world implications.

[15] Women's relationships with time in the context of fertility and pregnancy, and the temporality of female embodiment, carries with it a whole different range of associations in a contemporary context (Oliver 2010). Foster 1996 discusses the mapping of women's time through a calendar of menstrual cycles; and Delaney 1976: 267–273 explores biological rhythms in men and women, as well as the cultural gendering of cyclical bodily rhythms.

reflect an independent concept.[16] When the processes of women's bodies are used as temporal metaphors, the dynamics of these processes inform the way in which people understand the various concepts involved – time, gender, and female bodies. Women's bodies were not only "good to think with," but they changed people's fundamental conceptions of time by being used as central metaphorical systems.

Ironically, the particular instances in which women's bodies are invoked – eschatological, historical, and calendrical time – are ones in which actual women usually did not fully participate. As I explore in the Conclusion, this figurative invocation of women's bodies as a way of marking time involves the elision of actual women from the social, political, and religious events being metaphorically described. According to rabbinic theology and law, for example, women typically do not play an active role in hastening the redemption, nor are they permitted to serve as witnesses for observing the moon. The use of such metaphors of the female body, I argue, might even contribute to the exclusion of real women from partaking in the processes described. At the same time, I also suggest that the texts that make use of such metaphors indirectly preserve women's experiences and voices.

Ultimately, I argue that to gain a more complete understanding of ancient Jewish conceptions of time, we must pay careful attention to the language used to describe and define time. As we shall see, the male authors of ancient Jewish texts frequently looked toward women (or their imagination and legislation of women's bodies) in developing their temporal lexicon. The chapter thus concludes with reflections on the implications of this argument for our understanding not only of ancient Jewish conceptions of time or ancient Jewish conceptions of women but also the relationship between the two. Our ancient Jewish authors could hardly imagine time without appealing to their understanding of women and of mothers in particular, and they regularly imagined women in temporal terms.

Examining these metaphors related to time also sheds light on the question of ancient Jewish temporality. The scholarly attempt to distinguish between religious traditions that cultivated cyclical and linear temporalities has given way to understanding the multiple ways in which time operated within a given society.[17] Metaphors of women's bodies describe a range of temporal

[16] Lévi-Strauss wrote, in the context of his work on totemism, that animals are "bon à penser" (in contrast to "bon à manger") and later applied this idea to women as well (Lévi-Strauss 1964: 89; 1983: 61–63), on which see Brown 1988: 153–159; Kittay 1988; and Clark 2005.

[17] Eliade 1959: 68–113; Barr 1962; Momigliano 1966; Neher 1976; Steensgaard 1993; Rubenstein 1997; and Goldberg 2000, 2004. Likewise, the debate about cyclical and linear time is present in the literature on women's time (e.g., Kristeva 1981).

processes: women's menstrual cycles highlighted the cyclicality of the calendar, while the nine months of pregnancy followed by labor and birth was employed to demonstrate the linear progression of teleological history. These various metaphors found within ancient Jewish texts thus reveal not only that cyclical and linear conceptions of time and temporality coexisted unproblematically within a single framework but also that a single metaphorical system was employed to develop, articulate, and emphasize its different aspects.

1 Eschatological Time

Most frequently, ancient sources employ metaphors of women's bodies to convey ideas about redemptive and eschatological time. The association of disaster, and especially war, with laboring and birthing metaphors appears in the Hebrew Bible and is developed further in literature composed after the destruction of the Second Temple. The use of feminine bodily imagery to describe eschatological time seems to stem from a more overarching and basic metaphor of the earth as a mother that, as Karina Martin Hogan has demonstrated, pervades 4 Ezra and is already found in biblical and other extra-biblical sources.[18] Among the various ideas that the metaphor in its earliest iterations conveys is about the progress of time approaching the eschaton and the features of this final era. The idiom "the birth pangs of the messiah," used for the first time in rabbinic sources to refer to the catastrophic time anticipated before the appearance of the messiah, is anchored in this long metaphorical tradition.[19]

The pains of labor and the birthing process served as an apt metaphor for tapping into the fear associated with an anticipated time of judgment and redemption because childbirth in antiquity was a terrifying event. Maternal and infant mortality rates were high. For infants, estimated mortality rates reached 5 percent within the first four weeks after delivery.[20] About the

[18] Hogan 2011.

[19] חבלו של משיח in Mek. Šim. *Wa-yassa'* 6 (245 Lauterbach); b. Sanh. 98b; b. Šabb. 118a and b. Pesaḥ 118a. MS Munich 6 of b. Pesaḥ 118a inserts instead ושעבוד מלכיות; there is more variation in the MSS of b. Ketub. 111a, where the following variants are used: חבלא דמשיחא; חבלי דמשיחא; חבלי משיח, etc.

[20] Death of mothers in childbirth is discussed in rabbinic sources, e.g., m. Šabb. 2:6 and t. Šabb. 2:10 (Hauptman 2013). Rousselle 1991 and French 1986 estimate rates of death during childbirth to be approximately 5 to 10 percent. Such fears were addressed in part through magical practices aimed at protecting women's wombs. Childbirth amulets, e.g., *PGM* VII.260–271, protecting mothers and born and unborn children have been published by Bonner 1950: 92–93; Isbell 1975: 56–57, 152; Schiffman and Swartz 1992: 32, 41, 46–47, 56, 60, 77; Levene 2003: 40–43, 93–98; and Montgomery 2011: 259–260. See also the studies of Barb 1953; Alexander 1986: 349 Aubert 1989; Darr 1994: 210–213; Thierry de Crussol des

death of women, Plutarch writes, "the house of him who has married and later lost his wife is not only incomplete but also crippled."[21] Soranus devotes an entire section of his *Gynecology* to difficult labors and complications that arise in childbirth. Gregory of Nyssa, in his tractate *On Virginity*, tries to persuade young women to embrace an ascetic life by warning them of the potential for tragedy in marriage and childbirth:

> Assume that the moment of childbirth is at hand; it is not the birth of the child, but the presence of death that is thought of, and the death of the mother anticipated. Often, the sad prophecy is fulfilled and before the birth is celebrated, before any of the anticipated goods are tasted, joy is exchanged for lamentation.[22]

Gregory writes of the loss of a mother at childbirth with such emotion and psychological depth that it is often assumed that he lost his own wife in this way. He continues: "Still burning with affection, still at the peak of desire, without having experienced the sweetest things of life, one is all at once bereft of everything as if in a nightmare."[23] For Gregory, "the very sweetness of their life is the fomenting of their grief . . . Instead of a bridal chamber, death provides a tomb."[24]

Childbirth was thus a time not only of pain and new life but also often of death, tragedy and uncertainty. When ancient texts draw on this metaphor, they do so in part to recall that moments of crisis, including exile and the eschaton, would bring with them great – even unimaginable – loss along with redemption. It is not only the seemingly unbearable pain that is relieved with the birth of a child but also the potential (and often actual) loss of life itself on which the metaphor draws. While the sources regard pregnancy as a natural event and its timing as typically inevitable, they also recognize that pregnancy is prone to disruption – miscarriage, still birth, premature birth, maternal mortality. These events are regarded as simultaneously natural and inevitable parts of the process, and yet also as disruptions of the natural order of the world.

Épesse 2002; and Folmer 2011: 224. Complications in pregnancy and childbirth were also discussed in ancient medical literature (Marganne, 1981; Hanson 1987, 1991; and Parker 1999). The Hippocratic Corpus, which continued to circulate in late antiquity, includes extensive discussions of gynecological matters, as does Soranus's *Gynecology* and Galen's *On the Natural Faculties*.

21 Plutarch, *Moralia* 288–289.

22 Gregory of Nyssa, *On Virginity* 3 (trans. Callahan 1967: 15–16). Cf. John Chrysostom, *Seventeenth Homily on Genesis* 144c, who describes the pain of childbirth as being mitigated by the joy and benefit of the emerging child.

23 Callahan 1967: 16.

24 Callahan 1967: 14, 16.

Several biblical texts use imagery of a laboring woman to emphasize the pain and suffering associated with moments of crisis: exile, destruction, judgment (e.g., Jer 4:31, 6:24, 13:21, 22:23, 30:6–7; 49:24, 50:43; Isa 21:3, 26:17).[25] Laboring and birthing imagery is also invoked, however, to describe the specifically *temporal* dimensions of redemption, a theme on which texts composed after the destruction of the Second Temple in 70 CE expand as the concept of an end-times continued to develop. In Isa 42:14, the text speaks in God's voice: "I have been quiet for an eternity. I have been still and restrained myself, I will groan like a woman giving birth (כיולדה אפעה); I will blow and gasp." As Katheryn Pfisterer Darr has shown, this is the only instance in the entire Hebrew Bible in which God is compared to a woman in labor and in which the pain of labor has been used to portray a sense of power and strength rather than anguish.[26] God's uncontrollable crying in the throes of childbirth will break the eternity of silence. Sarah Dille writes of the temporal aspect of this passage: "The fruition of YHWH's period of apparent silence is as *inevitable* as labor. What has been germinating and gestating will come to birth . . . Birth comes after a time at the right time, and comes inevitably."[27] These are all ideas that the text tries to capture in its image of God as a laboring woman. The passage then evokes the idea that at the necessary time, even God, who has withdrawn himself, will emerge from silence and cry out. Of course, the intensity, pain, and visceral auditory response of God are essential parts of the experience: God's wailing, blowing, and gasping are all simultaneously emphasized. Three elements are thus combined in this passage to communicate the

[25] The biblical passages are examined in detail in Dille 2004: esp. 41–73; Bergmann 2008; and Kalmanofsky 2008. 1QH[a] 11:8–13 (DJD 40, 144), from the Hodayot at Qumran, uses an extended birthing metaphor about two mothers and two children, one of whom is born successfully and the other not, to describe the approaching eschaton; the discomfort of labor stands in for indescribable pain during a crisis. The metaphor begins: "I am in distress like a woman giving birth to her first-born (אשת לדה מבכריה)" (1QH[a] 11:8). While the first mother's labor pains ("deathly breakers" – this phrase is a pun on משברי מוות, because the word משבר denotes both a birthstool as well as a breaker; see also Jonah 2:4) give way to the delivery of a male son, the text details how the second mother progresses deeper into pain. The text and translation of 1QH[a] are found in Bergmann 2008: 172–173; see also García Martínez and Tigchelaar 2000: 1.164–165; and Stegemann, Schuller, and Newsom 2009. In general, the association of birth with pain might evoke Gen. 3:16, in which labor pains are presented as a punishment for Eve's sin.

[26] Darr 1987, also on the problems with the meaning of the verse; the metaphor of a laboring woman to invoke God's power is created in conjunction with the metaphor of God as warrior with which it is paired in the verse. Cf. Gruber 1992: 3–15 and Brettler 1998.

[27] Dille 2004: 70.

intensity of the moment: the anguish of labor, the power needed to sustain such suffering, and its occurrence at a specific time. The pregnancy and birth imagery highlights the intensity of the experience, as well as its inevitable timing.

Dille further juxtaposes the inevitability of God breaking his silence with a passage from Hos 13:13 about God's dissatisfaction with Israel and his subsequent judgment of the people.[28] The verse describes what happens when the natural process of birth is disrupted and time, as a result, becomes unreliable: "the pangs of childbirth (חבלי יולדה) come for him, but he is an unwise son, for there is no time at the breaking forth of babes (כי עת לא יעמד במשבר בנים)."[29] In this passage, the *son* fails to emerge at the proper time in the laboring process. The unnatural aspect of halting labor is also used in other biblical contexts, in which the laboring *mother* lacks the necessary strength to birth her child.[30] The metaphor of childbirth represents a time that is expected and assumed to be inevitable and yet does not materialize. If the baby fails to arrive or its emergence from the womb is interrupted at the anticipated time, nature has been undone.

In addition to sources that describe prolonged or stalled processes of birth, still other biblical texts employ these metaphors for the opposite effect, to refer to a rapid conclusion of a process that was expected to be much longer. Consider, for example, the ways in which Isa 66:7–12 utilizes the theme of time in its description of a metaphorical birth:

> Before she labored, she delivered (בטרם תחיל ילדה);
> Before her pang came, she bore a son (בטרם יבוא חבל לה והמליטה זכר).
> Who ever heard the like?
> Who ever witnessed such events?

28 This verse presents translational difficulties, on which see Andersen and Freedman 1980 on Hos. 13:13.

29 NRSV. This is not a typical form of עת, which is usually followed by an infinitive, as it is in Mic. 5:2 (Andersen and Freedman 1980: 638–639). Other biblical passages describe the miraculous – and nature-defying – features of the end of the age. An opposite evocation of unnatural timing in birth as an indication of approaching end-times appears in 4 Ezra 6:21: "Infants a year old shall speak with their voices, and women with child shall give birth to premature children at three or four months, and these shall live and dance." Even those born *not in their proper time* will flourish. 4 Ezra 5:8 speaks of a time when "menstruous women shall bring forth monsters"; 2 Bar 73:7 foretells the cessation of pain during childbirth as a feature of redemption; 2 Bar 10:13–16 mentions that the barren rejoice during times of mourning over the Temple's destruction.

30 In 2 Kgs 19:3, too, the natural order of birth is not followed, but this time the emphasis is not on the failure of the child to emerge but on the mother's lack of strength to birth her offspring. Cf. Isa 37:3 and the insightful study of the metaphor and its ancient Near Eastern context by Darr 1994: 205–224.

Can a land pass through travail in a single day (היוחל ארץ ביום אחד)?[31]
Or is a nation born all at once?
Yet Zion travailed (כי חלה)
And at once bore her children!
Shall I who bring on labor not bring about birth? Says the Lord.
Shall I who cause birth shut the womb? Said your God.
Rejoice with Jerusalem and be glad for her,
All you who love her!
Join in Jubilation,
All who mourn over her –
That you may suck from her breast
Consolation to the full,
That you may draw from her bosom
Glory to the full.[32]

This passage both evokes the metaphor of labor pangs and uses the inverted sequence – birth before labor – to emphasize Zion's swift and relatively painless delivery, the discomfort miraculously short or nonexistent. The redemption of Zion is regarded as miraculous because it abbreviates a process that is, in the natural order, much longer and more painful. Here, in other words, the birth metaphor is used not to highlight pain but to emphasize the absence of pain, and the possibility that a process that is expected to be lengthy can, if God so wills it, be remarkably brief.

Mic 4–5 relies on an extended birth metaphor to describe a painful temporal process with an anticipated, if not always precisely known, end. At first, labor pains are evoked to describe the experience of Israel's torturous exile and subsequent salvation from its enemies: "For agony has gripped you, like a woman in childbirth! Writhe and bring forth, O daughter Zion, like a woman in childbirth (חולי וגחי בת-ציון כיולדה)! For now you must leave the city and dwell in the country, and you will reach Babylon. There you shall be rescued, there the Lord will redeem you from the hands of your enemies" (Mic 4:9–10).[33] The labor terminology parallels similar language in Job 38:8, in which the text describes the creation of the world: "Who closed the sea

[31] The term used in this passage for "travail" is one reserved for childbirth contexts. The root חיל is used three times in these first two verses and is the verbal form of the noun for labor pains.

[32] NJPSV.

[33] NRSV translation with some modification. Cf. Andersen and Freedman 2000: 441–447. The LXX renders 4:10 somewhat differently – "be in pain and be manly (ἀνδρίζου), and draw near, Daughter of Zion – like a woman in childbirth!" – such that withstanding the pain of childbirth is regarded as a manly act, an interesting shift in the passage's use of gender.

behind doors when it gushed forth out of the womb (בגיחו מרחם יצא)?"[34] In the Job passage, the sea is created at the beginning of time by slithering out from a womb, while in the passage from Micah the daughter of Zion is likened to a woman in labor at the beginning of a redemptive end point. God contains the sea behind a barrier, preventing it from overwhelming the world, whereas the daughter of Zion's writhing in pain ends with God's redeeming and setting her free. Then, the *timing* of the labor and birth process becomes central in Mic 5:2–3: "from you shall come forth for me one who is to rule in Israel, whose origin is from of old, from ancient days. Therefore he shall give them up until the time when she who is in labor has brought forth (עד־עת יולדה ילדה)." Here, quite literally the "time" (עת) is what is important: only once the woman has birthed – that is, she (and by implication the people of Israel) has withstood the torments of labor – will the messiah figure begin paying attention to the people.

Many other ancient sources also employ the metaphor of childbirth to express notions of time. In 1 Thess 5:1–3, the metaphor of labor pangs is conjured specifically in a discussion of time to emphasize the temporal *unpredictability* of the moment. The passage begins with the words "now concerning the times and seasons (Περὶ δὲ τῶν χρόνων καὶ τῶν καιρῶν)."[35] The day of the Lord and its destruction, Paul writes, will be unexpected; that is, it will occur at a surprising time, just as "labor pains come upon a pregnant woman (ὥσπερ ἡ ὠδὶν τῇ ἐν γαστρὶ ἐχούσῃ)" – it will "surprise you like a thief" (cf. also Matt 24:42–44). Paul believes that he lives on the cusp of the eschaton, and he warns that at any point the "labor pains" of this impending period will begin, as they might for a woman who nears the end of her pregnancy. These labor pains thus indicate the fast-approaching end-times. Nonetheless, the precise onset of labor itself – even once a woman is full term – is unpredictable and can occur at any time. One must be ready for this time, Paul warns, at all moments.

In his epistle to the Romans, Paul again evokes the process of birth. He juxtaposes the suffering of the present with "the glory about to be revealed to us" (Rom 8:18); the creation has been forced to wait, Paul insists, but is eager to be set free through the glory of God's children. The metaphor with which

[34] A following verse, Job 38:10, also includes reference to breakers and other terms related to the birth metaphor, and the passage seems to reference Job's opening lament in 3:10–11 with its similar use of words and images ("Because it did not block my mother's womb [כי לא סגר דלתי בטני], and hide trouble from my eyes. Why did I die at birth, expire as I came forth from the womb [מבטן יצאתי ואגוע]?"). Cf. Job 1:21, though the terminology employed differs, and Hos 13:13, which also parallels the breaking of the waters of creation and childbirth (Andersen and Freedman 1980: 638–639).

[35] NRSV.

Paul describes the creation's anxious, tortured wait is through birthing imagery: "We know that the whole creation has been groaning in labor pains (συστενάζει καὶ συνωδίνει) until now; and not only the creation, but we ourselves, who have the first fruits of the Spirit, groan inwardly while we wait for adoption, the redemption of our bodies" (Rom 8:22–23).[36] Here, Paul evokes the bodily pains of labor (applied to the creation) to describe the state of being of those who seek the redemption of their bodies through baptism. The physical metaphor here serves two purposes: to articulate the idea of a belabored wait similar to that of labor and to highlight the tormented state from which one is awaiting redemption.

Strikingly, in later sources composed after the Second Temple's destruction in 70 CE, the metaphors of labor and birth become even more entwined with matters of time and, especially, with eschatological time. In a short passage in 2 Bar, another first-century apocalypse that struggles, as does the passage previously presented from 4 Ezra, with unmet expectations for redemption after the Temple's destruction, a childbirth metaphor is used to teach that it takes time for the world to be ready to herald the end of days, just as a baby requires a period of gestation before birth. After Baruch prays to God to bring about the final judgment and facilitate the end of times, God replies to Baruch to assure him, on the one hand, that God will fulfill his promise, but on the other hand to stress that enough time is needed before the end can come. God demonstrates the need for patience and for history to run its course with a series of analogies:

> Baruch, Baruch, why are you disturbed? Who starts on a journey and does not complete it? Or who will be comforted making a sea voyage unless he can reach a harbor? Or he who promises to give a present to somebody – is it not a theft, unless it is fulfilled? Or he who sows the earth – does he not lose everything unless he reaps its harvest in its own time? Or he who plants a vineyard – does the planter expect to receive fruit from it, unless it grows until its appointed time? *Or a woman who has conceived – does she not surely kill the child when she bears untimely?*[37]

[36] NRSV; on which see Sutter Rehmann 1995, 2004. This verse should be understood in the context of the preceding few chapters, in which Paul discusses the reign of sin and death in humanity's bodies of flesh (see, e.g., Rom 6:6) and the transformation of one's body after baptism. Cf. Gal 4:19–20, in which Paul describes himself to be suffering from the distress of labor pains: "My little children, for whom I am again in the pain of childbirth (ὠδίνω) until Christ is formed in you" (NRSV). Here is an example of the use of this metaphor in more casual speech, applying the theological idea to a social-historical moment.

[37] 2 Bar 22:1–8, trans. Klijn 1983: 1.629. Henze 2011: 278–293 explores additional temporal themes in 2 Bar.

The duration of pregnancy appears here as one example to illustrate the necessity for a process to run its course before its purpose can be realized.[38] Just as a woman's womb must wait for the child within it to mature and grow, so too must the nation await the historical moment ripe for its redemption. Were the redemption to arrive prematurely, it would not be effective.

Imagery of labor and birth is also evoked in Mark 13:7–8, in which the onset of the eschatological era is described in terms of labor pains.[39] In this passage, the metaphor serves to warn of the suffering that will precede redemption. After Jesus foretells of the Temple's destruction, he sits together with Peter, James, John, and Andrew at the Mount of Olives to disclose privately details of his prediction. They ask him: "Tell us, when will this be, and what will be the sign that all these things are about to be accomplished?" (Mark 13:4). Jesus warns his disciples not to allow others to lead them astray, and then says:

> When you hear of wars and rumors of wars, do not be alarmed; this must take place, but the end is still to come. For nation will rise against nation, and kingdom against kingdom; there will be earthquakes in various places; there will be famines. These are but the beginning of the birth pangs (ἀρχὴ ὠδίνων ταῦτα). (Mark 13:7–8, NRSV)

While in 4 Ezra 4 the metaphor of birth was used to emphasize the inevitability of the end and the attempt to hasten its coming, and the Hebrew Bible sources evoked labor primarily to stress the anguish associated with an approaching redemptive or eschatological period, here the gospel uses the birthing metaphor to somewhat different effect. Just as labor, which is painful, must precede the glory of birth and in fact is a sign that the infant's arrival is imminent, the wars and destruction foretold in this prophecy signal the beginning not only of cataclysm but also, eventually, of redemption. The metaphor of labor pains is evoked to highlight the pain of war but also to argue that such disaster should be seen positively, as a sign of a redemptive future, like a birth. Jesus adds words of encouragement: "the one who endures to the end will be saved" (Mark 13:13). Jesus concludes by emphasizing that the precise time of redemption is still unknown: "But about that day or hour no one knows, neither the angels in heaven, nor the Son, but only the Father. Beware, keep alert; for you do not know when the time will come" (Mark 13:32–33). The dating of this passage is debated.[40] If it was

[38] Likewise, Flannery 2012 argues that the pregnant body is evoked by Uriel in 4 Ezra as a pedagogical tool to teach Israel the virtues of patience.

[39] Parallel in Matt 24:3–8. Cf. Pitre 2005: 223–253.

[40] Frey 2011: 465–468.

written before 70 CE, it can refer to an earlier moment of struggle during the revolt or even before the war; if one prefers a post-destruction context, then its author, like those of 4 Ezra and 2 Bar, was coming to terms with the destruction and absence of the Temple. Reimagining the war or destruction as the beginning of the new age that Jesus had predicted all along brought a measure of comfort and optimism. The destruction, while traumatic, thus also became a symbol of the coming salvation – just as labor pains signal an impending birth.[41]

The Gospel of John makes use of the metaphor of a laboring woman in a different redemptive context, again to articulate an idea about time – current suffering in contrast with ultimate joy. When Jesus explains to his disciples that he is "going to him who sent me" before his arrest, he enigmatically declares that in "a little while . . . you will no longer see me, and then after a little while you will see me" (John 16:5, 16). His disciples are puzzled by what he means. Jesus thus explains:

> Very truly, I tell you, you will weep and mourn, but the world will rejoice; you will have pain, but your pain will turn into joy. When a woman is in labor, she has pain, because her hour has come (ἡ γυνὴ ὅταν τίκτῃ λύπην ἔχει, ὅτι ἦλθεν ἡ ὥρα αὐτῆς), but when her child is born, she no longer remembers the anguish because of the joy of having brought a human being into the world. So you have pain now, but I will see you again, and your hearts will rejoice, and no one will take your joy from you. On that day you will ask nothing of me. (John 16:20–23, NRSV)

All of the persecution that Jesus' disciples will surely face (and that John's readers currently do) is likened to a woman's temporary labor pains, which will not only end but will turn into pure rejoicing when Jesus will be reunited with his disciples in a redemptive moment (either at the resurrection or Jesus'

[41] Cf. Rev 12, in which a celestial image ("a great portent") appears in heaven and is described as "pregnant and crying out in birth pangs, in the agony of giving birth (καὶ ἐν γαστρὶ ἔχουσα καὶ κράζει ὠδίνουσα καὶ βασανιζομένη τεκεῖν)" (Rev 12:1–2). A second portent appears, of a dragon waiting to devour the woman's child. Once the child has been born, the dragon pursues the woman, who has been given wings to fly into the wilderness "to her place where she is nourished for a time, and times, and half a time" (Rev 12:14 NRSV). Here again the metaphor of labor is intertwined with a reflection about an anticipated time to come. The image of a pregnant woman seems to be inspired by Isa 26:17–27:1, in which Israel is described as calling out to God, remarking that they have sought God "like a woman with child, who writhes and cries out in her pangs when she is near her time" but ends up with "no victories on earth, and no one is born to inhabit the world" and is perhaps also related to Isa 66:7 and 1QH[a] 11:8–13 (DJD 40, 144), which describes the painful labor of a mother birthing a redemptive figure. See Lévi 1922; Yarbro Collins 2001: 57–155, esp. 67–69 and 104–107; Flusser 2009: 285–287; Pagels 2012: 5, 29–30, 181 n. 15; and Duff 2003.

second coming), one in which there is a metaphorical (re)birth. The passage invokes the childbirth metaphor to speak about the timing of sorrow and redemption on a communal and cosmic level.[42] The author of the gospel also sets up a parallel between the woman's "hour" of labor in this passage and Jesus' "hour" to be glorified in John 12:23 ("The hour has come for the Son of Man to be glorified" Ἐλήλυθεν ἡ ὥρα ἵνα δοξασθῇ ὁ υἱὸς τοῦ ἀνθρώπου).[43] The use of this time-marker in these two passages emphasizes the temporal focus of an anticipated redemption.

Several rabbinic sources, drawing on the same themes present in the Hebrew Bible and further developed in New Testament writings, characterize the period of catastrophes that will precede the messianic era specifically as "the birth pangs of the messiah (חבלו של משיח and חבלי משיח)."[44] The earliest source, the Mekhilta de-Rabbi Ishmael, commenting on Moses' instructions that the Israelites collect two portions of Manna on Friday in anticipation of the Sabbath, explains that one of the rewards for observing the Sabbath is protection from three catastrophic events including "the birth pangs of the messiah."[45] That is, those who are vigilant in their Sabbath observance will be spared the destruction expected in the time before the messiah's arrival. This midrash is likely alluding to another midrash in the Mekhilta that characterizes the Sabbath as a "taste of the world to come."[46] If one sanctifies the time of the earthly Sabbath of this world, the logic goes, one will be safe during the messianic transition to the redeemed world that is to come.

The phrase is invoked more frequently in the Babylonian Talmud. The most developed use appears in tractate Sanhedrin, in a pericope devoted to a discussion of redemption and the end of days.[47] Citing Mic 5:3, "therefore he shall give them up until the time when she who is in labor has brought forth,"

[42] The NRSV: 1845 commentary makes a similar observation.

[43] NRSV.

[44] The use of the term חבל draws on language from Jer 13:21, 22:23, 49:24; Isa. 13:8, 26:17, and 66:7 (other biblical references to the pains of a woman giving birth do not make use of this word) and is given further expression in 1QH[a] 11:8–13, on which see Flusser 1983: 130–133; 2009: 285–287. The Mekhilta is the earliest source to employ this specific phrase ("the birth pangs *of the messiah*") to refer to the messianic era (as opposed to the more general "birth pangs" in biblical sources). There is syntactic ambiguity in the phrase חבלי משיח, "the birth pangs of the messiah." It probably refers to the pangs felt by a woman who gives birth (to the pangs suffered by those in the world that are giving birth to the messiah, the pangs that signal his arrival), but syntactically it could also refer to the pangs felt *by* the messiah, who oversees the birth of a new age.

[45] Mek. *Wa-yassa'* 6 on Ex 16:25 (245 Lauterbach).

[46] Mek. *Shabbeta* 1 on Ex 31:13 (495 Lauterbach); cf. b. Ber. 57b.

[47] b. Sanh. 98b. Cf. b. Pesaḥ. 118a; b. Šabb. 118a, in which a person who eats three meals on the Sabbath is spared the "travails of the messiah," which is likely based on the passage in the Mekhilta discussed earlier and b. Ketub. 111a.

Rav proclaims that the messiah, son of David, will not arrive until the Romans hold Israel for a period of nine months – the precise duration of pregnancy, at the end of which labor begins. Ulla and Rabbah (some MSS: Rava) both respond that while they wish for the messiah to come, they do not want to see the messiah themselves or to be present during these precarious times. Abaye asks Rabbah if the reason he wants to avoid the messiah's arrival is that he fears the birth pangs associated with the messianic redemption. Attempting to calm Rabbah, Abaye explains that to be spared from suffering during that period, all he must do is engage in study and benevolence.[48]

The texts surveyed here use metaphors associated with women's bodies to articulate ideas about eschatological time, and especially the element of anticipating an expected eschaton or messianic redemption that will bring along with it not only eventual peace but also, beforehand, temporary devastation and chaos. The inevitability of an approaching time, the inability to hasten or delay that time, the chaos and pain that precede this time – all of these ideas find expression through the evocation of the stages of women's pregnancies, labors, and births.[49] The birthing metaphor works on a number of other levels as well. An entire pregnancy is uncomfortable and mirrors the times of increasing frustration before the climax of war. Labor is painful and represents the extreme moments of social and political chaos that are said to precede the messianic age in the process of redemption. Labor itself is a process that usually starts gradually and increases in pain and intensity as it progresses to the moment of birth. Though the ancient texts do not elaborate in such detail

[48] Later rabbinic sources associate times of suffering with metaphors of childbirth even when those metaphors do not appear in the biblical base texts being explicated. For instance, Midr. Pss. 20 (c. 1000 CE) reads the metaphors of labor and birth into a biblical text that uses the generic phrase "*time of trouble* [lit. ביום צרה]" (Ps 20:2; cf. Ps 91:15; cf. also Deut. Rab. *Wa-etḥanan* 2). In the context of the biblical text, Psalm 20 refers to a general time of trouble and assures the readers that God will answer them during desperate moments. The midrash draws an analogy between the cries of Israel and the cries of a laboring woman to stress the theme of desperation felt both by Israel and by God during the Temple's destruction. A story is told about a pregnant woman who had an argument with her mother. When the woman began laboring, her mother went up to the attic while her daughter cried in agony downstairs; on hearing her daughter's cries, the mother, too, began wailing sympathetically. Questioned about the usefulness of her cries, the mother answers: with my daughter in pain, "how can I tolerate her cries other than by wailing along with her, as the pain of my daughter is my own?" This is similar, the midrash explains, to the destruction of the Temple, when a cry of agony rang out in the entire world, and God (who had been angry at Israel) nonetheless wept along with his people over their distress (cf. Isa 42:14). The midrash plays on a passage from Isaiah that highlights God's sympathetic pain: "In all their [Israel's] troubles he [God] was troubled (בכל צָרָתָם לו צר)" (Isa 63:9; cf. Jer 49:24).

[49] Death is another example of a life event that is inevitable but unpredictable in its timing.

about this crescendo – pregnancy/upheaval, pre-labor/beginnings of discomfort, labor/chaos and pain, transition/agony – it very well may be why the metaphor is used with such realistic effect. There is also often an accompanying fear of stillbirth and maternal death, just as redemption is eagerly anticipated but can be accompanied by doubt or uncertainty.

2 Historical Time

Various aspects of historical time, too, are described in ancient Jewish sources with reference to women's bodies. To illustrate the *longue durée* of historical time and the succession of generations through history, 4 Ezra develops an extended metaphor related to a woman's fertility and childbearing over the course of her lifespan. Ezra is frustrated with how distant redemption still seems and asks the angel why all of humanity was not created simultaneously so that the final judgment would occur sooner: "Could you not have created at one time those who have been and those who are and those who will be, that you might show your judgment the sooner?" (5:43). The angel tells Ezra that the world cannot hold all of its creations at one time and provides the following explanation: "Ask a woman's womb, and say to it, If you bear ten children, why one after another? Request it therefore to produce ten at one time" (5:46). Ezra answers: "Of course it cannot, but only each in its own time" (5:47). Ezra is told that the world cannot contain all of its creations and that therefore generations of people populate the world at different times in history, just as a woman cannot carry all of her children in her womb simultaneously and must, instead, generate one child at a time.

The passage then points out that a woman's fertility does not last indefinitely and, moreover, observes that with each generation there seems to be a decline (in size and in quality), similar to a woman's fertility, which ebbs as she ages:

> He [Uriel] said to me, "Even so have I given the womb of the earth to those who from time to time are sown in it. For as an infant does not bring forth, and a woman who has become old does not bring forth any longer, so have I organized the world which I created." Then I [Ezra] inquired and said, "Since you have now given me the opportunity, let me speak before you. Is our mother, of whom you have told me, still young? Or is she now approaching old age?" He replied to me: "Ask a woman who bears children, and she will tell you. Say to her, 'Why are those whom you have borne recently not like those whom you bore before, but smaller in stature?' And she herself will answer you, 'Those born in the strength of youth are different from those born during the time of old age, when the womb is failing.' Therefore you also should consider that you and your contemporaries are

> smaller in stature than those who were before you. And those who come after you will be smaller than you, as born of a creation which already is aging and passing the strength of youth." (4 Ezra 5:48–55)[50]

Again, several aspects of time are articulated through metaphors related to bearing children: first, a woman usually only bears a single child (a woman cannot bear ten children in her womb at the same time, though she admittedly can carry two or three simultaneously), and thus each child has its own time to be born as each generation succeeds the previous one. Second, an infant cannot bear children, for only a mature woman can produce children. Third, a woman is more fertile and begets larger and stronger children when she is young. Fourth, when a woman approaches old age, her fertility diminishes and her progeny become smaller and weaker just as each generation is less impressive than the one that succeeded it. Fifth, eventually a woman's fertility expires and she cannot bear more children just as eventually there is no renewal of life on earth and the end of days begins. This is the way the world has been created, we are told. In this metaphor, it is not the process of birth but rather the arc of fertility of a woman's body and her maturation into and out of the maternal role during her own life cycle that is utilized to illuminate the natural order of the world and the passage of linear time and generations within it. The tangible metaphors of a woman's reproductive capabilities in general and that of an aging woman's body in particular are used to describe the less tangible passing of time through generations, and the decline and deterioration of generations.

The linear duration of time is expressed in 4 Ezra through the metaphor of the arc of female fertility over the course of a lifetime. In rabbinic literature, another aspect of historical time also gains expression through the details of female reproductive mechanics. Gwynn Kessler has demonstrated that rabbinic sources build on biblical precedents in their application of metaphors of embryonic conception to cosmogonic origins such that the world's creation – which includes the beginning of historical time – was conceived in much the same way as human life begins within a mother's womb.[51] For

50 The temporality of a woman's body is also referenced in m. Nid. 5:7, in which a metaphor of a fig is used to describe aging: an unripe fig is like a baby girl, a ripening fig refers to the days of her girlhood, and a fully ripe fig is a grown woman. The metaphor of a fig tree in relation to barren women is also employed by Augustine at the beginning of Sermon 60 on the New Testament (Luke 13:6).

51 Kessler 2013. A similar association between creation and procreation is attributed to Empedocles by Aëtius: "Why are seven-month fetuses viable? Empedocles [said that] when the human race was born from the earth the day was as long as a ten-month period is now, because of the slow progress of the sun. And as time progressed the day was as long as a seven-month period is now. This is why there are ten-month and seven-month fetuses,

example, Genesis Rabbah and Leviticus Rabbah apply a passage from Job 10:10 – "Did you not pour me out like milk and curdle me like cheese?" – to their descriptions of the formation of an embryo within a woman's womb and the creation of the world.[52] In these midrashim, the entire world and each individual human embryo was/is created through a process that is similar to the curdling of milk when a drop of resin is placed into it. Another passage from Leviticus Rabbah likens a human embryo's nine months of gestation in a woman's womb with God's containment of the primordial sea within the cosmic womb during creation.[53] These ideas climax with the Tanhuma's statement that "the creation of an embryo is like the creation of the world."[54] Here, then, is a set of examples in which the imagery of procreation is applied to creation: what occurs within a woman's body is used to describe the process by which the world itself was created by God. The first moment of human conception in a woman's body mirrors the very beginning of historical time through the shared metaphor of the curdling of milk with a drop of resin that is applied to each process. What is more, Kessler argues, the analogy between the creation of the world and the creation of individual embryos also emphasizes that God is constantly acting as creator – God recreates the world and human life each and every day. Kessler writes: "By reading creation with (pro)creation, we see rabbinic portrayals of cosmogony as an ongoing, constant process."[55] Here, then, the moment of embryonic conception within a woman's womb becomes a metaphor both for the first moment of the world's creation (the beginning of historical time) and also for God's repeated creation thereafter, at all times (the past and future).

3 Calendrical Time

In rabbinic sources, female bodies are also evoked in the context of *calendrical* time, a development that corresponds well with the replacement of the biblical masculine noun for moon (ירח) with the feminine noun (לבנה), a term rarely used in the Hebrew Bible, highlighting another connection that

the nature of the cosmos having contrived things so that the fetus should grow in one day" (Aëtius 5.18.1, *Dox. Gr.* 427, on which see Wilford 1968). Empedocles also refers to a clepsydra as a womb, connecting the mechanism of a water clock to a woman's uterus (King 1873: 116).

[52] Gen. Rab. 4:7; 14:5; Lev. Rab. 14:9.

[53] Lev. Rab. 14:4; in connection with Job 38:8–11. Cf. Lev. Rab. 14:1 on the creation of the embryo in comparison with the creation of the first human being.

[54] Tanḥ. *Pekudei* 3.

[55] Kessler 2013: 138.

develops between women and the calendar.[56] This new discourse does not anchor itself in more ancient precedents; it seems to be a rabbinic innovation. Even though debates about competing calendars were of central concern in the sectarian disputes during the end of the Second Temple period[57], it is only in rabbinic sources about the calendar and its observation and calculation (as far as I am aware) that the discussion draws on the stages and processes of women's bodies and the terminology associated with female bodily practices.

In a famous *mishnah* about declaring a new month, the waxing of the moon is likened to a pregnant woman's abdomen.[58] When Rabban Gamliel accepts the testimony of false witnesses (that they observed a new moon when in fact it becomes clear the following day that their testimony was incorrect), his critic, Rabbi Dosa, declares: "They are false witnesses: how can they testify that a woman has given birth, when, on the very next day, her stomach is still up there between her teeth!" The metaphor of a pregnant woman is used not only because declaring a birth to have taken place when in fact a woman is still pregnant is a sure sign of false testimony. It is also an appropriate metaphor because the visual imagery of a pregnant woman's body evokes that of the moon's crescent, and because the last night of a long month (i.e., one that contains thirty days rather than twenty-nine) is called ליל עבור, a pregnant night, and is mentioned within the story – an artful intertextual pun.

Similarly, tractate Niddah places in parallel observations of the onset of menstruation and of the new moon.[59] In a mishnaic rule attributed to Rabbi Eliezer, it is asserted that "in the case of four classes of women it suffices [for them to reckon] their [period of uncleanness from] the time [of their discovering of the flow]..."[60] In other words, certain women must only consider themselves impure from the moment when they see blood, rather

[56] The term לבנה appears infrequently in the Hebrew Bible, e.g., Isa 24:23, 30:26; Song 6:10, and always in parallel with חמה, rather than with שמש. It appears, however, throughout rabbinic sources and essentially replaces ירח. On the terms, see Rendsburg 2003a: 26; 2003b: 110; on the sex of the sun and the moon, see Lévi-Strauss 1983: 211–221.

[57] E.g., Stern 2001; Talmon 2005, 2007; Fraade 2009–10.

[58] m. Roš Haš. 2:8. On the mishnaic story, see Assis 2009 and Simon-Shoshan 2012: 167–193. A passage in rabbinic sources (b. B. Bat. 16a–b) also explains God's precision with time through a reference to the birthing habits of goats and hinds, e.g.: "This hind has a narrow womb. When she crouches for delivery, I [God] prepare a serpent that bites her at the opening of the womb, and she is delivered of her offspring; and were it one second too soon or too late, she would die. I do not confuse one moment with another" (trans. Soncino; cf. b. ʿErub. 54b).

[59] In Latin, "month" and "menstruation" derive from the same word: *mensis – menses*.

[60] m. Nid. 1:3.

than retroactively, as is the case for all other types of women according to the tractate's preceding *mishnah*. The witnessing of blood determines the beginning of these women's status of impurity. In their first comment on this *mishnah*, the Palestinian and Babylonian Talmuds cite a text from Tosefta Niddah: "It was taught: Rabbi Eliezer said to Rabbi Joshua, '. . . people do not ask one who has not seen the moon to come and tender evidence but only one who has seen it.'"[61] Here, the rabbinic text consciously makes the connection between detecting menses and observing a new moon as they construct a legal parallel: only on seeing blood (just as with a new moon) can one be certain of its appearance, at which point a woman's time of impurity begins.

In his study of female imagery in rabbinic calendar literature, Ron Feldman has pointed out that rabbinic terminology "concerning key moments of calendrical time, especially lunation, drew heavily on feminine biological processes of pregnancy and birth."[62] Feldman identifies a number of legal terms employed both in calendrical contexts and in a variety of texts related to women's rituals, including menstruation and marriage. First, the term "in its [proper] time – בזמנו" refers both (1) to observing the new moon "in its proper time," on the evening before the thirtieth day, thus completing a month of twenty-nine days (a thirty-day month occurs when the new crescent is not seen "in its [proper] time") in discussions of the calendar and (2) to observing menstrual blood during the time of menstruation, as in the phrases "it is her time to observe [menstrual blood] – זמנה לראות" and "in her [proper] time – בזמנה" in texts about menstrual purity (as opposed to seeing blood "not at her [proper] time," for example, before a girl has menstruated for the first time, or at other times when menstruation is not expected).[63] Even more striking is the phrase employed in the Babylonian Talmud, "in her [proper] time or not in her [proper] time – בין בזמנה בין שלא בזמנה," to refer to the time a woman immerses herself in a ritual bath, which is parallel to language used in relation to the observation of the moon, "either in its [proper] time or not in its [proper] time – בין בזמנן בין שלא בזמנן."[64] The term זמן appears all over rabbinic sources, of course,[65] but this particular

[61] t. Nid. 1:5; y. Nid. 2:1, 1b; b. Nid. 7b.

[62] Feldman 2010. On rabbinic practices of determining new months, cf. Stern 2012 and Leicht 2014.

[63] E.g., m. Nid. 10:1; t. Nid. 1:5; 9:6; y. Ber. 2.6, 5b; y. Nid. 1.2, 49a; 1.4, 49c; 1.6, 49c; b. Ketub. 6a; b. Nid. 5a; 9b; 10b; 64b; 65a, all cited by Feldman 2010. Note especially that the term "in her [proper] time" is linked to "observing" (לראות) – as in, "in her [proper] time to observe" – in m. Nid. 10:1, as are the two terms in m. Roš Haš. 2:8–10.

[64] Compare b. Nid. 67b with m. Roš Haš. 2:9 and b. Sanh. 10b, discussed by Feldman 2010.

[65] See Stern 2003: 26–30 and Kaye 2012: 174–176, 199–200.

construction is less ubiquitous. Feldman suggests that it is more likely that the formulation was first used with reference to the observation of female biological phenomena and then applied to calendrical matters. Second, the term מעובר is used to describe a woman who is "pregnant" and a month that "contains a thirtieth day" – that is, the month is metaphorically "pregnant" with an extra day (each month in the rabbinic calendar contains either twenty-nine or thirty days).[66] It is also used to describe the extra, intercalated month of Adar Sheni that is added during a leap year.[67] Third, the verb מקודש is used to mean "sanctification" of the new moon and also "sanctification" for marriage (מקודשת), and the terms "rebirth – תולדה"/"born – נולד" refer to the "birth" of the new moon (technically the mean conjunction when the moon becomes visible) and a woman's delivery of a child.[68] Shemaryahu Talmon has pointed out that the technical phrase for a new moon (*molad*) draws on the idea of a "birth," thus associating human birth by a woman to calendrical contexts in which the moon is "born."[69] The term *molad* may refer more specifically to "horoscope"; this technical definition maintains the connection to the day of a birth, even if an astrological context is assumed.[70] In each of these cases, women's bodies are associated with and compared to the monthly phases of the moon, the length of the months, the process of intercalation, and thus calendrical time more generally.

Feldman suggests that language that referred originally to female biological processes was adapted in the discourse surrounding the calendar during the rabbinic period. This is most obvious in the case of the term מעובר, which refers to a woman who is pregnant and is then extended to include the moon, which is metaphorically pregnant with an extra day, or a leap year, which

[66] E.g., b. Roš Haš. 20a–b; b. Ta'anit 18a; b. Ketub. 112a; Pesiq. Rab. 15, 78a; Pirqe R. El. 7 (41–42 Friedlander), and perhaps also in texts from Qumran, including 1Q27 1 i 5–7; 1QH[a] 20:11; 1QS 9:26–10:17; 4Q299 frg. 5; 4Q417 1 i 11; 4Q418 123 ii 2–6 and 4Q416, on which see Talmon 1999: 36–38.

[67] E.g., b. Sanh. 11b, 12b. I do not know whether the term was first used with reference to the additional day in a full month of thirty days, or to the intercalated month in a leap year. Interestingly, the use of the idea of "pregnant" to indicate a leap year seems to be particular to Jewish sources; the Syriac (*shunta kbīshta*) and Arabic (*sana kabīsa*) designations for leap years are linguistically unrelated, as is the Latin term *saltus lunae*, which refers to a "jumping" moon, from which the English term "leap" comes.

[68] E.g., m. Roš Haš. 2:7; b. Roš Haš 20b. The term "sanctification" in the context of marriage is discussed by Satlow 2001: 76–77 and Labovitz 2009: 76–79.

[69] Talmon 1999, including Tg. Ps-J. on Gen. 1:14 and Tg. Ket. on 1 Chr. 12:32; Talmon more tentatively suggested that the term is used in some Qumranic (e.g., 1Q27 1 i 5–7) and rabbinic sources in conjunction with turns of phrase that the Hebrew Bible uses to describe the motif of the "barren wife" (33–38).

[70] Morgenstern 2000: 141 translates *molad* as "astrological sign under which one was born" in light of Syriac and Aramaic parallels.

includes an intercalated month. The terminology of observation, too, might originate with the observation of menstrual blood and then transfer to other forms of observation such as the observation of the moon's cycles. For Feldman, this terminological overlapping and the parallels that develop out of these similarities indicate rabbinic efforts to control nature, women, and time – control over the untamed female body and the untamed natural monthly phases of the moon.

While Feldman stresses the parallels between rabbinic discourses on time and on women to show how the rabbis aim to control both, I build on Feldman's suggestions to argue further that these discourses are not merely parallel but also that the rabbis *use* their understanding of women and their bodies to explain and define time. For the rabbis, the distant, feared, and intangible eschaton is best understood with reference to childbirth and its complicated emotional and physical dimensions, and the cycles of the distant moon are most readily conceived through the metaphor of menstruation and its regular rhythms. I agree that power dynamics are also involved, as Feldman persuasively suggests; what is fascinating for our purposes is that women's bodies are a central conceptual tool for the rabbinic imagination of time. In other words, out of the many possible terms the rabbis might have used or metaphors they might have employed or comparisons they might have drawn to other concepts or laws as they devised their calendrical system, the rabbis chose women's maternal bodies, pregnancy, and the laws of menstruation.

4 Pregnant with Meaning: Feminine Discourse and Women's Voices

In all of the examples discussed earlier, the schedule of pregnancy, the temporal inevitability of labor and birth, the season of female fertility, and the cycles of menstruation are employed metaphorically in descriptions of the eschatological end-times, historical time, or calendrical time.[71] Women's bodies are marked *by* time more than they mark

[71] We might add, as well, "pedagogical time." Ancient Jewish and Christian sources at times refer to lactation as a metaphor for education and pedagogy, though the focus in these sources is usually the nursing infant or child rather than the lactating mother. The process, from nursing an infant through weaning a more independent toddler who can eat on his or her own and eventually also sustain him- or herself, becomes applied to the process of educating children and people. Here, again, a maternal bodily metaphor is used to describe pedagogical time. E.g., Heb 5:12–14; 1 Pet. 2:2–3; 1 Cor 3:2; Dio Chrysostom, *Or.* 33.7.44, discussed by Bynum 1985; Corrington 1989; Muers 2010.

time; all of the menstruation, pregnancy, fertility, and birth metaphors highlight the inability to change the timing of events and expectations, even as they happen within bodies, the heavens, and communities. The act of birthing a child is itself generally more active than menstruation, yet both processes are equally passive with regard to their inevitable timing, over which no one has full control (which is precisely the point).[72]

The relationship between time and women's bodies in all of these instances consistently remains on the level of discourse and metaphor. Women's bodies are used to illustrate theological concepts and ideas; these are not actual pregnant bellies or labor pains, but metaphorical ones. Women, after all, were banned by rabbinic law from serving as witnesses of the new moon and from sanctifying it, and they seemingly play no active role in the redemption as it is imagined in any of the sources cited earlier.[73] Moreover, the earliest rabbinic texts already exempt, and thus also exclude, women from observing those rituals that belong to the category labeled by rabbinic sources as "positive, time-bound commandments" (and it is these very rituals that come to define *men's* daily rhythms in contrast to women's time).[74] In the same literature that associates a timelessness to women and their ritual lives, women's bodies are also regarded as primary models for understanding time and temporality. The rhetorical intertwining of women's bodily metaphors to describe temporal processes such as the end-times and the calendar evoke the language of female bodily time while excluding real women from those times.

It has been argued that when women and their bodies are evoked metaphorically – to stand in for concepts such as justice, or Israel, or the Torah – actual women are necessarily barred from participating in the abstraction because the feminine image stands in for one of the male partners. For example, when the Torah is described as a female erotic object within a heterosexual paradigm, women do not have access to the Torah because it is

[72] There were, of course, ancient labor-inducing techniques as well as methods used to try to postpone labor or accelerate a long and difficult labor, which were attempts to control an unpredictable process. On ancient childbirth techniques, see Hippocrates, *Mul.* 1.68 = 8.142.13–144.16, reproduced in Hanson 1991: 91–92.

[73] There are isolated exceptions in which women do play some role in the redemption, but their presence in these contexts highlights their absence elsewhere. On the theme, see Himmelfarb 2002.

[74] E.g., m. Qidd. 1:7; t. Qidd. 1:10; Mek. *Pisḥa* 17 on Exod. 13:9 (68 Horovitz and Rabin); Mek. Šim. to Exod 13:9 (41 Epstein and Melamed); y. Qidd. 1:7, 61c; y. Ber. 2:3, 4c; b. Qidd. 33b–35a and b. Sukkah 28a–b. On men and women's time in rabbinic sources, see also Gribetz 2013: 124–248.

men who are presented as those drawn to the (female) Torah.[75] With the case of the sanctification of the new moon, and even with the process of redemption, the mapping of these times onto women's bodies ironically highlights their forced distance from the unfolding process. These metaphorical uses of women's bodies stand in marked contrast to the bodily cycles of real women, whose times were also marked, in part, by their bodies – for instance, menstruation and childbirth determined a woman's time of purity and impurity.[76]

Despite the marginalization of actual women, we might still detect women's voices and experiences behind our sources. Scholars assume that men authored all the sources discussed throughout the chapter.[77] By what means did men in antiquity know about the childbirth process? Were they present at the births of their children (according to biblical passages as well as Roman medical texts, most roles were filled by midwives and other women during standard deliveries)?[78] Did they hear secondhand from their wives and daughters about the process, rather than witness it directly? Even if men were not typically present for their wives' childbirths, those who lived in agrarian settings frequently witnessed – and cared for – animals giving birth. In fact, Eran Viezel has argued that biblical descriptions of a baby born with outstretched arms, found in Gen. 25:25–26 and Gen. 38:28–30, are based on men's experience with the birth process of cows, sheep, goats, donkeys, and camels, who typically emerge out of their mothers with their front hooves before their bodies, rather than those of women, whose babies are not born in this way.[79] In Greco-Roman contexts, medical texts – such as Pliny's *Historia Naturalis*, Soranus' *Gynecology*, and discussions by Galen and Celsus – also describe labor and the birthing process; these texts, which men might have read or the content of which might have been familiar to them via oral transmission, provided yet another source of knowledge about birth to those who were not present in the birthing room. However these male authors learned of

[75] Fraade 1986: 275 and Boyarin 1993: 134–166 explore the tension between Torah study and marriage. Labovitz 2007: 31–34 discusses the exclusion of women from Torah study.

[76] On this topic, see Gribetz 2013: 124–248.

[77] Though on the potential for female authorship, cf. Brenner and van Dijk-Hemmes 1993; Gruber 2007.

[78] Biblical accounts usually only mention midwives and other women, while men were informed of their children's births, cf. Gen. 35:17; 38:28; Exod. 1:15–21; 1 Sam. 4:20; Jer. 20:15; see Philip 2006 and Viezel 2011.

[79] Viezel 2011.

the pain and anticipation of the birthing process, authoring these metaphors required an act of imagination on the part of their male authors. The writing of these texts necessitated that men imagine themselves to be in the position of women in labor – to envision themselves as going through, or anticipating, the emotional (anxiety) and physical (painful) experiences of labor and birth, to write from the perspective of a laboring woman. Reading texts authored by men that enter imaginatively into and give voice to women's experiences, we have the opportunity to access what these male authors valued and regarded as important about these experiences.

Despite this filtering of women's voices and experiences through the voices of men, as they seem to be here, it is also the case that women's voices are preserved indirectly and claim an important role in the literature. These metaphorical invocations become passages that, in a certain sense, access and channel women's experiences, even as each individual woman's experience of them is also unique. These texts convey emotionally sensitive and nuanced understandings of some women's maternal bodily experiences and their various temporal dimensions, which are then used to understand and articulate other ideas about time. Real women may be marginalized through metaphor, but some of their feminine, embodied, motherly experiences also take center stage in those very same metaphors.

Bibliography

Aaron, D. H. 2001. *Biblical Ambiguities: Metaphor, Semantics, and Divine Imagery*. Brill Reference Library of Judaism 4. Leiden: Brill.

Alexander, P. S. 1986. "Incantations and Books of Magic." In E. Schürer (ed.), *The History of the Jewish People in the Age of Jesus Christ, 175 B.C. – A.D. 135*, vol. 3.1. Rev. ed. by G. Vermes, F. Millar, and M. Goodman. Edinburgh: T & T Clark, 342–379.

Andersen, F. I. and D. N. Freedman. 1980. *Hosea: A New Translation with Introduction and Commentary*. Anchor Bible 24. New York: Doubleday.

Andersen, F. I. and D. N. Freedman. 2000. *Micah: A New Translation with Introduction and Commentary*. Anchor Bible 24E. New Haven: Yale University Press.

Assis, A. 2009. "Two Kings, One Crown, and Raban Gamliel's Court: Between Strategies of Justifying Authority and Signification of Time." *Jerusalem Studies in Hebrew Literature* 23: 53–75 (in Hebrew).

Aubert, J.-J. 1989. "Threatened Wombs: Aspects of Ancient Uterine Magic." *Greek, Roman, and Byzantine Studies* 30: 421–449.

Barb, A. A. 1953. "*Diva Matrix:* A Faked Gnostic Intaglio in the Possession of P.P. Rubens and the Iconology of a Symbol." *Journal of the Warburg and Courtauld Institutes* 16: 193–238.

Barr, J. 1962. *Biblical Words for Time*. London: SCM Press.

Bergmann, C. D. 2008. *Childbirth as a Metaphor for Crisis: Evidence from the Ancient Near East, the Hebrew Bible, and 1QH XI, 1–18*. Beihefte zur Zeitschrift für die alttestamentliche Wissenschaft 382. Berlin: de Gruyter.

Bonner, C. 1950. *Studies in Magical Amulets Chiefly Graeco-Egyptian*. Ann Arbor: University of Michigan Press.

Boyarin, D. 1993. *Carnal Israel: Reading Sex in Talmudic Culture*. Berkeley: University of California Press.

Brenner, A. and F. van Dijk-Hemmes. 1993. *On Gendering Texts: Female and Male Voices in the Hebrew Bible*. Biblical Interpretation Series 1. Leiden: Brill.

Brettler, M. Z. 1998. "Incompatible Metaphors for YHWH in Isaiah 40–66." *Journal for the Study of the Old Testament* 78: 97–120.

Brown, P. 1988. *The Body and Society: Men, Women, and Sexual Renunciation in Early Christianity*. New York: Columbia University Press.

Buell, D. K. 1999. *Making Christians: Clement of Alexandria and the Rhetoric of Legitimacy*. Princeton: Princeton University Press.

Bynum, C. W. 1985. "Fast, Feast, and Flesh: The Religious Significance of Food to Medieval Women." *Representations* 11: 1–25.

Callahan, V. W. 1967. *Saint Gregory of Nyssa: Ascetical Works*. Washington, DC: The Catholic University of America Press.

Clark, E. 2005. "Thinking with Women: The Uses of the Appeal to 'Woman' in Pre-Nicene Christian Propaganda Literature." In W. V. Harris (ed.), *The Spread of Christianity in the First Four Centuries: Essays in Explanation*. Leiden: Brill, 43–51.

Collins, J. J. 2009. "The Beginning of the End of the World in the Hebrew Bible." In J. J. Ahn and S. L. Cook (eds.), *Thus Says the Lord: Essays on the Former and Latter Prophets in Honor of Robert R. Wilson*. The Library of Hebrew Bible/Old Testament Studies 502. New York: T&T Clark, 137–155.

Corrington, G. P. 1989. "The Milk of Salvation: Redemption by the Mother in Late Antiquity and Early Christianity." *Harvard Theological Review* 82/4: 393–420.

Darr, K. P. 1987. "Like Warrior, Like Woman: Destruction and Deliverance in Isaiah 42:10–17." *Catholic Biblical Quarterly* 49: 560–571.

Darr, K. P. 1994. *Isaiah's Vision and the Family of God*. Louisville, KY: Westminster John Knox Press.

Delaney, J. 1976. *The Curse: A Cultural History of Menstruation*. New York: Dutton.

Dille, S. 2004. *Mixing Metaphors: God as Father and Mother in Deutero-Isaiah*. Journal for the Study of the Old Testament Supplement Series 398. London: T & T Clark.

Duff, P. B. 2003. "Wolves in Sheep's Clothing: Literary Opposition and Social Tension in the Revelation of John." In D. L. Barr (ed.), *Reading the Book of Revelation: A Resource for Students*. Resources for Biblical Study 44. Atlanta: Society of Biblical Literature, 65–79.

Eliade, M. 1959. *Cosmos and History: The Myth of Eternal Return*. Trans. W. R. Trask. New York: Harper Torchbooks.

Feldman, R. H. 2010. "Controlling Women and Controlling Time: The Use of Female Imagery in Rabbinic Calendar Literature." Association for Jewish Studies Conference Paper, December 2010, Boston, MA.

Flannery, F. 2012. "'Go, ask a woman's womb': Birth and the Maternal Body as Sources of Revelation and Wisdom in *4 Ezra*." *Journal for the Study of the Pseudepigrapha* 21/3: 243–258.

Flusser, D. 1983. "The Reflection of Jewish Messianic Figures in Early Christianity." In Z. Baras (ed.), *Messianism and Eschatology: A Collection of Essays*. Jerusalem: Zalman Shazar Center, 103–134.

Flusser, D. 2009. *Judaism in the Second Temple Period: The Jewish Sages and Their Literature*. Trans. A. Yadin. Grand Rapids, MI: Eerdmans.

Folmer, M. L. 2011. "A Jewish Childbirth Amulet from the Bibliotheca Rosenthaliana." In W. T. van Peursen and J. W. Dyk (eds.), *Tradition and Innovation in Biblical Interpretation*. Studia Semitica Neerlandica 57. Leiden: Brill, 223–241.

Foster, J. 1996. "Menstrual Time: The Sociocognitive Mapping of 'The Menstrual Cycle.'" *Sociological Forum* 11/3: 523–547.

Fraade, S. 1986. "Ascetical Aspects of Ancient Judaism." In A. Green (ed.), *Jewish Spirituality: From the Bible Through the Middle Ages*. New York: Crossroad Press.

Fraade, S. 2009–10. "Theory, Practice, and Polemic in Ancient Jewish Calendars." *Diné Israel: Studies in Halakhah and Jewish Law* 26/27: 147–181.

French, V. 1986. "Midwives and Maternity Care in the Roman World." *Helios* 13/2: 69–84.

Frey, J. 2011. "Temple and Identity in Early Christianity and in the Johannine Community: Reflections on the 'Parting of the Ways.'" In D. R. Schwartz and Z. Weiss (eds.), *Was 70 CE a Watershed in Jewish History? On Jews and Judaism Before and After the Destruction of the Second Temple*. Ancient Judaism and Early Christianity 78. Leiden: Brill, 447–508.

García Martínez, F. and E. J. C. Tigchelaar. 2000. *The Dead Sea Scrolls Study Edition*. 2nd ed. 2 vols. Leiden: Brill.

Goldberg, S. A. 2000. *La Clepsydre: Essai sur la pluralité des temps dans le judaïsme*. Paris: Albin Michel.

Goldberg, S. A. 2004. *La Clepsydre II: Temps de Jérusalem, temps de Babylone*. Paris: Albin Michel.

Gribetz, S. K. 2013. *Conceptions of Time and Rhythms of Daily Life in Rabbinic Literature, 200–600 C.E.* PhD Thesis, Princeton University.

Gruber, M. I. 1992. *The Motherhood of God and Other Studies*. Atlanta: Scholars Press.

Gruber, M. I. 2007. "Women's Voices in the Book of Micah." *lectio difficilior* 1/2007. www.lectio.unibe.ch/07_1/pdf/mayer_gruber_womans_voices.pdf (accessed August 20, 2015).

Hanson, A. E. 1987. "The Eight Months' Child and the Etiquette of Birth: 'Obsit Omen!'" *Bulletin of the History of Medicine* 61: 589–602.

Hanson, A. E. 1991. "Continuity and Change: Three Case Studies in Hippocratic Gynecological Therapy and Theory." In S. B. Pomeroy (ed.), *Women's History and Ancient History*. Chapel Hill: University of North Carolina Press, 73–110.

Hauptman, J. 2013. "A Synchronic and Diachronic Reading of Mishnah Shabbat 2:6: On the Topic of Why Women Die in Childbirth." In R. S. Boustan et al. (eds.), *Envisioning Judaism: Studies in Honor of Peter Schäfer on the Occasion of His Seventieth Birthday*, vol. 1. Tübingen: Mohr Siebeck, 475–486.

Henze, M. 2011. *Jewish Apocalypticism in Late First Century Israel: Reading Second Baruch in Context*. Texts and Studies in Ancient Judaism 142. Tübingen: Mohr Siebeck.

Himmelfarb, M. 2002. "The Mother of the Messiah in the Talmud Yerushalmi and Sefer Zerubbabel." In P. Schäfer (ed.), *The Talmud Yerushalmi and Graeco-Roman Culture III*. Texts and Studies in Ancient Judaism 93. Tübingen: Mohr Siebeck, 369–389.

Hogan, K. M. 2011. "Mother Earth as a Conceptual Metaphor in 4 Ezra." *Catholic Biblical Quarterly* 73/1: 72–91.

Hogan, K. M. 2014. "Mother Zion, Mother Earth: Maternal Personification in 4 Ezra and 2 Baruch." Paper delivered at the "Conceiving Maternity and Motherhood in Jewish, Christian, and Muslim Contexts" Conference at Fordham University, New York.

Iricinschi, E. 2013. "Interroga matricem mulieris: The Secret Life of the Womb in *4 Ezra* and Sethian Cosmology." In R. S. Boustan et al. (eds.), *Envisioning Judaism: Studies in Honor of Peter Schäfer on the Occasion of His Seventieth Birthday*, vol. 2. Tübingen: Mohr Siebeck, 751–770.

Isbell, C. D. 1975. *Corpus of Aramaic Incantation Bowls*. Missoula, MT: Society of Biblical Literature.

Kalmanofsky, A. 2008. "Israel's Baby: The Horror of Childbirth in the Biblical Prophets." *Biblical Interpretation* 16: 60–82.

Kaye, L. 2012. *Law and Temporality in Bavli Mo'ed*. PhD Thesis, New York University.

Kessler, G. 2013. "Constant Creation: (Pro)Creation in Palestinian Rabbinic Midrashim." In L. Jenott and S. K. Gribetz (eds.), *Jewish and Christian Cosmogony in Late Antiquity*. Texts and Studies in Ancient Judaism 155. Tübingen: Mohr Siebeck, 126–138.

King, C. W. 1873. *Early Christian Numismatics, and Other Antiquarian Tracts*. London: Bell & Daldy.

Kittay, E. F. 1988. "Woman as Metaphor." *Hypatia* 3/2: 63–86.

Klijn, A. F. J. 1983. "2 (Syriac Apocalypse of) Baruch." In J. H. Charlesworth, *The Old Testament Pseudepigrapha*, vol. 1. New Haven: Yale University Press.

Kristeva, J. 1981. "Women's Time." Trans. S. Jardine and H. Blake. *Signs* 7/1: 13–35.

Labovitz, G. S. 2007. "The Scholarly Life – The Laboring Wife: Gender, Torah and the Family Economy in Rabbinic Culture." *Nashim* 13: 8–48.

Labovitz, G. S. 2009. *Marriage and Metaphor: Constructions of Gender in Rabbinic Literature*. Lanham, MD: Lexington Books.

Lakoff, G. and M. Johnson 1980. *Metaphors We Live By*. Chicago: University of Chicago Press.

Lakoff, G. and M. Turner. 1989. *More Than Cool Reason: A Field Guide to Poetic Metaphor*. Chicago: University of Chicago Press.

Leicht, R. 2014. "Observing the Moon: Astronomical and Cosmological Aspects in the Rabbinic New Moon Procedure." In S. Stern and C. Burnett (eds.), *Time, Astronomy, and Calendars in the Jewish Tradition*. Time, Astronomies, and Calendars 3. Leiden: Brill, 27–39.

Levene, D. 2003. *A Corpus of Magic Bowls: Incantation Texts in Jewish Aramaic from Later Antiquity*. London: Routledge.

Lévi, I. 1922. "Le ravissement du Messie à sa naissance." *Revue des Études Juives* 74: 113–126, reprinted in E. Patlagean (ed.), 1994. Le Ravissement du Messie à sa naissance et autres essais. Paris: Peeters, 228–241.

Lévi-Strauss, C. 1964. *Totemism*. Trans. R. Needham. London: The Merlin Press.

Lévi-Strauss, C. 1983. *Structural Anthropology*, vol. 2. Chicago: University of Chicago Press.

Long, K. P. de 2012. "'Ask a Woman': Childbearing and Ezra's Transformation in 4 Ezra." *Journal for the Study of the Pseudepigrapha* 22/2: 114–145.

Marganne, M. H. 1981. *Inventaire analytique des papyrus grecs de medicine*. Geneva: Librairie Droz.

McVey, K. E. 2001. "Ephrem the Syrian's Use of Female Metaphors to Describe the Deity." *Zeitschrift für Antikes Christentum* 5: 261–288.

McVey, K. E. 2003. "Images of Joy in Ephrem's Hymns on Paradise: Returning to the Womb and the Breast." *Canadian Society for Syriac Studies* 3: 59–77.

Metzger, B. M. 1983. "4 Ezra." In J. H. Charlesworth (ed.), *The Old Testament Pseudepigrapha, vol. 1*. New Haven, CT: Yale University Press.

Meyers, C., T. Craven, and R. S. Kraemer (eds.). 2000. *Women in Scripture: A Dictionary of Named and Unnamed Women in the Hebrew Bible, the Apocryphal/Deuterocanonical Books, and the New Testament*. Grand Rapids, MI: Houghton Mifflin.

Momigliano, A. 1966. "Time in Ancient Historiography." *History and Theory* 6: 1–23.

Montgomery, J. A. 2011. *Aramaic Incantation Texts from Nippur*. Cambridge: Cambridge University Press.

Morgenstern, M. 2000. "The Meaning of בית מולדים in the Qumran Wisdom Texts." *Journal of Jewish Studies* 51: 141–144.

Muers, R. 2010. "The Ethics of Breast-Feeding: A Feminist Theological Exploration." *Journal of Feminist Studies in Religion* 26/1: 7–24.

Neher, A. 1976. "The View of Time and History in Jewish Culture." In L. Gardet (ed.), *Cultures and Time*. Paris: The Unesco Press, 149–168.

Oliver, K. 2010. "Motherhood, Sexuality, and Pregnant Embodiment: Twenty-Five Years of Gestation." *Hypatia* 25.4: 760–777.

Pagels, E. 2012. *Revelations: Visions, Prophecy, and Politics in the Book of Revelation*. New York: Viking.

Parker, H. N. 1999. "Greek Embryological Calendars and a Fragment from the Lost Work of Damaste, on the Care of Pregnant Women and of Infants." *Classical Quarterly* 49/2: 515–534.

Philip, T. S. 2006. *Menstruation and Childbirth in the Bible: Fertility and Impurity*. Studies in Biblical Literature 88. New York: Lang.

Pitre, B. 2005. *Jesus, the Tribulation, and the End of the Exile: Restoration, Eschatology, and the Origin of the Atonement*. Wissenschaftliche Untersuchungen zum Neuen Testament II/204. Tübingen: Mohr Siebeck.

Rendsburg, G. A. 2003a. "A Comprehensive Guide to Israelian Hebrew: Grammar and Lexicon." *Orient* 38: 5–35.

Rendsburg, G. A. 2003b. "The Geographical and Historical Background of the Mishnaic Hebrew Lexicon." *Orient* 38: 105–115.

Rousselle, A. 1991. "La politique des corps: Entre procreation et continence à Rome." In G. Duby and M. Perrot (eds.), *Histoire des femmes en Occident*. Paris: Plon.

Rubenstein, J. 1997. "Mythic Time and the Festival Cycle." *The Journal of Jewish Thought and Philosophy* 6: 157–183.
Satlow, M. 2001. *Jewish Marriage in Antiquity*. Princeton: Princeton University Press.
Schiffman, L. H. 2006. "Messianism and Apocalypticism in Rabbinic Literature." In S. T. Katz (ed.), *The Cambridge History of Judaism*, vol. 4. Cambridge: Cambridge University Press, 1053–1072.
Schiffman, L. H. and M. D. Swartz. 1992. *Hebrew and Aramaic Incantation Texts from the Cairo Geniza: Selected Texts from Taylor Schechter Box K1*. Semitic Texts and Studies 1. Sheffield: JSOT Press.
Simon-Shoshan, M. 2012. *Stories of the Law: Narrative Discourse and the Construction of Authority in the Mishnah*. New York: Oxford University Press.
Steensgaard, P. 1993. "Time in Judaism." In A. N. Balslev et al. (eds.), *Religion and Time*. Studies in the History of Religions 54. Leiden: Brill, 63–108.
Stegemann, H., E. Schuller, and C. Newsom. 2009. *Qumran Cave 1. III: 1QHodayot[a], with incorporation of 1QHodayot[b] and 4QHodayot[a–f]*. Discoveries in the Judaean Desert 40. Oxford: Clarendon Press.
Stern, S. 2001. *Calendar and Community: A History of the Jewish Calendar, 2nd Century BCE – 10th Century CE*. Oxford: Oxford University Press.
Stern, S. 2003. *Time and Process in Ancient Judaism*. Oxford: Oxford University Press.
Stern, S. 2012. "The Rabbinic New Moon Procedure: Context and Significance." In J. Ben-Dov, W. Horowitz, and J. M. Steele (eds.), *Living the Lunar Calendar*. Oxford: Oxbow, 211–230.
Studelska, J. 2012. "The Last Days of Pregnancy: A Place of In-Between." *Mothering: The Home for Natural Family Living* blog, www.mothering.com/community/a/the-last-days-of-pregnancy-a-place-of-in-between (accessed August 20, 2015).
Sutter Rehmann, L. 1995. *Geh, frage die Gebärerin: Feministisch-befreiungstheologische Untersuchungen zum Gebärmotiv in der Apokalyptik*. Gütersloh: Chr. Kaiser.
Sutter Rehmann, L. 2004. "To Turn the Groaning into Labor: Romans 8: 22–23." In A.-J. Levine (ed.), *A Feminist Companion to Paul: Authentic Pauline Writings*. London: Continuum, 74–84.
Talmon, S. 1999. "Anti-Lunar-Calendar Polemics in Covenanters' Writings." In M. Becker and W. Fenske (eds.), *Das Ende der Tage und die Gegenwart des Heils: Begegnungen mit dem Neuen Testament und seiner Umwelt. FS H.-W. Kuhn*. Arbeiten zur Geschichte des antiken Judentums und des Urchristentums 44. Leiden: Brill, 29–40.
Talmon, S. 2005. "Anti-Lunar-Calendar Polemic in the Covenanters' Writings." *Meghillot* 3: 69–84 (in Hebrew).
Talmon, S. 2007. "What's in a Calendar? Calendar Conformity and Calendar Controversy in Ancient Judaism: The Case of the 'Community of the Renewed Covenant.'" In J. H. Charlesworth (ed.), *The Bible and the Dead Sea Scrolls*, vol. 2. Waco, TX: Baylor University Press, 25–58.
Thierry de Crussol des Épesse, B. 2002. "Le sang et les intailles magiques gynécologiques gréco-égyptiennes: Un exemple d'une greffe antique de la magie sur la science." In R. Gyselen (ed.), *Charmes et sortilèges: Magie et magiciens*. Bures-sur-Yvette: Groupe pour l'étude de la civilisation du Moyen-Orient, 275–281.
Tromp, N. 1969. *Primitive Conceptions of Death and the Netherworld in the Old Testament*. Rome: Pontifical Biblical Institute.

Viezel, E. 2011. "The Influence of Realia on Biblical Depictions of Childbirth." *Vetus Testamentum* 61: 685–689.

Wilford, F. A. 1968. "Embryological Analogies in Empedocles' Cosmogony." *Phronesis* 13/2: 108–118.

Yarbro Collins, A. 2001. *The Combat Myth in the Book of Revelation*. Eugene, OR: Wipf & Stock.

Young, I. M. 1990. *Throwing Like a Girl and Other Essays in Feminist Philosophy and Social Theory*. Bloomington: Indiana University Press.

11

THE BEGINNING OF SABBATH AND FESTIVALS IN ANCIENT JEWISH SOURCES

Lutz Doering

1 Introduction

The beginning of the Jewish Sabbath and festivals is a specific case of how the beginning of the day is reckoned and observed in Judaism. It is an issue of both ritual marking and Jewish Law, or *halakhah*, since both Sabbath and festivals require specific behavior and the abstention from certain activities. As to the general question of whether the day began with morning or evening, there is reason to assume that Jews in the Second Temple period, in particular in Palestine but arguably also in parts of the Diaspora, reckoned the day from evening to evening. To be sure, the reception of scripture implies some interaction with earlier reckoning from morning, as can be assumed for the sacrificial legislation in the Pentateuch,[1] but it can be shown that the textual corpus for which a start of the day with sunrise has most vigorously been claimed, the Dead Sea Scrolls, have adjusted the respective legislation to a reckoning from evening, possibly out of a tendency for consistency.[2] Already the Hebrew Bible, outside sacrificial legislation in the strict sense, emphasizes the role of "evening" (*'erev*) in ritual contexts, as in the fast of Yom Kippur, which is to be held "on the ninth day of the [sc. seventh] month at evening, from evening to evening" (Lev 23:32). This points to a reckoning of the day from evening to evening[3] but also raises the question of whether "evening" belongs to the previous day or the next. Further biblical texts suggest that the day was reckoned from evening in the postexilic period;[4] in

[1] E.g. Lev 7:15, 22:30; cf. b. Hul. 83a. See Milgrom 1991–2010: 417–418, 420.

[2] Cf. Birenboim 1998: 241–244; Kister 1999: 317–372, here: 339, 390–391 n. 207. *Contra* Talmon 1958, 1960, 2003, 2004. I engage with Talmon's claims later, particularly at n. 49.

[3] Cf. Stern 2001: 112 n. 44; Niehr 1989: 363. See further Exod 12:18; Lev 15:5–11; Isa 30:29.

[4] Cf. Neh 13:19; and Niehr 1989: 363.

fact, the "overwhelming majority of evidence from Second Temple times points to a reckoning of the day from sunset."[5]

The present chapter takes its departure from this "overwhelming majority" and focuses on certain ambiguities concerning the precise time around sunset in which the day was taken to begin. It first looks at constructions of "evening" in ancient Jewish sources, especially at interpretations of the phrase "between the two evenings" that is given as the timing of two sacrifices in the Bible, the Passover sacrifice and the second daily offering (§ 2). Next, we discuss texts dealing with a public announcement of the approach and end of the Sabbath that provide a framework for common Sabbath observance (§ 3). The chapter further investigates traditions in which a period of time prior to the beginning of the Sabbath is already invested with Sabbath-conforming behavior (§ 4) and confronts these with rabbinic texts that suggest that Jews used to do "labor" until very shortly before the onset of the day; these texts suggest that the Sabbath began at the end of evening twilight, which proved difficult to define (§ 5).[6] Finally, the role of Sabbath evening meals is investigated, which initially began before the onset of the Sabbath and during which the participants experienced the transition from weekday to holy day (§ 6).

2 Between the Two "Evenings": *bein ha-'arbayim*

The relevant texts in the Pentateuch stipulate that both the second of the two daily sacrifices (*tamid*) and the slaughtering of the Passover animal are to be conducted *bein ha-'arbayim*, literally, "between the two evenings."[7] The original meaning of the phrase is uncertain. Even though *'arbayim* may originally have been an adverbial form,[8] the preceding *bein* "between"

[5] Ben-Dov and Saulnier 2008: 158. In § 5, we address the question of the precise start of day around "sunset." In his subsequent work, Saulnier claims that "there is solid evidence to suggest that in some traditions of the 364DY [364-day year] calendar the day was reckoned to start at sunrise" (Saulnier 2012: 216), although he does not adduce any new positive evidence for this claim but simply rehearses S. Talmon's relevant suggestions (see n. 49). His new hypothesis that the *mishmarot* texts from Qumran (4Q320, 4Q321, and 4Q321a) correlate a reckoning of the day from morning with a lunar reckoning from evening is unconvincing in my view, but an engagement with it is beyond the scope of the present chapter.

[6] This explains why Roman-style sundials, which have been found on the Temple Mount in Jerusalem and elsewhere in Judea (cf. Ben-Dov 2011: 222–223 and fig. 6–7), would have been of little help for establishing the precise onset of Sabbath and festivals.

[7] Exod 29:39, 41; Num 28:4, 8 (*tamid*); Exod 12:6; Lev 23:5; Num 9:3, 5 (Passover); 9:11 (Second Passover). Cf. further Exod 16:12 (meat in the desert); 30:8 (setting up of lamps by Aaron).

[8] Cf. Niehr 1989: 362. One may compare *ṣohorayim*, which denotes a point in time ("noon").

turns it into a dual. This may imply that one of the two "evenings" belongs with one day and the other one with the following day. More specifically, *bein ha-'arbayim* was initially taken as occurring at some point during twilight.[9] Reflecting such an understanding, the Israelite Samaritans up to the present day slaughter the Passover shortly after sunset, except when Nisan 15 is a Sabbath.[10] What sense did Jews in antiquity make of this phrase and how did they construe the respective "evenings"? It may be useful to start with the rendering of the phrase in the Septuagint. Most of the relevant passages[11] translate it with πρὸς ἑσπέραν "toward evening," while a couple of others use τὸ δειλινόν "at even";[12] neither of these reveals a particular interest in giving a precise time bracket or interpreting the peculiar wording of the Hebrew phrase. However, Lev 23:5, referring to Passover, deploys the rendition ἀνὰ μέσον τῶν ἑσπερινῶν "midway between the eventides." This remains close to the Hebrew text.[13]

Philo of Alexandria tells his readers that the second *tamid* was brought δείλης ἑσπέρας, literally, "in the afternoon, in the evening," or perhaps "in the early evening" (*Spec.* 1:169). Philo may here depend on the usual Septuagint timing for the *tamid*, "toward evening." While this will certainly have been before darkness, the precise timing of the sacrifice remains unclear. For Philo, the morning *tamid* is brought "for the benefactions of the day-time," the early evening one for those "of the night" (ibid.). Against the practical trend in late Second Temple Judaism (see presently), Philo maintains a close connection between the second *tamid* and nighttime. As to the time for the slaughtering of the Passover animals, Philo knows a wider margin: "from midday to evening" (ἀπὸ μεσημβρίας ἄχρι ἑσπέρας; *Spec.* 2:145). As we shall see, Philo appears to share a notion of "between the evenings" close to what we later find in rabbinic literature. Philo is tacit about the sequence between the Passover and the early evening *tamid*.[14]

9 Cf. Milgrom 1991–2010: 1968–1969.

10 The Passover is slaughtered about two minutes after sunset, taken to be midway between the first evening, starting when the sun becomes yellowish, and the second evening, ending with nightfall; cf. Jeremias 1932: 80. When Nisan 15 is a Sabbath, the sacrifice is brought forward to c. 30 min after noon.

11 LXX Num 28:4, 8; Exod 12:6; 16:12; Num 9:3, 11 (lacking in 9:5).

12 LXX Exod 29:39, 41; cf. ὀψέ "late in the day, at even": Exod 30:8.

13 Cf. Harlé and Pralon 1988: 188, about translation technique in LXX Lev: "Ce littéralisme est propre au *Lévitique*."

14 Philo's Alexandrian predecessor, the Torah interpreter Aristobulus, according to a fragment deriving from the third-century Christian writer Anatolius and preserved by Eusebius, states, "the day of the Passover (τῆς τῶν διαβατηρίων ἡμέρας) has been assigned to the 14th of the month after evening (μεθ' ἑσπέραν)" (Eusebius, *Hist. Eccl.* 7.32.18). What Aristobulus wishes to show is that the sun and the (full) moon stand in opposition during this

In his rewriting of the biblical law, Josephus maintains that the *tamid* was brought "at both the opening and the closing of the day" (ἀρχομένης τε ἡμέρας καὶ ληγούσης; *A.J.* 3:237), with "day" here ostensibly referring to daytime. However, he is more specific about the timing of the second *tamid* when he describes concrete practice: during the siege under Pompey in 63 BCE, the priests continued to offer the *tamid* "twice a day, in the morning and around the ninth hour" (περὶ ἐνάτην ὥραν; *A.J.* 14:65). The Passover is slaughtered "in the same way" as the Egyptian Passover, commemorating the latter, "because on that day God passed over them" (*A.J.* 2:313; cf. 3:248). While in his rewriting of Exod 12, Josephus makes no mention of the time of slaughtering, he states in his description of a Passover during the Jewish War that the animals would have been killed "from the ninth hour to the eleventh" (*B.J.* 6:423).[15] This earlier time bracket probably served to accommodate the large number of animals that were slaughtered in the Temple precincts.[16] Does the mid-afternoon time for the *tamid* similarly reflect practical concerns?

The Mishnah (m. Pesaḥ. 5:1), redacted in the early third century CE, views the *tamid* slaughtered at eight and a half hours and offered at nine and a half, which comes close to the time given by Josephus. On Passover eve, the *tamid* is brought forward by an hour: slaughtered at seven and a half and offered at eight and a half, with the Passover animal following, that is, from about the ninth hour (around 3 p.m.),[17] on either a regular weekday or a Sabbath.[18] However, if Passover coincides with Sabbath eve, the *tamid* is brought forward by a further hour: slaughtered at six and a half and offered at seven and a half, followed by the Passover animal. Clearly such early timing of both *tamid* and Passover addresses largely practical concerns, that is, to leave enough time for preparing the slaughtered Passover animal in advance of the Sabbath.[19] Conceptually, this timing is justified by a particular

time. As is known, the visibility of the full moon begins with sunset; thus, "after evening" likely means "after sunset" here. Beyond the observation that Aristobulus, according to this fragment, reckons some time after sunset with Nisan 14, we cannot say anything specific about his envisioned timing for any part of the celebration.

15 Cf. Colautti 2002: 25, 79, 121–122, and on the dating of this Passover: 115–116.

16 Cf. Safrai and Safrai 2009: 209–210; Tabory 2000: 87–88. This applies despite the possibility that Josephus's numbers are exaggerated; cf. Colautti 2002: 115–116 with n. 81.

17 The ninth temporal hour coincides with 3 p.m. at the time of the vernal and autumnal equinoxes.

18 As to the sequence of *tamid* and Passover, Werman 2011: 303–304 suggests (against Yadin) that it is also preserved in the Temple Scroll and that 11QT[a] 17:6–7 merely requires that the Passover be brought before the afternoon grain and drink offering.

19 The animal should preferably be completely roasted but at any rate had to be in the oven before Sabbath; cf. m. Pesaḥ. 6:1; t. Pesaḥ. 7:1; Jeremias 1932: 83–84.

understanding of *bein ha-'arbayim*: the "two evenings" are taken as the beginning and the end of the afternoon. This reasoning is developed in the Mekhilta de-Rabbi Ishmael (probably redacted in the second half of the third century CE),[20] where the period for slaughtering the Passover animal is said to be from the sixth hour onward, when the day turns toward evening; this is corroborated with a quotation from Jer 6:4, according to which "noon" is the time from which "the day declines, the shadows of evening lengthen."[21] In this context, the Mekhilta also provides an attempt at harmonizing the timing "between the two evenings" with the times given in the second – yet different – main biblical Passover passage, Deut 16: "in the evening, at sunset, the time you departed from Egypt" (Deut 16:6). The beginning of the departure is assumed with the sixth hour of the day, for which Exod 12:41 is referred to, saying that all companies of the Lord went out "on this day itself." According to R. Shim'on ben Yohai, the times as given in Deut 16:6 must be reversed: the time of departure corresponds to the slaughtering, sunset to roasting, and evening to eating the Passover animal.[22]

A particular interpretation of "between the two evenings" can be found in the Book of Jubilees.[23] The section on the Passover in this text from the middle of the second century BCE, originally written in Hebrew yet fully transmitted only in Ethiopic and partly also in Latin,[24] starts with the command (Jub. 49:1):

> Remember the commandments which the Lord gave you regarding the passover so that you may celebrate it at its time[25] on the fourteenth of the first month, that you may sacrifice it before evening (*za-'enbala yemsay*), and so that they may eat it at night (*ba-lēlit*) on the evening (*'ama mesēt*) of the fifteenth from the time of sunset.

The slaughtering must take place in the afternoon of the fourteenth, while the roasted meat is consumed after nightfall on the fifteenth. It is clear that

20 Cf. Stemberger 2011: 282.

21 Mek. *Pisḥa Bo'* 5 (17–18 Horovitz and Rabin); cf. Siphra *'Emor pereq* 11:1 (100a Weiss).

22 Mek. ibid.; Midr. Tannaim Deut 16:6 (92 Hoffmann); the passage was later also added as an alternative explanation (*davar 'aḥer*) to Siphre Deut. 133 (190 Finkelstein); cf. Epstein 1957: 721. In the main text, Siphre Deut. (ibid.) quotes both R. Eliezer and R. Aqiva, who interpret the phrase "in the evening, at sunset" (Deut 16:6) as follows: "in the evening you slaughter, at sunset you eat," with "sunset" apparently indicating the earliest time for eating; cf. Shemesh 1996: 7 n. 19. Here, "evening" must refer to the hours of the afternoon.

23 On Jubilees' understanding of "between the two evenings" cf. also Werman 2011: 301–304. On the use of passages from Exod 12, Lev 23, Num 9, and Deut 16 in Jub. 49, see Halpern-Amaru 2007.

24 The Ethiopic text and the English translation used here follow VanderKam 1989.

25 See Num 9:2–3; cf. Num 28:2.

this requires a beginning of the new day some time in the evening. While this passage develops the timing "in the evening," "at the time of sunset" from Deut 16:4, 6,[26] another passage explicitly explains the phrase "between the two evenings" and specifies further the times for slaughtering and consuming the meat (Jub. 49:10–12):

> The Israelites are to come and celebrate the passover on its specific day – on the fourteenth of the first month – between the evenings (*ba-me'kala mesyātāt*), from the third part of the day (*'em-sālestā la-'elat*) until the third part of the night (*'eska sālestā la-lēlit*). For two parts of the day (*keflē makfaltā la-'elat*) have been given for light (*la-berhān*) and its third part (*wa-sālestā*) for the evening (*la-mesēt*). (11) This is what the Lord commanded you – to celebrate it between the evenings (*ba-me'kala mesētāt*). (12) It is not to be sacrificed at any hour of the daylight but in the hour of the boundary of the evening (*gizē wassana mesēt*). They will eat it during the evening hour(s) (*ba-gizē mesēt*) until the third part of the night. Any of its meat that is left over from the third part of the night and beyond is to be burned.

To this, we can add a third passage (Jub. 49:19):

> At the time when the house is built in the Lord's name in the land which they will possess, they are to go there and sacrifice the passover in the evening when the sun sets, in the third part of the day.[27]

The period of daylight is thus divided into three parts. The third part of the day has been given to the evening, and it is during this time that the animal has to be slaughtered, at "the boundary of the evening." Apparently, "between the evenings"[28] is here interpreted to mean the bracket of time between the beginning of the "evening" part of the day and the "evening hour(s)," reckoned with the night, during which the meat is eaten. This bracket of time is "when the sun sets." This is also coherent with 49:1, according to which the meat is consumed only "*from* the time of sunset." While the period of sacrifice starts around the eighth hour,[29] Jubilees maintains a strong connection with the setting of the sun and may thus distance itself from the view, variously affirmed,

[26] Cf. Halpern-Amaru 2007: *82–*84.

[27] Kugel 2012: 198–200 reckons Jub. 49:10–12 among the passages introduced by an interpolator. There is no room to discuss this issue here; suffice it to note that I do not consider Kugel's criteria for identifying the hand of the interpolator compelling. In the present case, Kugel has to admit (202) that 49:19, not from the interpolator, agrees with the "third part of the day" as time for slaughtering the Passover.

[28] Note however that the Latin translation does not render the term accordingly but simply uses variations of "evening": *ad uesperam* (Jub. 49:10), *in uespertino* (49:11), and *in uespertina* (49:12).

[29] That is, c. 2 p.m. around the time of the vernal equinox.

as we have seen, by Philo, Josephus, and the rabbinic tradition, that "between the evenings" starts from midday and that the slaughtering may be practically brought forward to the early hours of the afternoon.[30] Yet, Jubilees accepts that "between the evenings" is not strictly limited to the time around sunset, as we observed for the Samaritan tradition. It appears, on the one hand, to harmonize the scriptural heritage with its double emphasis on evening/sunset (Deut) and "between the two evenings" (Exod, Lev, Num) and, on the other hand, to negotiate between scriptural heritage and Second Temple practice, which knew wider margins of "between the two evenings." It does so with a unique, "literal" reading of "between the two evenings" that refers one of these "evenings" to the last "daylight" part of the day, the other one to the "night" part of the following day.

It is debated whether the "third part of the night" refers to the first third (inclusively counted)[31] or the last third (exclusively counted).[32] I deem the former solution preferable, in which the periods are adjacent to one another: Jubilees here deploys a conceptual reinterpretation of "between the two evenings" that takes the first evening as the *last third of the day,* during which the animals are slaughtered, and the second evening as the *first third of the night,*[33] during which their meat is eaten.[34] Contrary to the scriptural deployment of the phrase "between the two evenings," which relates only to the sacrifice, Jubilees extends the meaning to encompass both the slaughtering and the consumption. A similar concept might be found in the dictum of one Ben Bathyra in the Mekhilta: "Give an evening to slaughtering it and an evening to eating it."[35] The limitation of meat consumption to the *first third*

30 Cf. Kugel 2012: 200; in contrast, Werman 2011: 304 plays down the differences regarding the time of slaughter.

31 E.g. Werman 2011: 302, 305; Kugel 2012: 199–200.

32 E.g. Albeck 1930: 13; Halpern-Amaru 2007: *86–*87. That this count could not be inclusive is clear because it otherwise would designate the very end of the night – unlikely in view of the comparative texts mentioned later.

33 The assumption of adjacent periods (both inclusively reckoned) is more satisfactory than coupling the third part of the night as the "evening of the night" (exclusively reckoned) with the "evening of the day" (inclusively reckoned), as does Halpern-Amaru 2007: *86–*87.

34 The Latin translation seems to corroborate this by speaking of meat consumption "*in* the third part of the night" (*tertia noctis*) rather than "until the third part" (Jub. 49:12). The statement in the same verse about meat "that is left over from the third part of the night and beyond" may then either be taken as "from *the end* of the third part of the night and beyond" or, if the Latin is correct that leaves out "and beyond," simply as "from the third part," i.e. from the period of consumption. *Pace* VanderKam 1989: 319 (trans.), Latin *hoc* here may be pleonastic (*quod* . . . *hoc* . . . "what . . . that . . .") rather than reflecting the misconstruction of a Hebrew equivalent of "and beyond."

35 Mek. *Pisḥa Bo'* 5 (18 Horovitz and Rabin). Alternatively, Ben Bathyra might combine the different rabbinic views regarding "evening" discussed earlier.

of the night reflects stringency beyond Exod 12:10, which merely says that nothing must be left over *until morning*. In rabbinic texts, we find a debate between the majority of the sages, who hold that it was permissible to eat the Passover animal until the early dawn, and R. Eliezer, who gives midnight as a limit. The shorter margin was eventually accepted to prevent transgression ("to make a fence around the Torah"; see m. 'Abot 1:1).[36]

3 Public Announcement of the Approach and the End of the Sabbath

According to several sources, a series of trumpet blasts marked the approach of the Sabbath. The first one to describe this is Josephus (*B.J.* 4:582)[37]; he says that the zealots under John of Giscala, enclosed in the Temple area, built four towers. One of them

> was erected above the roof of the priests' chambers (παστοφορίων), at the point where it was custom for one of the priests to stand and to give notice, by sound of trumpet, in the afternoon (δείλης) of the approach (εἰσιοῦσαν), and on the following evening (περὶ ἑσπέραν) of the close (τελεσθεῖσαν), of every seventh day (ἑκάστην ἑβδομάδα), announcing to the people the respective hours for ceasing work and for resuming their labors.

It is interesting to find a trumpet signal also after the end of the Sabbath. This suggests that Saturday night was deemed a time in which "labors" (ἔργα) would be resumed.[38] What is remarkable is the element of social coordination

[36] Mek. *Pisḥa Bo'* 6 (19 Horovitz and Rabin); Siphre Deut. 133 (190 Finkelstein); m. Pesaḥ. 10:9; t. Pesaḥ. 1:34 [= 2:22]; y. Ber. 1:1, 3a; b. Pesaḥ. 120b; Tg. Ps.-J. on Exod 12:8. R. Eliezer builds his argument on a *gezerah shawah* regarding the term "night" (לילה) for the eating of Passover (Exod 12:8) and God's slaying the firstborn (Exod 12:12; cf. 11:4; 12: 29–31). This points to an apotropaic understanding of the annual Passover that was not generally shared; cf. Shemesh 1996: 5–9. The Temple Scroll intriguingly does not give a temporal limit within the night (11QT[a] 17:8–9).

[37] Trans. Thackeray, LCL (spelling adapted).

[38] Mark 16:1–2 suggests that trade resumed on Saturday night: the women "bought" (ἠγόρασαν) the spices "after the Sabbath had passed" (διαγενομένου τοῦ σαββάτου), and they came to Jesus's tomb "early in the morning of the first day of the week." While the evidence for shops opening in the evening is limited in rabbinic texts (but see m. Taan. 1:6 regarding shops opening on Monday evening during a sequence of public fasts), there are references to "waiting for darkness at the (Sabbath) limit" (*maḥshikhin 'al ha-teḥum*) to conduct business, bring in fruit, or care for the needs of a bride or a dead body, after the Sabbath has ended (m. Šabb. 23:3–4; t. Šabb. 17[18]:10–13; b. 'Erub. 39a; b. Šabb. 150a–151a). Some purposes of such waiting are forbidden, others permitted: according to m. Šabb. 23:3; t. Šabb. 17[18]:13, Abba Sha'ul set up the rule that for anything one may say on Shabbat one may also wait for the darkness at the Sabbath limit.

by means of the signal marking the beginning and the end of the rest. The connection with the Temple and the priests creates a common public context for observing the set time and keeping Shabbat, independent of the different sects and groups within ancient Judaism that otherwise follow their own legal and practical traditions about the Sabbath. The details Josephus gives for the locations of the towers are not fully clear. However, the *pastophoria*, priestly chambers mentioned by this name also in the Greek scriptures,[39] were likely situated at the western side of the Temple enclosure.[40]

Even so, the precise times given by Josephus are ambiguous. Josephus speaks of a signal marking out the approach and the end of the ἑβδομάς, a term frequently used in the Hellenistic Jewish writers for "Sabbath." The temporal expression used for the approach is δείλη, "afternoon," that concerning the end of the Sabbath, περὶ ἑσπέραν, "around evening." As to the latter, we may assume that the signal would have been issued at or, perhaps more appropriately, after nightfall. As to the former, there is reason to assume that Josephus refers to a signal issued some time *before* the actual beginning of the seventh day, to allow the people to cease work and prepare for the Sabbath rest.

A similar understanding of the trumpet signals is suggested by several references in rabbinic texts. Thus, m. Sukkah 5:5, discussing the number of trumpet blasts on various days and festivals in the year, states:[41]

> And on Sabbath eve they used to add a further six, three to cause the people to cease labor and three in order to separate between holy and profane.

The Tosefta here clearly comments on this mishnah and fills in a number of details (t. Sukkah 4:11–12):[42]

> "Three to cause the people to cease from labor" – how so? The officer of the synagogue (*ḥazzan ha-knesset*) takes trumpets and ascends to a rooftop at the highest place in town. He starts to blow: those close to town cease (from labor), those close to the (Sabbath) limit enter and come into the limit; they

[39] See LXX 1 Chron 9:26; 23:28; 26:16; 28:12; 2 Chron 31:11; 1 Esdr 8:58; 9:1; 1 Macc 4:38, 57; Isa 22:15; Jer 42:4; Ezek 40:17, 38. Where Hebrew text is extant, it is often לשכה (but no building is referred to in Isa 22:15 MT); in 1 Chron 26:16, LXX appears to translate לשכה, whereas MT has שַׁלֶּכֶת.

[40] The text from Josephus can be compared with an archaeological discovery that came to light during the excavations at the Temple Mount. Not far from the southwestern corner, an inscription connected to a corner piece of the top ridge of the Herodian Temple enclosure was found, for which the following restoration has been proposed: *le-beit ha-teqi'ah le-hakh*[*riz*] "For the house of trumpeting, to announce." See Mazar 1978: 30, 34–35.

[41] The translation follows Danby 1958.

[42] Trans. LD, based on MS Erfurt. Cf. for the issue Elbogen 1911: II–VII; Gilat 1992: 316–317.

> would not enter immediately but wait until all come, and then enter all at the same time. When does he (himself) enter? After he would have filled a barrel of water, fried a fish and lit the (Sabbath) lamp for himself. (12) "Three to separate between the holy and the profane" – how so? The officer of the synagogue takes a trumpet and ascends to a rooftop at the highest place in town. He starts to blow: a person transfers a dish from the stove, stores a hot kettle away and lights the lamp for himself. When he stops blowing, he does not even store away a hot kettle but sets it down on the earth, he does not even put a lamp in his hand onto the lamp-stand but sets it down on the earth.

Although an "officer of the synagogue" is occasionally mentioned for the Temple Mount (e.g. m. Yoma 7:1), the Tosefta reads as if a procedure similar to the one Josephus reports for Jerusalem would take place in any (larger?) town. It is doubtful that this would reflect historical reality during Second Temple times; more likely, the Tosefta might here be seen as "democratizing" a Jerusalem institution. Whether this corresponds to actual practice during the rabbinic period is not fully clear; some scattered references to *shofar* signals on Sabbath eve may suggest this.[43] Be that as it may, the three trumpet blasts inviting people to cease from labor were staggered and came in intervals. At the beginning of the second set of three blasts, domestic preparations still seem possible, whereas by the last blast they need to be interrupted. However, the order, particularly the position of kindling the Sabbath light, is not entirely clear here. The Babylonian Talmud gives different details for the process (b. Šabb. 35b):[44]

> Our Rabbis taught: Six blasts were blown on the eve of the Sabbath. The first, to cause the people to cease labor on the fields; the second, to cause the city and shops to cease (labor); the third, for the lights to be kindled: that is R. Nathan's view. R. Judah the Nasi said: The third is for the *tefillin* to be removed. Then there was an interval for as long as it takes to bake a small fish, or to make a loaf stick to the oven, and then a *teqi'ah,* a *teru'ah* and a *teqi'ah* were blown, and one began the Sabbath rest.

We note that the time of the kindling of Sabbath lights is still debated.[45] While the first three blasts here call for a staggered cessation of work, the actual Sabbath

[43] Cf. Gilat 1992: 316–317, citing references from both Babylon and Eretz Israel. Note, however, that "trumpets," as mentioned in the Tosefta, are the privilege of the Temple (cf. y. Roš Haš. 3:4, 58d), though in b. Sukkah 34a trumpets and *shofars* are viewed as interchangeable.

[44] ET of b. Šabb. follows the Soncino translation, with some adaptations.

[45] The Babylonian Talmud ibid. adduces a further variant of the tradition, attributed to the Amora Samuel (*de-be Shmu'el*), but in Talmud editions mistakenly assigned to the school of *R. Ishmael* (Epstein 2000: 212–214): the first blast ceases work on the fields, the second closes the shops, and the third is the signal to remove hot water and pots, as well as light the lamp. The remainder is the same as in the first variant.

rest is preceded by the second set of three blasts "blown in rapid succession."[46] The interval between the first and second set of three blasts is given as the cooking time for small items of food. This is significant since that is what Jewish women would be doing on Friday afternoon: preparing the Sabbath meal. The stoves and ovens were still hot and would be used until close to the beginning of the Sabbath rest. These rabbinic texts, then, distinguish between the time given to preparation for the Sabbath and the beginning of Sabbath rest proper; the latter probably occurred rather close to the actual beginning of the day.[47] This distinction between time given to preparation and actual Sabbath rest should also be considered in other testimonies, such as Augustus's exemption of Jews in Asia Minor from being called to court or from standing security to appear in court (ἐγγύας ὁμολογεῖν, *vadimonia facere*) on Sabbath and on Friday after the ninth hour, that is, after around 3 p.m. This does not suggest that Jews in the Diaspora would have started their actual Sabbath rest at that time.[48]

4 Early Start of Sabbath Observance

In contrast to the texts discussed so far, there is evidence that Jews at certain times and places started their actual Sabbath rest somewhat in advance of the day. Thus, the Damascus Document (CD-A 10:14–17), a medieval copy of a Second Temple text also found among the Qumran scrolls, states about the onset of Sabbatical rest:

> No-one may do, on the (15) sixth day, labor from the time that the orb of the sun will be (16) distant from the gate (by) its fullness; for this is what he said: "Guard the (17) day of the Sabbath to sanctify it."

This text takes the observable distance of the sun from "the gate" as the indicator of the time from which no labor should be done on Friday evening.[49] There is a long-standing debate over what is referred to by

[46] So the comment in the Soncino translation *ad loc.* (166 n. 2).

[47] Cf. Safrai and Safrai 2008: 29.

[48] Josephus, *A.J.* 16:162–165 (edict by Augustus); Doering 1999: 301–302. *Contra* Gilat 1992: 317, who holds that the Asian Jews started to observe Sabbath several hours before sunset.

[49] This passage presents a *prima facie* obstacle to the theory of the late Shemaryahu Talmon that a solar calendar would feature a reckoning of the day from morning, not evening. Talmon 1958: 193 attempted to resolve this difficulty as follows: "the copyist" of CD-A, "whether a Karaite or a Rabbanite, altered the ancient wording" to bring it in line with the beginning of Shabbat in the Middle Ages. According to the later version of this theory, the original reading would have been: "no one may do, on the day of the *Sabbath*, labor from the time that the orb of the sun will be distant from the gate (by) its fullness" (Talmon 1960: 394–395; cf. 2004: 88–92). Talmon reasoned that this would be the orb of the rising sun over the city gate on Saturday morning. However, such a late start of Sabbatical rest relative to the

"gate" here: it might be either the city gate[50] or a celestial gate.[51] I have elsewhere argued that the second solution, first proposed by Louis Ginzberg, is more convincing and that the first one suffers from two problems: first, it is unclear *which* gate would be in view, since the group associated with the Damascus Document Sabbath code may have lived in various places, not all of which necessarily had walls and gates; second, it remains open *at what distance* the observer is taken to stand from the relevant gate, since it will obviously make a difference if one is five, fifty, or five hundred yards away.[52] The reference to Neh 13:19, sometimes invoked here,[53] is problematic, since here the phrase "when the gates of Jerusalem threw shadows" does not depend on a *precise* position of the setting sun, which is, however, needed in our case.[54] It is therefore preferable to think of the celestial gate into which the sun sinks at the horizon. In recent years, this solution has gained plausibility by a better understanding of the notion of heavenly gates in the

beginning of daylight would sit uncomfortably with the sect's otherwise strict Sabbath regulations, as attested by CD-A 10:14–11:18. In addition, the theory of scribal change remains speculative. *Pace* Talmon 2003: 83–87, I do not think that CD-A 10:19, with its reference to "issues of labor or work to do *next morning*" (*le-mashkim*), points to an end of Sabbatical rest in the morning. This may simply be a reference to issues one would want to, or could only, pursue the morning after, such as going on a journey, traveling by ship, or any activity that would require continuation during daylight. More generally, the order of day and night in 1QS 10:10 or of the activities in 1QS 10:13–14 (cf. Talmon 1958: 189; Saulnier 2012: 216–217) is indecisive for the question of the beginning of the day since it may follow the cycle of activities from awakening to bedtime, perhaps – thus far can be conceded – with an emphasis on daylight. Most importantly, however, the 364-day calendar that is presupposed in many texts from Qumran is – *contra* Talmon – not a solar, but a schematic calendar, so it does not *structurally* prioritize the day over the night; it is only Jubilees that shows a solar emphasis (Jub. 6:32; cf. Ben-Dov and Saulnier 2008: 125, 136–138), and there can be no doubt that Jubilees affirmed a beginning of the day with evening (see § 2).

50 E.g. Gilat 1992: 258–259, who thinks of a western gate in Jerusalem (see n. 52) or the city in which the sectarians reside.

51 Thus the majority of earlier scholars; see the list in Doering 1999: 134 n. 80. Schiffman 1975: 84 n. 1 remains undecided.

52 Doering 1999: 133–138. As to Gilat's suggestion (n. 50) that Jerusalem is in view, it should be noted that the so-called first wall, which, coming from the southwestern hill, bent toward the Temple Mount roughly at today's Jaffa Gate, did not have a western gate. Another possibility would be a gate on the Temple Mount; here the potential difference in distances would be reduced, though still be an issue.

53 Especially by Talmon 2003: 91; 2004: 21–23. On Talmon's peculiar understanding of the passage see n. 49. It should be stated that Neh 13:19 speaks of preparation for the Sabbath by closing the city gates in the later afternoon and not necessarily of the time when obligatory Sabbath rest began; *contra* Ta-Shma 1983: 323.

54 To be sure, the Talmudim tell of situations where people are advised to take the position of the sun over specific landmarks as a marker of time, but here the position of these people is known to those involved as well; see e.g. b. Šabb. 35b; y. Ber. 4:1, 7c.

Astronomical Book of Enoch.[55] While lacking the complexity of the latter, the passage in CD-A 10:14–17 might well refer to the position of the sun above the heavenly gate it enters at sunset. To be sure, there may be weather conditions or topographical features such as mountains that make it impossible to see the sun stand above the horizon in the west; but this problem could be resolved by referring to corresponding time markers in such conditions, for which we have evidence in other sources.[56] In any event, the Damascus Document requires abstention from work from a few minutes before sunset and about half an hour before the end of what is now called "civil twilight," which may serve us here as an approximate marker, from which onward one *may* speak of "darkness": by then, first magnitude stars have become visible and daylight has receded considerably.[57]

This text, then, suggests that the *precise* start of the Sabbath was not of primary concern, as already a small portion of Friday is subject to the behavioral mode relevant for the day, that is, Sabbatical rest. A similar concept is found in rabbinic texts. In the Siphra, it is initially derived from extending the fast on Yom Kippur beyond the limits of this day, both before and after.[58] As the passage in the Siphra clarifies, this extension also applies to the Sabbath and further holidays, and it means that "you add from the profane to the holy." The aspect of an additional rest is initially called *tosephet mela'khah* "addition to (the ban on) labor" (Siphra ibid.) but later becomes known under the name *tosephet shabbat* "addition to the Sabbath." In tannaitic midrashim, the exegetical justification for this is found in the dual wording of the Sabbath commandment in the two versions of the Decalogue: "'Remember' [Exod 20:8] and 'Guard' [Deut 5:12] – remember before it, and guard after it; hence they said: one adds from the profane to the holy."[59] We note that the passage in the Damascus Document relates the phrase "Guard the Sabbath, to sanctify it" to the stretch of time

[55] 1 En. 72:1–5 features a double set of six gates; although no Aramaic text has been preserved for the section, the concept of gates through which the sun enters and exits is integral to the synchronistic system in the Aramaic fragments of the Astronomical Book from Qumran. On the gates cosmology of the Astronomical Book cf. now Ratzon 2015.

[56] See the advice given for cloudy days in b. Šabb. 35b.

[57] The end of "civil twilight" is reached when the center of the sun is 6° below the horizon. The sun's position of one diameter above the horizon implies a height of 0.75° above the horizon, but one has to take the "refraction" of the sun into account. A calculation for the cardinal points and the latitude of Jerusalem yields 29min 50s for the equinoxes and between 33min and 34min 20s for the solstices. I wish to thank Dr. Björn Voss, LWL-Museum für Naturkunde, Münster, for the relevant information.

[58] Siphra *'Emor pereq* 14:7–9 (102a Weiss); b. Yoma 81a.

[59] Mek. *Ba-ḥodesh (Yitro)* 7 (229 Horovitz and Rabin), Mek. Šim. *Yitro* (148 Epstein and Melamed) (both on Exod 20:8) and Midr. Tannaim on Deut 5:12 (21 Hoffmann).

before the Sabbath. There is thus a similar exegetical sensitivity to that of the later rabbis, though not the same exegetical solution: the rabbinic version appears to have further reflected on the implications of the *difference* between the Exodus and the Deuteronomy Decalogue and considered the *end* of the Sabbath as well, in line with its more comprehensive *midrash halakhah*.

However, it is unclear how much time would have been added before and after the Sabbath according to the rabbinic texts quoted. Moreover, it is also unclear how widely this view was initially accepted. For example, in y. Šeb. 1:1, 33a, it is clearly stated that one is entitled to work on both Sabbath eve and New Year's eve "until the sun sets." To be sure, the Talmudim and midrashim state that some rabbis began Shabbat early, though this was an issue of personal style and piety, not a legal requirement.[60] In addition, Talmudic sources continue to mention the idea of *tosephet shabbat*, although they do not normally state how much time one must add, and they also provide evidence that the addition practiced by some was actually minimal at best.[61] Only later, in medieval Ashkenaz, do we find fixed public – and much longer – periods of *tosephet shabbat*, reflecting that the sun sets so late in summer that it would be difficult to wait with the lighting of lamps and the Sabbath meal.[62]

5 Ambiguities Surrounding the Beginning of the Sabbath

In contrast to the sources discussed earlier, the Mishnah does not mention *tosephet shabbat* at all.[63] Instead, in m. Šabb. 2:7 we read:

> There are three things that a person must say in his house on the eve of the Sabbath with (approaching) darkness (*'im ḥashekhah*): Have you tithed? Have you made an *'eruv*?[64] Light the lamp. If it is doubtful whether it is darkness or not, one does not tithe the certain (non-tithed) items (*wadai*) and dip the vessels and light the lamps. But one tithes uncertain items (*demai*), makes an *'eruv* and stores hot water.

[60] Cf. y. Ber. 4:1, 7c; Ber. R. 10:8 [84–85 Theodor and Albeck] (R. Haninah ben Dosa). Cf. also the paraenetical warning – certainly without any legal implications – that no blessing will be on work done after the *minḥah* on Friday (b. Pesaḥ 50b).

[61] Cf. Gilat 1992: 318–320, though assuming that the old practice, which would have been strict in his view, was relaxed.

[62] See Ta-Shma 1983.

[63] Cf. also Safrai and Safrai 2008: 31–36.

[64] A symbolic meal deposited before Shabbat, whereby a courtyard shared by two or more parties becomes a common domain into which each participating party is allowed to carry objects.

This corroborates what we have suggested previously with respect to the trumpet blasts, namely, that it would have been quite common to continue with one's preparatory "labors" until quite close to the Sabbath. From this mishnah, it emerges that *'im ḥashekhah* "with (approaching) darkness" is slightly before the time when "it is doubtful whether it is darkness or not"; during the latter, only actions not prohibited by the Torah on Sabbath are allowed, such as tithing *demai*, making an *'eruv* or storing hot water, whereas actions prohibited by the Torah, such as lighting lamps on Sabbath, are already banned.[65] This in turn implies that according to this text, the legally binding beginning of the Sabbath was taken to be "darkness": it is only from then that forbidden labor *had* to stop.

A similar situation emerges from the debates between the Houses of Hillel and Shammai on the starting of processes continuing on their own, such as soaking dyes in water or drying flax in an oven: according to the House of Shammai, such processes may only be started before the Sabbath if they are also completed *mi-be-'od yom* "while it is still day" (m. Šabb. 1:5–6, 8). The House of Hillel, however, permits all of these, provided they are initiated *'im ha-shemesh* "during sunlight" (m. Šabb. 1:8). Both schools agree that the beams of the oil press may be loaded with weights *'im ḥashekhah* "with (approaching) darkness" but that meat, an onion, or eggs may not be fried at that time unless they are completely done before the Sabbath (t. Šabb. 1: 20–21; cf. m. Šabb. 1:9–10).[66]

In sum, early rabbinic texts suggest that the Sabbath was begun at "darkness." Around sunset, "with (approaching) darkness," was the last opportunity for certain activities. In the period between sunset and darkness, "biblically" prohibited activities were no longer practiced. This intermediary period of time is called *bein ha-shemashot*, "twilight" (literally "between the suns"), in some early rabbinic texts.[67] The problem with this period is that it is difficult to define. When precisely the Sabbath as well as other days started

65 Cf. the rule recorded at t. Šabb. 2:9.

66 Safrai and Safrai 2008: 104 give practical reasons for the agreement of the schools (and the respective deviation of one of them from its usual approach) in these two latter rulings. Cf. also t. Šabb. 1:23 with further activities, such as channeling water to a garden, putting ointment on an eye or frankincense on coals on Sabbath eve "with (approaching) darkness." Cf. also the prohibition that an artisan go out with his tool *samukh la-ḥashekhah* "close to darkness": m. Šabb. 1:3; t. Šabb. 1:8.

67 See m. Nid. 1:7; 6:14; 10:2; m. Zabim 1:6; t. Zabim 1:13; m. Šabb. 19:5; Siphra *Sheraṣim parashah* 11:2 (58c Weiss); m. Ker. 4:2; t. Ker. 2:15; Mek. *Shabbeta Wa-yaqhel* 1 (346 Horovitz and Rabin); Siphra *Ḥovah parashah* 5:2 (20a–b Weiss); *Ḥovah parashah* 12:1 (26c Weiss); Siphre Zut. Num 9:3 (258 Horovitz). Milgrom 1991–2010: 1969 refers to M. Jastrow's explanation of the term as "between the two services, between the rulership of the day and that of the night."

was hard to establish experientially. Early rabbinic texts and traditions provide a few, albeit limited, pointers as to delimiting this period. Thus, in t. Ber. 1:1, the "visibility of the stars" (*ṣe't ha-kokhavim*) is taken as a "sign" for the point from which one may say the evening *Shema' Yiśra'el*, which is identified with the time of the Sabbath eve meal (see § 6). The Talmudim refer to early rabbinic traditions that classify the color of the sky: when the eastern sky is red it is certainly daylight (*yom*); when it is silver-colored it is twilight (*bein ha-shemashot*); and when it has become black, the zenith as well as the horizon, it is night (*layla*).[68] The Talmudim then try to establish further criteria for differentiating between daylight, twilight, and nightfall, such as the number and size of visible stars or the distance one manages to walk until nightfall. However, the issue remains debated, so that the Jerusalem Talmud invokes the coming of Elijah, who will say, "That is twilight," and the Babylonian Talmud provides two definitions, one earlier for Sabbath observance and one later for the allowance of *terumah* consumption (after purification), assigning to each issue the more stringent option.[69] In sum, the precise transition to the Sabbath remains a problem for the practitioner, and it is not sufficiently clarified in the early rabbinic (tannaitic) period.

6 The Role of Meals

In this context, the celebration of meals at the onset of the Sabbath may have had a mitigating role. Importantly, in this respect, in the early tannaitic period there would have been no prayer assembly in the synagogue marking the beginning of the Sabbath. Instead, the seventh day would have been welcomed during a common meal,[70] which took place somewhat *later* than on other days, when dinners, especially of the elite, were held as early as around the ninth hour.[71] According t. Ber. 1:1, R. Meir identifies the time from which the evening *Shema' Yiśra'el* may be recited with the time when the meal inaugurating the Sabbath is usually held, whereas the (other) Sages identify it with the time when the priests are entitled to eat *terumah* (cf. m. Ber. 1:1). The latter is clearly after sunset: the priests must eat *terumah* in a status of ritual purity; but to acquire this status they have to await sunset after their purification (cf. e.g. Lev 15:1–11, 16, 18). As we have seen, however, there are discussions about the precise time of eating *terumah*.

[68] So y. Ber. 1:1, 2b. Cf. the variant in b. Šabb. 34b.

[69] See y. Ber. 1:1, 2a–c; b. Šabb. 34b–35b. Cf. also b. Pesaḥ. 94a.

[70] Cf. Elbogen 1911: VII–IX; Gilat 1992: 328.

[71] See e.g. b. Pesaḥ. 107b (King Agrippa), further t. Ber. 5:1–2 (and see later). For Greco-Roman practice cf. Klinghardt 1996: 45.

At any rate, the Jerusalem Talmud states that according to tannaitic teaching, the two time markers point roughly to the same period of time.[72]

Moreover, the context of m. Šabb. 2:7 quoted earlier (§ 5) clearly points to a domestic setting, in which the meal would have been prepared: the "tithe" mentioned in this situation can only refer to ingredients of the meal, and the *'eruv* is a symbolic meal deposited between neighbors in their shared courtyard before Shabbat. These preparatory activities, including the lighting of the lamps, are the domain of women, but this mishnah suggests that the *paterfamilias* would have monitored them. The lighting of the lamps, already mentioned by Seneca (first century CE) as a Jewish practice on the Sabbath,[73] has the dual character of a custom and a commandment. It is not absolutely necessary to light the lamp, as the continuation in m. Šabb. 2:7 shows: if one is too close to darkness, the lighting is discouraged because of the possibility that the Sabbath may have already begun. In addition, the use of lamps would have been a powerful custom illuminating the unusually late meal at the beginning of Shabbat and aptly expressing the joy of the day.[74] On the other hand, the lighting of the lamp was considered a commandment, applicable to women, which could be "transgressed."[75]

According to the Tosefta, rabbis of the third tannaitic generation (of Usha) engage in debate about the continuation of a meal on Sabbath eve (t. Ber. 5:2):

> *Ma'aśeh* about Rn. Shim'on ben Gamli'el, R. Judah and R. Yose, who were reclining at Akko and the day became holy over them. Rn. Shim'on ben Gamli'el said to R. Yose, Let us interrupt for the Sabbath. He (R. Yose) said to him, Every day you prefer my words over R. Judah's, and now you are preferring R. Judah's words over mine ... He (Rn. Shim'on) said to him, If so, then let us not interrupt, so that they may not thus establish the *halakhah* in Israel. And they said that they did not move away thence until they established the *halakhah* according to R. Yose.

72 Cf. y. Ber. 1:1, 2a. However, the Jerusalem Talmud goes on to problematize this: whereas the priests' eating of *terumah* happens "while it is still daylight and the stars are there" (i.e., shortly after sunset), it is claimed that the Sabbath meal takes place about one or two hours later. This betrays a significant shift over against the earlier tannaitic practice: for the later (amoraic) sages in Palestine, the evening prayer in the synagogue precedes the meal, which is therefore held much later than in tannaitic times, except perhaps for scattered hamlets, where people prefer to go home while it is still daylight (ibid.).

73 Seneca, *Ep.* 95.47.

74 Although the House of Shammai tended to disallow processes to continue on their own on Shabbat (see before n. 66), they accepted the use of Sabbath lamps as established custom.

75 Cf. m. Šabb. 2:6: lack of care with lighting the Sabbath lamps is one of the "transgressions" on account of which women die in childbed.

The difference between R. Judah and R. Yose is succinctly expressed in the parallel to this section in y. Pesaḥ. 10:1, 37b and b. Pesaḥ. 100a, where this *chreia* or *ma'aśeh* is prefaced with the following statement: "One interrupts for the Sabbaths – words of R. Judah. R. Yose says, One does not interrupt." What is established in this short narrative about a legal example is that meals may continue from Friday afternoon into the beginning of the Sabbath, and that one does not have to interrupt the meal for the Sabbath. R. Judah wants to distinguish more clearly between a Friday meal and the Sabbath; thus, he demands that guests who have reclined for a meal when the Sabbath approaches go to the *bet ha-midrash* "with (approaching) darkness" (*'im ḥashekhah*) and, upon their return, bless the "sanctity of the day" (*qedushat ha-yom*) over a glass of wine (t. Ber. 5:3). However, according to t. Ber. 5:4, reflecting the position of R. Yose, one speaks, after the meal, first the benediction over the meal, referring to the Sabbath, and then blesses the "sanctity of the day."[76] In t. Ber. 3:7, R. Eleazar b. Ṣadoq transmits, in his father's name,[77] a brief prayer for Sabbath night as well as a benediction over the cup, "Blessed be the One who has sanctified the Sabbath day." Since the consumption of wine requires a benediction of its own, already the Houses of Hillel and Shammai are said to have debated whether the benediction over the day or over the wine comes first.[78] In sum, the earlier tradition locates the *qiddush*, the blessing over wine and day, at the end of a meal during which the Sabbath had begun or, in rabbinic terminology, "become holy" (*qadash*).[79]

The celebration of a meal starting on Friday but honoring the Sabbath is also attested for the Jewish Diaspora by traditions connected with the elusive term *cena pura*. The evidence is circumstantial, but it can be inferred from the use of the term *cena pura* for *Friday*[80] that it derived from the practice of a Jewish meal, presumably in honor of the Sabbath, on (late) Friday afternoon.[81] According to William Horbury, Latin-speaking Jews used the

76 Cf. also the tannaitic midrash Mekhilta de-Rabbi Ishmael, usually assumed to have been redacted around the time of the Tosefta, which states (Mek. *Baḥodesh [Yitro]* 7): "they sanctify over the wine at its beginning," and then asks, "I have only the holiness of the day – whence the holiness of the night?" The midrash answers with reference to Exod 31:14, "and you shall keep the Sabbath," which is taken to be composed of night *and* day.

77 This R. Eleazar b. Ṣadoq might be a contemporary of the other rabbis or a grandfather of the same name; cf. Stemberger 2011: 86, 93.

78 Cf. m. Ber. 8:1; m. Pesaḥ. 10:2; t. Ber. 5:25 [6:1]; t. Pesaḥ. 10:2–3. The halakhah follows the House of Hillel, according to whom the blessing over the wine comes first.

79 E.g. t. Ber. 5:2, 3.

80 E.g. in Old Latin manuscripts at Judith 8:6; Matt 27:62; Mark 15:42; Luke 23:54 for either παρασκευή, or προσάββατον, or both. Cf. Horbury 2006: 108–111.

81 Cf. Horbury 2006: 104.

term *cena pura* for the Friday evening dinner and then for Sabbath eve.[82] These Jews will have celebrated the onset of the Sabbath; yet, the name was applied to Friday.

For the Western Diaspora at the turn from the first to the second century, Plutarch mentions that Jews invite one another to drink wine on the Sabbath, presumably on Friday night; he interprets this as a form of Dionysian cult.[83] Already earlier, and in Palestine, the Book of Jubilees repeatedly points out that the Israelites are to "eat, drink and keep Sabbath" on the seventh day (*Jub.* 2:31; cf. 2:21; 50:9, 10). As Horbury has argued, this sequence may point to the practice of inaugurating the Sabbath with a festive meal, although I find it difficult to accept Horbury's view that during daylight on the Sabbath hardly anything would have been eaten:[84] Jubilees explicitly forbids fasting on the Sabbath (50:12).

Similarly, the end of the Sabbath, and thus the beginning of the first day of the week, would have elapsed during a meal that was started on late Sabbath afternoon and continued until darkness, without marking out the precise time of the beginning to the new day.[85] The *havdalah*, the ritual culminating in the separation between holy and profane after the end of Shabbat, was clearly part of a meal in the time of the Mishnah, as the introduction to m. Ber. 8:1 explicitly states: "these are the controversies between the House of Shammai and the House of Hillel on the meal (*ba-se'udah*)." The "Houses" in the late Second Temple period are said to have discussed the order of blessings during this ritual (m. Ber. 8:5).

In sum, before it became customary to pray in the synagogue at the onset and close of the Sabbath,[86] Jews spent both the beginning and the end of the Sabbath days at dinner. These meals provided a framework for Sabbath-conforming demeanor from before and until after the Sabbath, and participating in them absolved Jews from the need to determine the precise transition from one day to the other. Importantly, the rituals referring to the distinction

[82] Cf. Horbury 2006: 109–111, discussing examples in Tertullian and Augustus.

[83] Plutarch, *Quaest. conv.* 4.6.2 [*Mor.* 671e–672a] (trans. Clement and Hoffleit, LCL): "I believe that even the feast of the Sabbath is not completely unrelated to Dionysus. Many even now call the Bacchant *Sabi* and utter that cry when celebrating the god ... The Jews themselves testify to a connection with Dionysus when they keep the Sabbath by inviting each other to drink and to enjoy wine; when more important business interferes with this custom, they regularly take at least a sip of neat wine."

[84] Horbury 2006: 117–118. Some Jews seem to have fasted on the Sabbath, but this appears to me a minority position (Doering 1999: 105–107).

[85] Cf. Elbogen 1911: XI–XV.

[86] As well as to attend sermons or lectures in the synagogue; cf. the anecdote of the woman eager to attend R. Meir's sermons on Friday nights: y. Soṭah 1:4, 16d; Lev. Rab. 9:9 (1: 191–193 Margoulies).

between sacred and profane time (*qiddush, havdalah*) were integrated into the grace after meals.

7 Summary

This chapter has focused on rituals and observations surrounding the transition from one day to the next, particularly from weekday to Sabbath or festival. It has emerged that "evening" is an ambiguous time and could variously refer to parts of the afternoon and the earliest part of the night. Also, the delimitation of "twilight," and thus the precise transition to Sabbath and festival, was difficult to establish and remained debated for a long time. One solution to these difficulties was to add notional time from the previous day, thereby creating a "buffer" that would prevent transgression of behavior specifically demanded for the day, such as the Sabbath, Yom Kippur, and other holidays. Related exegetical approaches in the Damascus Document and in rabbinic texts about the *tosephet mela'khah* take the phrase "Guard the Sabbath day" to mean such forward extension. However, there is evidence in rabbinic literature that many would have conducted "labor" until shortly before nightfall and therefore would not have accepted, or would have limited to a minimum, any addition of rest before the Sabbath proper. A useful framework was the celebration of a meal during which the new day began, which was customary in the early tannaitic period. Crucially, the meal would have been prepared and members of the household would have gathered some time before the day came to an end, thereby already allowing some control over appropriate festal behavior during the uncertain period of transition from weekday to holiday. The earliest tannaitic references to *qiddush* and *havdalah* connect the blessing over the day closely with a meal at home, and there is also evidence of meals on Friday afternoons and evenings in the Jewish Diaspora. Meals, therefore, provided an opportunity for rendering the legally relevant ambiguities of time practically irrelevant by a controlled domestic celebration.

Bibliography

Albeck, C. 1930. "Das Buch der Jubiläen und die Halacha." *Bericht der Hochschule für die Wissenschaft des Judentums in Berlin* 47: 1–60.

Ben-Dov, J. 2011. "The Qumran Dial: Artifact, Text, and Context." In J. Frey, C. Claußen, and N. Kessler (eds.), *Qumran und die Archäologie: Texte und Kontexte*. Wissenschaftliche Untersuchungen zum Neuen Testament 278. Tübingen, 211–237.

Ben-Dov, J. and S. Saulnier. 2008. "Qumran Calendars: A Survey of Scholarship 1980–2007." *Currents in Biblical Research* 7: 124–168.

Birenboim, H. 1998. "The Laws of the Well-Being Sacrifice in the *Miqṣat Ma'aśeh* [*sic*] *Ha-Torah* Scroll." *Tarbiz* 67: 241–244 (in Hebrew).

Colautti, F. M. 2002. *Passover in the Works of Josephus*. Journal for the Study of Judaism in the Persian, Hellenistic, and Roman Periods, Supplement Series 75. Leiden: Brill, 2002.

Danby, H. (trans.). 1958. *The Mishnah*. Reprint. Oxford: Oxford University Press.

Doering, L. 1999. *Schabbat: Sabbathalacha und -praxis im antiken Judentum und Urchristentum*. Texts and Studies in Ancient Judaism 78. Tübingen: Mohr Siebeck.

Elbogen, J. 1911. "Eingang und Ausgang des Sabbats nach talmudischen Quellen." In M. Brann and idem (eds.), *Festschrift zu I. Lewy's siebzigstem Geburtstag*. 2 vols. Breslau: Marcus, 176–187 [German part] (reprint Jerusalem, 1971/72).

Epstein, J. N. 1957. *Introduction to Tannaitic Literature: Mishna, Tosephta and Yerushalmi*. Ed. E. Z. Melamed. Jerusalem: Magnes Press; Tel Aviv: Dvir (in Hebrew).

Epstein, J. N. 2000. *Introduction to the Mishnaic Text*. 2 vols. 3rd ed. Jerusalem: Magnes Press; Tel Aviv: Dvir (in Hebrew).

Gilat, Y. D. 1992. *Studies in the Development of the Halakha*. Ramat Gan: Bar Ilan University (in Hebrew).

Halpern-Amaru, B. 2007. "The Use of the Bible in Jubilees 49: The Time and Date of the Pesaḥ Celebration." *Meghillot* 5–6: *81–*100.

Harlé, P. and D. Pralon. 1988. *Le Lévitique*. La Bible d'Alexandrie 3. Paris: Cerf.

Horbury, W. 2006. "*Cena pura* and Lord's Supper." In idem, *Herodian Judaism and New Testament Study*. Wissenschaftliche Untersuchungen zum Neuen Testament 193. Tübingen: Mohr Siebeck, 104–140.

Jeremias, J. 1932. *Die Passahfeier der Samaritaner und ihre Bedeutung für das Verständnis der alttestamentlichen Passahüberlieferung*. Beihefte zur Zeitschrift für die neutestamentliche Wissenschaft 59. Giessen: Töpelmann.

Kister, M. 1999. "Studies in 4QMiqṣat Ma'aśe Ha-Torah and Related Texts: Law, Theology, Language and Calendar." *Tarbiz* 68: 317–372 (in Hebrew).

Klinghardt, M. 1996. *Gemeinschaftsmahl und Mahlgemeinschaft: Soziologie und Liturgie frühchristlicher Mahlfeiern*. Texte und Arbeiten zum neutestamentlichen Zeitalter 13. Tübingen: Francke.

Kugel, J. 2012. *A Walk through* Jubilees*: Studies in the* Book of Jubilees *and the World of Its Creation*. Journal for the Study of Judaism in the Persian, Hellenistic, and Roman Periods, Supplement Series 156. Leiden: Brill.

Mazar, M. 1978. "Jerusalem in the Time of the House of Herod in Light of the Excavations South and South-West of the Temple Mount." *Cathedra* 8: 29–41 (in Hebrew).

Milgrom, J. 1991–2010. *Leviticus*. Anchor Bible 3. 3 vols. New York: Doubleday; New Haven, CT: Yale University Press.

Niehr, H. 1989. "עֶרֶב *'æræḇ*. " *Theologisches Wörterbuch zum Alten Testament* 6: 359–366.

Ratzon, E. 2015. "The Gates Cosmology of the Astronomical Book of Enoch." *Dead Sea Discoveries* 22: 93–111.

Safrai, S. and Z. Safrai. 2008. *Tractate Shabbat*. Mishnat Eretz Israel, Moed A–B. Jerusalem: Lipshitz College Publishing House (in Hebrew).

Safrai, S. and Z. Safrai. 2009. *Tractate Psachim*. Mishnat Eretz Israel, Moed D. Jerusalem: Lipshitz College Publishing House (in Hebrew).

Saulnier, S. 2012. *Calendrical Variations in Second Temple Judaism: New Perspectives on the "Date of the Last Supper" Debate*. Journal for the Study of Judaism in the Persian, Hellenistic, and Roman Periods, Supplement Series 159. Leiden: Brill.

Schiffman, L. H. 1975. *The Halakhah at Qumran*. Studies in Judaism in Late Antiquity 16. Leiden: Brill.

Shemesh, A. 1996. "Why this Passover?" *AJS Review* 21, 2:1–17 (in Hebrew).

Stemberger, G. 2011. *Einleitung in Talmud und Midrasch*. 9th ed. Munich: Beck.

Stern, S. 2001. *Calendar and Community: A History of the Jewish Calendar, Second Century BCE–Tenth Century CE*. Oxford: Oxford University Press.

Tabory, J. 2000. *Jewish Festivals in the Time of the Mishnah and Talmud*. 3rd rev. and enlarged ed. Jerusalem: Magnes Press (in Hebrew).

Talmon, S. 1958. "The Calendar Reckoning of the Sect from the Judaean Desert." In C. Rabin and Y. Yadin (eds.), *Aspects of the Dead Sea Scrolls*. Scripta Hierosolymitana 4. Jerusalem: Magnes Press, 162–199.

Talmon, S. 1960. "Concerning the Calendar of the Judaean Desert Sect." *Tarbiz* 29: 394–395 (in Hebrew).

Talmon, S. 2003. "Sabbath Observance according to the Damascus Fragments: Evening to Evening or Morning to Morning?" *Meghillot* 1: 71–93 (in Hebrew).

Talmon, S. 2004. "Reckoning the Sabbath in the First and Early Second Temple Period – From the Evening of the Morning?" In G. J. Blidstein (ed.), *Sabbath: Idea, History, Reality*. Beer Sheva: Ben Gurion University of the Negev Press, 9–32.

Ta-Shma, I. M. 1983. "The 'Addition' to the Sabbath." *Tarbiz* 52: 309–324 (in Hebrew).

VanderKam, J. C. 1989. *The Book of Jubilees*. 2 vols. Corpus Scriptorum Chistianorum Orientalium 510, 511. Leuven: Peeters.

Werman, C. 2011. "The Festivals of the Year." In eadem and A. Shemesh, *Revealing the Hidden: Exegesis and Halakha in the Qumran Scrolls*. Jerusalem: Bialik Institute, 274–421 (in Hebrew).

12

SEASONING THE BIBLE AND BIBLIFYING TIME THROUGH FIXED LITURGICAL READING SYSTEMS (LECTIONARIES)

Daniel Stökl Ben Ezra

Judaism and Christianity have impregnated popular cyclical time with sophisticated narratives. Since late antiquity until our very days, public readings and theatrical or choral performances juxtapose scripture and the seasons of the year and build a cyclically repeated narrative of carefully selected scriptural passages that were not necessarily serially connected in the Bible. Sacred time, sacred space, and sacred narrative have been – in varying ways according to times, places, and groups – woven into a multidimensional mythological re-enaction of religious ideals.[1] The selection of specific biblical texts reflects denominational predilections for certain passages and establishes a kind of "canon within the canon," a part of the Bible that plays a role in the religious life that is stronger than other passages.[2]

Beyond the simple fact of selection, these passages are interconnected, especially those juxtaposed on the same or adjacent liturgical occasions. The foundations of this ecosystem that anchors the canon of the canon in the calendar and vice versa the calendar in the canon of the canon are first and foremost the fixed liturgical reading systems that started to develop in late antiquity.[3] Of course, not only are the narratives very different (in the inter-religious as well as in the intrareligious, interdenominational comparison) but so too are their ways of constructing time and being constructed by time.

The present chapter attempts to investigate some of these aspects in late antique Jewish rites in comparison with some Christian celebrations. Unlike

[1] Bell 2009: 102–108.

[2] For an investigation of the concept of a "canon in the canon" in the Reformation, see Lønning 1972.

[3] On Christian lectionaries, see Amphoux and Bouhot 1996. On Jewish reading systems, see Stemberger 1999. For a convenient freely accessible database of Jewish and Christian lectionaries, see Stökl Ben Ezra 2015 (see further n. 21).

other comparative studies of liturgical readings, this investigation does not so much focus on the possible influence of readings taking place in the *same* time, but rather the choice and organization of similar readings that take place at *different*, albeit complementary seasons. From a methodological standpoint, it is important to emphasize that the readings are not individual readings but part and parcel of larger blocks. The main point of the chapter is the focus on a series of events rather than single events. It is clear that two entities can be related if they share many characteristics.

I focus on two series of Jewish liturgical events: the Shabbatot (Sabbaths) in spring before Purim and Pesach (Passover) and the series of Shabbatot before and after the commemoration day of the destruction of the Temple in the summer (Tish'ah Be-'Av). Both are different in form and content and represent different stages. I argue that the second Jewish event series responds to the challenge of Christian lectionaries with their powerful and coherent narrative that had emerged in the meantime. The second series of Shabbatot uses similar readings as contemporary Syriac Christians used during Lent, but it is purposefully established in a season that is rather neglected in Christianity. It is organized around the commemoration of the destruction of the Temple, a potentially highly difficult historical event that could have undermined Jewish identity and played into Christian hands and converts it from a *skandalon* into a cornerstone and into a response to Christian polemics, employing many readings used by Christians around the commemoration of an equally problematic event: the death of Jesus Christ.

1 Time Control: From the Temple to Lectionaries

Before the destruction of the Temple, Jews were in no dire need of a unified lectionary to organize time. The Temple cult took care of that. Some communities may already have had some sort of *lectio continua* of the Torah and here and there chosen readings from other books – whether from what became *nevi'im* (Prophets) or *ketuvim* (Writings) or still something else. Some communities may also have had eclectic readings proper for certain festivals. Especially for the particularly long meetings on Yom Kippur (where texts were needed for the liturgy), ready-made texts such as the biblical texts or something close to the sources of Mishnah Yoma and the Epistle of Barnabas 7 may well have served communities that did not have the poets to write new texts or teachers to expound twenty-four hours.[4] So the Temple functioned

[4] Stökl Ben Ezra 2011: 169–170.

in a way like a standard clock to which all others could decide to relate positively or negatively. We have no indications, however, that any *lectio continua* (*stricto sensu*) was coordinated between different communities in rhythm, frequency, or length apart from the fact that it mostly took place on the Sabbath.

With the destruction of the Temple, this standard clock stopped. Certainly, some festivals were celebrated in Judea, Galilee, Babylonia, Egypt, Asia Minor, North Africa, and other places. Some such as the weekly Sabbath, Purim, Passover, the week of unleavened bread, Yom Kippur (the Day of Atonement), and Sukkot (Tabernacles) may have been celebrated by (some) Diaspora Jews before the destruction of the Temple.[5] Other festivals may have been newly created or recreated.[6] In any case, without the possibility of referring to the Temple service, as far as it may have been or as "distorted" as it may have been in the eyes of others, there was a completely different need for a new coordination device. We do not have to imagine a device similar to unified time in central Europe as was needed with the invention of railway services. In his fine study, Sacha Stern has clearly shown that different local calendars were in use in different communities.[7] We do not even have to imagine that everybody celebrated the same festivals on the same dates. Approximation in time and a common pool of events suffice to create a network of collective identities.

Rabbinic literature makes a strong claim for providing such an overall structure transposing their interpretation(s) of different biblical narratives into cyclical time while leaving some questions open for differing views. It took quite some time, several centuries, before this claim was more or less accepted by a significant proportion of Jews in Palestine, in Babylonia, and even more so in the Western Diaspora.[8] What was needed were tools and frameworks to implement Rabbinic ideology (or rather ideologies) in the life of the general Jewish population. One was of course studying and learning rabbinic texts in schools, but this could touch only a limited number. Another framework was the synagogue and, within its ritual, the reading of biblical passages, as explained according to rabbinic ideology. Again, diversity in synagogue worship may have been considerable, especially in the first half of the first millennium.[9]

[5] Doering 2013; Stökl Ben Ezra 2013.

[6] See Leonhard 2006 and Rouwhorst 2001 with regard to Pentecost.

[7] Stern 2001: 154.

[8] Hezser 1997; Schwartz 2001.

[9] Levine 2013.

2 The Shabbatot before Pesach

Reading, explaining, and studying texts in meetings was nothing specifically Jewish (or Christian) but of course part and parcel of a symposium. Reading, explaining, and studying the same texts on the same specific calendar date every year, however, was a different qualitative turn. The oldest passages about the regular study of biblical texts in Jewish meetings on the Sabbath are well known.[10] We do not know a lot about liturgical readings in Jewish and Christian services in the first centuries beyond the fact that they took place and that they were not fixed, as is clear from Justin Martyr's description.[11] Clearly, *incipit, desinit*, selection, and length of passages were not prescribed. From Origen's homilies that explain whole books chapter after chapter, we can infer that *lectio continua* followed by an explanation existed in these communities for the majority of the books that later became canonical, including books that one rarely finds in Christian lectionaries today, such as the descriptions of sacrifices in Leviticus.

The earliest source about fixed liturgical readings comes from rabbinic sources, roughly contemporaneous to Tertullian and Origen: m. Meg. 3:4–6:[12]

> (4) If the first day of the month Adar falls on the Sabbath, they read the section [in the Law] "Shekels" (Exod 30:11ff.).[13] If it falls in the middle of the week they read it earlier on the Sabbath that goes before, and on the next Sabbath they break off (מפסיקים) [from the reading of the four portions prescribed for the month of Adar]. On the second [Sabbath of the month they read the section] "Remember what Amalek did" (Deut 25:17ff.); on the third, the section of "The Red Heifer" (Num 19:1ff.); on the fourth, the section "This month shall he unto you . . ." (Exod 12:1ff.)
>
> On the fifth they revert to the set order (חוזרין לכסדרן). At all these times they break off [from the set order in the reading of the Law]: on the first days of the months, at the [Feast of the] Dedication [Hanukkah], at Purim, on days of fasting, and at Ma'amads[14] and on the Day of Atonement.

10 Stemberger 1999.

11 And on the day called Sunday, all who live in cities or in the country gather together to one place, and the memoirs of the apostles or the writings of the prophets are read, as long as time permits; then, when the reader has ceased, the president verbally instructs, and exhorts to the imitation of these good things (Justin, 1. Apology 67:3, ANF).

12 Trans. Danby 1933, my line breaks.

13 Usually, Rabbinic sources give only the *incipit*, not the *desinit*. Therefore, "ff." is used here to indicate an unspecified number of verses following.

14 m. Ta'an. 4:2 explains the "Ma'amad" as a group consisting of Israelites and priests. When the latter went to Jerusalem to perform their service at the Temple liturgy whenever their priestly watch had its turn, the former would assemble for a liturgical meeting in their cities.

> (5) At Passover they read the section "The Set Feasts" (Lev 23:1ff.) in the Law of the Priests; at Pentecost, [the section] "Seven weeks" (Deut 16:9ff.); at the New Year, "In the seventh month in the first day of the month . . ." (Lev 23:23ff.); on the Day of Atonement, "After the death . . ." (Lev 16:1ff.); on the first Festival-day of the Feast [of Tabernacles] they read the section "The Set Feasts" in the Law of the Priests, and, on all the other days of the Feast, about the offerings at the Feast (Num 29:17ff.).
>
> (6) At the [Feast of the] Dedication [they read the action] "The Princes" (Num 7:1ff.); at Purim, "Then came Amalek . . ." (Exod 17:8ff.); on the first days of the months, "And on the first days of your months . . ." (Num 28:11ff.); at the Ma'amads, from the story of Creation (Gen 1:1ff.); on the days of fasting, "The Blessing and the Curses" (Lev 26:3ff.). They make no break in the reading of the curses, but the one [reader] reads them all.
>
> On Mondays and Thursdays and on Sabbaths at the Afternoon Prayer they read according to the set order; and these are not taken into account. For it is written, And Moses declared unto the children of Israel the set feasts of the Lord (Lev 23:44) – the law prescribed for them is that they should read each one in its set time.

There are four important things to learn from this well-known passage: (1) At least in some communities, there was a *lectio continua* of the Torah.[15] The rabbinic movement may have been tiny at this time, and many congregations may have followed completely different rules about how many verses to read each week, and on which days to do so (if they had rules at all). Still, they published a text at the beginning of the third century that made a claim of uniformity. However, even if we presume a large ratio of uniformity, the liturgical reading of the Bible was not yet closely linked to the year, at least not in Palestine, because there, according to most scholars, a triennial Torah reading system was in operation.[16]

This fundamental dissimilarity between Palestinian and Babylonian Judaism with regard to the reading cycles is a crucial point for the consideration of the Rabbinic perception of time. According to most scholars, most

[15] While there is a discussion in the Babylonian Talmud whether "breaking off" refers to the order of Torah readings or to the order of Haftarot readings (readings from the prophets that accompany the Torah readings), modern scholarship accepts, quite unanimously, that this passage refers to an interruption of a *lectio continua* of the Torah. Fishbane 2002: xx–xxi, Naeh 1998: 174, and the literature mentioned there. Most scholars do not even discuss this possibility.

[16] Naeh 1998; Fleischer 1991.

Babylonian Jews read the Torah in an annual cycle in one year,[17] while Palestinian Jews read shorter passages and needed more than three years.[18] This so-called triennial cycle divided the Torah into between 154 and 175 passages.[19] Experiencing the biblical narrative in the triennial cycle must have been fundamentally different from the Jewish Babylonian world – and, as we shall see, from the Christian systems – since in the triennial system the year and the narrative are not related. The yearly festivals were the *only* institutions to anchor the biblical narratives in the yearly cycle. About sixty percent of the non-daily liturgical events would have been Shabbatot of the triennial rite and only about forty percent would have been annual events. Therefore, each yearly event that received fixed readings must be understood as a step toward the Babylonian Jewish and the Christian Mediterranean conceptions of time.

(2) The *lectio continua* of the Torah readings is interrupted by specifically selected Torah passages on fifteen liturgical events. This is the earliest tangible form of a fixed selection of readings in the pre-Constantinian time. The selection was clearly made by people who had a deep nostalgia for sacrifices rather than narrative. Accordingly, at Passover Leviticus 23 is read, not Exodus 12.

A Baraita from the Tosefta found also at the end of tractate Megillah in the Babylonian Talmud expresses this strategy well:[20]

> Our Rabbis taught: Moses laid down a rule for the Israelites that they should enquire and give expositions concerning the subject of the day – the laws of Passover on Passover, the laws of Pentecost on Pentecost, and the laws of Tabernacles on Tabernacles.

(3) The Mishnah deduces the rationale for the correlation of time and reading biblical passages from Lev 23:44, "And Moses declared unto the children of Israel the set feasts of the Lord." In the Masoretic Text, this is an introductory verse with nothing following that seems superfluous in its current context. Since according to Rabbinic hermeneutics, nothing is superfluous in the Bible, the editors of the Mishnah conclude that the verse demands reading the biblical passages of the festival calendar, each at its appointed festival: this is presented as the reason for the existence of this verse.

(4) The exceptions are Hanukkah and the four special Shabbatot before Passover and Purim, which are named according to keywords in the biblical

[17] Cf. e.g. b. Meg. 31b.
[18] b. Meg. 29b.
[19] Naeh 1998.
[20] t. Meg. 3:5.

passages prescribed for these days: *Sheqalim, Zakhor, Parah, Haḥodesh*. It is with regard to the latter that we can get a first glimpse of a narrative creating a new yearly season, during Adar and Nisan with a climax toward Pesach. The narrative does not, however, exactly follow the chronology of the Torah and some events are less closely related than others.

(a) The Torah passage Exodus 30:11ff. of the first Sabbath, *Sheqalim* (ThALES E2524),[21] recalls the one-time ransom in the time of Moses. According to the biblical narrative, Exodus 30 should of course happen *after* Exodus 12. In the Second Temple period, however, this ransom was interpreted as a yearly Temple tax of a half-shekel to be paid at this time of the year in the month of Adar (m. Sheq. 1:1). The reading of this passage therefore re-enacts a Temple ritual rather than the biblical narrative.

(b) For the second Shabbat, *Zakhor*, the reading of Exod 17:8ff. is prescribed, the battle between the Israelites and the Amalekites. This contradicts the biblical chronological order where the battle between the Israelites and the Amalekites would take place *after* Pesach (Exod 17 after Exod 12). This Shabbat is primarily linked to Purim and makes less sense in a series of Shabbatot leading to Pesach.

(c) The third Shabbat, *Parah*, requires the reading of the red cow purification ritual in Numbers 19. The Jerusalem Talmud (y. Meg. 3:5, 74b) justifies this by pointing to the need of purification before entering the Temple for the slaughtering of the Pesach sacrifice but complains that Shabbat *Parah* should have followed Shabbat *Ha-Ḥodesh*, since the Tabernacle was only erected on Nisan 1, and the ashes of the Red Heifer were burnt the following day.

(d) Exod 12:2ff. for Shabbat *Ha-Ḥodesh* (E2572) describes the preparations of the lamb four days before Pesach. This is the most closely related to Pesach. Interestingly, there is no reading yet for the Shabbat preceding Pesach, which will be called Shabbat Ha-Gadol ("the great Sabbath") in the Middle Ages.

Probably a little later than the Mishnah, the Tosefta, in a passage that parallels the one from the Mishnah, adds Haftarot, prophetic readings, for the four special Shabbatot (t. Meg. 3:5). These Haftarot are all relatively closely linked by their content to their corresponding Torah passages.[22] (A statement

[21] The article uses the identification scheme for lectionary data proposed in the lectionary database *Thesaurus Antiquorum Lectionariorum Synagogaeque* (www.lectionary.eu) = Stökl Ben Ezra 2015. References beginning E denote IDs for liturgical events (Shavu'ot Day 1) in a specific lectionary source. R refers to individual readings on one such event. Each lectionary has a unique ID.

[22] The Tosefta adds 2 Kgs 12:8ff. (R6669), linked to the Torah passage Exod 30:11ff. through the common reference to public donations – in the wilderness in the days of Moses and in the time of the first Temple to the Temple. For Shabbat *Zakhor* (E2548), the Tosefta adds

in the Babylonian Talmud [b. Meg. 29b] claims that that was the rule.) These Haftarot reinforce the narrative character of each single Shabbat by linking it to other biblical *narratives* (e.g., instead of further sacrificial details). Apart from the last, they do not, however, give the series an overall plot leading toward Pesach. We can see this even in later sermon collections such as Pesiqta de-Rav Kahana. Exodus 12 or Pesach is not referred to in the first three special Shabbatot homilies. After all, the general tendency of these prescribed lections is more toward a commemoration of the chronology of the Temple ritual than toward the establishment of a biblical narrative in cyclical time. Still, as stated earlier, each yearly event that received fixed readings must be understood as a step toward the Babylonian Jewish and the Christian Mediterranean conceptions of time.

3 Lent

The special Shabbatot commence about three weeks before Purim, which is seven weeks before Pesach. This period closely corresponds to Lent. According to Tertullian, the Montanists fasted a fortnight before Easter from as early as about 200 CE (*Jejun.* 15:2), but our first attestations for a full-fledged forty-day fast come only from the fourth century: the fifth canon of the Council of Nicea in 325 CE and a festal letter of Athanasius from 331 CE. In the Western tradition, the forty days of Lent consist of a little less than seven weeks. In the Eastern tradition, where the Sabbath is excluded from fasting, the pre-Easter Great Fast begins eight weeks before Easter.

Both periods are dedicated to studying. On the rabbinic side, thirty days before Pesach, the commandments of Pesach have to be learned. On the Christian side, Lent is the period of catechesis of the new Christians. Furthermore, as Israel Yuval has underlined,[23] the preparation periods of Jews and Christians have the opposite emotional connotation. While the

1 Sam 15:2ff. (R6671), linked to the Torah passage Deut 25:17–19 through the common reference to the destruction of the Amalekites. Its primary link to this period is via Purim as Haman the Agagite is understood to be an offspring of the Amalekite king Agag (1 Sam 15:8). The readings for this Shabbat still focus on Purim without reference to Pesach. For Shabbat *Parah* (E2596), Ezek 36:25ff. (R6673) is prescribed, a promise of eschatological purification, the return of the exiles and rebuilding of the ruined towns. It is linked to the Torah passage Num 19:1ff. through the common reference to purification sprinkling. For Shabbat *Ha-Ḥodesh* (E2572), the Tosefta prescribes Ezek 45:18ff. (R6675), linked to the Torah passage Exod 12:2–20 through the common reference to the Pesach ritual and its preparations. Four days before Pesach, the lamb has to be prepared.

[23] In an unpublished paper given at the Pontifical Gregorian University on October 23, 2013, "Did Rabbinic Judaism Emerge out of Christianity?" (p. 12 of the manuscript).

Christian Lent is a period of retreat and admonition, the weeks before Pesach are a time of joyful preparation. After Pesach/Easter again, both religions follow a period of seven weeks and again, the emotions are reversed. The Jewish ʻOmer is a time of mourning for a plague that killed 24,000 students of Rabbi Aqiva in the second century (b. Yebam. 62b), while the fifty days until Pentecost are the most joyful period in the Christian calendar.

With regard to the special Shabbat lections, it may be interesting to note that three emphasize atonement, a point usually less related to Passover. Exodus 30 regards the payment as atonement (כסף הכפורים). A Piyyut (liturgical poem) of the Jewish hymnographer Yannai confirms the importance of atonement and ransom with regard to *Sheqalim*. The Red Heifer is a purification ritual, and Ezek 45:20 interprets the Passover sacrifice as atonement for the Temple (וְכִפַּרְתֶּם).This proximity to the theological heart of the Christian narrative may or may not be a coincidence. Knohl and Naeh have proposed a pre-Christian link between the priestly ordination (*millu'im*) in the days before Pesach and the atonement (*kippurim*).[24] Christians were still very few in numbers in the second and the first half of the third centuries, but in some areas they may have been more numerous than Rabbinic Jews.

The Jerusalem Talmud to Megillah does not add much to this picture. The readings given are exceedingly close to the Mishnah. It quotes relatively few traditions from the Tosefta and gives few new indications of new readings of its own, especially when compared with the Babylonian Talmud. Still, it attests to two new alternative Torah readings for the festivals of Shavuʻot (Exod 19, the revelation at Sinai) and New Year (Gen 21, the expulsion of Hagar and Ishmael). The combination of Shavuʻot (Weeks) with the revelation of the Torah, taking place in the third month, is alluded to (but not made explicit) in the biblical narrative. It is explicit in Jubilees.[25] The relation of Gen 21 to New Year is a novelty.[26] Both readings, Exod 19 and Gen 21, emphasize the mythological narrative rather than the sacrificial nostalgia. The Shavuʻot reading creates therefore a continuation of the Pesach narrative.

[24] Knohl and Naeh 1992.

[25] Stökl Ben Ezra 2007: 280.

[26] We may speculate whether it is possible that the selection of Gen 21 might be the result of the impact of the triennial reading rite. If the cycle began after Pesach, Gen 20 for the 17th Sabbath would have frequently been read at the last Shabbat of Elul, just before Rosh Ha-Shanah. The Rosh Ha-Shanah reading also attests to the transition of the Aqeda narrative from Passover to Rosh Ha-Shanah.

In sum: toward the end of the third century, we can clearly perceive the inception of single events with fixed liturgical readings in Palestinian Rabbinic Judaism. Yet, a full-fledged lectionary year is still in its infancy. The readings do not yet serve to connect events into cohesive series. With the Jerusalem Talmud, we get a hint that this is about to change.

4 Systems of Fixed Readings in Early Christianity

Intriguingly, it is about the same time that Church Fathers from different parts of the empire attest to an explosion of liturgical readings fixed on a large series of events. In Northern Italy, Ambrose (340–397) tells in an epistle about the Last Week of Lent in the year 385:[27]

> You have heard, my children, the reading of the book of Job, which, according to the appointed order and season, is being gone through.

A little later, in North Africa, Augustine (354–430) says on Easter:[28]

> So what does it all mean? I am in the habit of speaking to you about it every year. But as this same reading is solemnly read every year, let the same sermon be solemnly preached every year.

The crown witness for the earliest development of Christian eclectic reading cycles, however, is the pilgrim Egeria, who visited Jerusalem in the 380s. She has described the rituals at the places she visited in a kind of diary, especially in the time between the Epiphany and Pentecost. Special services are held on the spot and on dates purportedly those of the event during which hymns, antiphons, and biblical readings take place. And the readings are *semper aptae diei et loco*:[29]

> (5) And on arriving at the Lazarium, so great a multitude assembles that not only the place itself, but also the fields around, are full of people. Hymns and antiphons suitable to the day and to the place are said, and likewise all the lessons are read.
>
> Then, before the dismissal, notice is given of Easter, that is, the priest ascends to a higher place and reads the passage that is written in the Gospel: When Jesus six days before the Passover had come to Bethany, and the rest. So, that passage having been read and notice given of Easter, the dismissal is made.

27 Ambrose, *Ep.* 20, trans. Nicene and Post-Nicene Fathers.
28 *Sermo* 246:3, trans. Rotelle 1990: 103.
29 Egeria 29.5–6, trans. McClure and Feltoe 1919: 64.

(6) This is done on that day because, as it is written in the Gospel, these events took place in Bethany six days before the Passover; there being six days from the Sabbath to the fifth weekday on which, after supper, the Lord was taken by night.

The imperial construction activity in and around Jerusalem contributed to the re-enactment of "the" biblical narrative according to Christian perspective. The ritual is not limited to retelling, singing, praying, and wandering around; it is not limited to ephemeral elements but includes buildings and tangible objects, as is evident in the veneration of relics (e.g., the Cross and the ring of Solomon on Good Friday) and the use of other objects such as palm leaves, to re-enact certain moments symbolically. The effect is not just pedagogical. The ritual becomes a mystagogical transformation transposing the participant to a different time at the same place. It is the threefold matching of formative narrative, "historical" place, and the "same" moment in the cyclical time that permits the transposition to the corresponding moment in linear time. Those that do not understand Greek can listen to translations into Syriac or Latin. Religious tourism may have played a significant role in creating the need for a detailed fixed liturgical year from the birth of Jesus via his presentation in the Temple, his mission, suffering, and death, as well as his resurrection and until his ascension.

While this need was particularly felt in Jerusalem, we have seen that other regions also developed fixed reading systems at about the same time. While returning pilgrims importing the idea of a liturgical year into their home communities may have played a certain role here, additional local and global reasons have to be sought out: many new festivals emerged probably because of the new role of Christianity as the official imperial cult that had to entertain a whole population in religious terms.

In any case, outside of Jerusalem, the creation of the mystagogical experience could only be based on narrative and time. Both can be joined in transportable literature related to time, that is, lectionaries. By transposing what they *can* transpose – a link of narrative and calendar – they are able to transfer the geographical sacredness that is more difficult to transport (though not impossible).

For those who did not want to reconstruct a full-fledged local copy of Jerusalem,[30] the easiest physical and intellectual means of transport were lectionaries. It may not be by chance that the witnesses for the earliest fixed reading cycles describe Jerusalem shortly after Egeria. I refer, of course,

[30] On this phenomenon, see Stroumsa 1999.

to the famous Old Armenian lectionary of which three medieval copies describe the rites for Jerusalem's sacred places for c. 417–429 CE. Masterly edited by Charles Renoux,[31] it has been studied many times. The fact that it is extant in medieval Armenian and was used in Armenia exemplifies the geographical transportability of its conception of linear time cyclically re-enacted by reading the narrative. A coherent narrative is constructed from Epiphany to Pentecost from the birth of Christ and his presentation in the Temple through the forty days of Lent, Lazarus Saturday, Palm Sunday, Holy Week, Good Friday, and Easter, followed by Pentecost with the Ascension. Such a liturgical year leaves more than fifty percent of the year relatively empty, just as Egeria. With one exception, nothing special takes place from roughly June to the end of December. This one exception is the eight-day festival of the Dedication of Holy Sepulcher and the Invention/Exaltation of the Cross in September.[32]

5 Three Shabbatot of Rebuke and Seven of Consolation in Pesiqta de-Rav Kahana

It may not be by chance that the most impressive creation of a liturgical season in the Jewish world fills exactly this period left relatively empty by the Christian narrative. I refer to the ten weeks around Tish'ah Be-'Av, the day commemorating the destruction of the Temple. The day itself may have existed for a long time, but the evidence is extremely slim.[33] (Attempts to find traces already in Philo[34] or in the Epistle to the Hebrews[35] are not convincing.) The very long period around it is unattested in the Mishnah, the Tosefta, and the Jerusalem Talmud and certainly of a late date. It appears for the first time in the late antique homiletic Midrash Pesiqta de-Rav Kahana, maybe from fifth- or sixth-century Palestine, and considerably later again in the other collection of sermons, Pesiqta Rabbati.

This season begins with Tammuz 17, the commemoration day for the breaching of the walls of Jerusalem, followed by three Shabbatot of Admonition/Rebuke until Tish'ah Be-'Av, the commemoration day for the destruction of the Temple. Between Tish'ah Be-'Av and Rosh Ha-Shanah are seven Shabbatot of Consolation. One might even add

[31] Renoux 1969, 1971.
[32] Schwartz 1987; Stökl Ben Ezra 2003.
[33] Alexander 2013.
[34] Cohen 2007.
[35] Gelardini 2007.

immediately subsequent series of events in Tishri, first Rosh Ha-Shanah, then the Fast of Gedaliah, Shabbat *Shuvah*, Yom Kippur, and Sukkot. This is an impressively long period lasting for a whole quarter of a year.

From the perspective of lectionary studies, one of the most interesting points is that the narrative, its cohesion, and its plot are constructed not via the Torah reading, which is commonly regarded as the more central one, but through the Haftara readings, usually secondary in importance. The Torah readings in this period follow the regular *lectio continua* of this season. It is the Haftara readings, picked from the beginning of Jeremiah and Isaiah for the Shabbatot of Rebuke (Jer 1:1–2:4; Jer 2:4–28; [4:1–2]; Isa 1:1–27), and from Second Isaiah for the Shabbatot of Consolation, that have become the main base text for the sermons. Unlike the series of four Shabbatot preceding Pesach, here we have a clear *season* of linked Shabbatot that creates a cohesive narrative, which in its importance for the formation of Jewish identity can only be compared with the sequence of Lent, Easter, and Pentecost in the Christian world.

The first pivot of this season is the commemoration day for the destruction of the Temple, a historical event that deprived Judaism of its geographical center and organ of calendric orientation. The national religious catastrophe is liturgically reinterpreted during the three Shabbatot of Admonition as the punishment for individual and collective misbehavior. Collective and individual ethical misbehavior causes catastrophes of a size that comes close to complete destruction of the entire Jewish people. This rationale is nothing new and can be found in much older sources. The innovation consists in its ritual implementation over an extremely long period that covers several months. The present generation is exhorted to correct conduct to avoid such a national catastrophe. At first, Jerusalem behaved correctly and was therefore divinely protected. Then it forsook God for no reason and consequently lost his protection as read in the opening chapter of Jeremiah:

> (16) And I will utter my judgments against them, for all their wickedness in forsaking me; they have made offerings to other gods, and worshiped the works of their own hands. (17) But you, (Jeremiah) gird up your loins; stand up and tell them everything that I command you. Do not break down before them, or I will break you before them. (18) And I for my part have made you today a fortified city, an iron pillar, and a bronze wall, against the whole land – against the kings of Judah, its princes, its priests, and the people of the land. (Jer 1:16–18)

The listener can choose to identify himself or herself either with the addressees or with the role of Jeremiah, the narrator. In the Haftara on the Sabbath before Tish'ah Be-'Av, the community is the remnant that has survived the catastrophe:

> (7) Your country lies desolate, your cities are burned with fire; in your very presence aliens devour your land; it is desolate, as overthrown by foreigners. (8) And daughter Zion is left like a booth in a vineyard, like a shelter in a cucumber field, like a besieged city. (9) If the LORD of hosts had not left us a few survivors, we would have been like Sodom, and become like Gomorrah. (Isa 1:7–9)

The very sense of liturgy is called into question, especially also the Temple cult with its incense and offerings:

> (12) When you come to appear before me, who asked this from your hand? Trample my courts no more; (13) bringing offerings is futile; incense is an abomination to me. New moon and sabbath and calling of convocation – I cannot endure solemn assemblies with iniquity. (14) Your new moons and your appointed festivals my soul hates; they have become a burden to me, I am weary of bearing them. (Isa 1:12–14)

Similar to the Christian pilgrims to Jerusalem, the ritual actions help create a mystagogical experience. On the day of Tish'ah Be-'Av, the fasting and mourning members of the community recite Lamentations and can therefore identify themselves with the Jerusalemites starving after the siege and devastated from the capture of the city and the loss of their beloved.[36] As a consequence of the preceding Shabbatot, the participants in the service are also likely to see their own misbehavior as responsible for national catastrophes such as this one. The destruction of the Temple on Tish'ah Be-'Av has been chosen as the primary identity marker that links the individual participant via the current community to the historical nation, the cyclical time to the linear.

The Haftarot for the ensuing seven Shabbatot of Consolation after Tish'ah Be-'Av come from Deutero-Isaiah mostly in the order of the biblical chapters. In Pesiqta de-Rav Kahana, the following Haftarot are indicated for the Shabbatot of Consolation:

1 Isa 40:1ff. (= Pesiq. Rab.);
2 Isa 49:14ff. (= Pesiq. Rab.);
3 Isa 54:11ff. (=Pesiq. Rab.);

[36] Stern 2004: 171–174.

4 Isa 41:12ff. (but Pesiq. Rab.: Isa 51:12ff.);
5 Isa 54:10ff. (but Pesiq. Rab.: Zech 9:9ff.);
6 Isa 60:1ff. (but Pesiq. Rab.: Zech 2:14ff.);
7 Isa 61:10ff. (= Pesiq. Rab.).

These passages convey the message that God is merciful to those repenting. God is said to love Israel and redemption is promised in a world in which the national center, the Temple, is still in ruins.

The choice to create this season in this way was undertaken in fifth-/sixth-century Galilee, a world ruled by Christians for whom the ruined Temple is one of the best arguments in their rationale of the *status quo*. Most Church Fathers would have agreed that the destruction of the Temple was a divine punishment for terrible misbehavior. They would also have agreed that repentance would eventually bring about redemption. They would disagree, of course, about two central features: first, what constituted the most central sin – the rejection of Jesus the Messiah; and second, what would express true repentance – the acceptance of Jesus as a Messiah.

The Haftarot counter these Christian claims. According to Isaiah, the sins have been forgiven. Israel is reinstalled as beloved. Its enemies will be defeated. While the Temple, so central in the commemoration day for its destruction, does not play a role in the Haftarot, it is important in the Piyyutim.[37] It is also present in the Midrashic homilies.[38] Still, it is the conduct of the individual and the congregation and the possibility to repent that are more important for asserting God's benevolence.

In this way, the series of ten Shabbatot can also be understood as a long foreplay to the cycle of the festivals of Tishri. The readings artfully allude to the end of the long cycle at Yom Kippur, more than two months later, when repentance will bring atonement and redemption. On the last Shabbat before Yom Kippur, for example, Isaiah 1 is read:

> (18) Come now, let us argue it out, says the LORD: though your sins are like scarlet, they shall be like snow; though they are red like crimson, they shall become like wool . . . (27) Zion shall be redeemed by justice, and those in her who repent, by righteousness. (Isa 1:18, 27)

This reading is attested also in the Babylonian Talmud as Haftarah for Tish'ah Be-'Av and also for the New Moon of 'Av in case it coincides with Shabbat. The red crimson becoming white is closely related to legends that such a red crimson was attached to the Temple on Yom Kippur and turned

[37] Stern 2004: 136, 153–154.
[38] Barth 1982, contra Stern 2004: 154.

white as soon as the scapegoat had been killed, indicating miraculously the forgiveness of sins.

Those who listen well in Tammuz, 'Av, and 'Elul can interiorize the message how sins lead to destruction and how repentance leads to forgiveness and redemption. The past is a lesson for the future, the Day of Judgment that begins on Rosh Ha-Shanah on Tishri 1. Both construct the conception of cyclical time and Jewish identity in past, future, and, most significantly, present. It is the Jewish Lent of Tishri.

6 The Burkitt Lectionary

Intriguingly, Christians also read many of the same passages in their services. The opening chapters of Jeremiah play a central role during Lent and in the teaching of the catechumens. Some of the Isaianic passages of consolation are read around Easter (UM1, Marono1) in the East, while in the West the birth of John the Baptist, Christmas, and Epiphany have attracted these readings (LR1, LG2). Ludwig Venetianer tried to link these Haftarot to Latin lectionaries.[39] Eric Werner (1959) corrected him in referring to the Armenian lectionary of Jerusalem and other Eastern sources. Most of the parallels are, however, quite general: Jews and Christians share the emphasis of reading Deutero-Isaiah and the opening chapters of Jeremiah and Isaiah. Stéphane Verhelst (1997) added further observations that the readings of the Shabbatot of Consolation with the Armenian and the Georgian lectionaries from Jerusalem consist of Isa 40:1ff. and Isa 60:1ff. These, however, are the only parallels with these important Jerusalem lectionaries, and his hypothesis of an early common Jewish-Christian substrate seems to me unnecessary.

There is one comparison that has not so far been undertaken that shows a quite impressive case of proximity. This is the famous source commonly known as "Burkitt lectionary."[40] Written in sixth-century Estrangelo, this list comes from a Syriac-speaking area with connections to Jerusalem. Its provenance is therefore intriguingly close to Pesiqta de-Rav Kahana in time and place. Its colophon indicates the purpose of the work:

> Here endeth the Table of the Lessons of Holy Scripture, by which anyone can know the Lessons read on all the Feasts, and on the commemorations of the martyrs and on the commemoration of our Fathers, the Bishops who have departed in the true faith, of the Holy Apostles, and the Lessons read on the Departure or on the commemoration of the departed.

[39] Venetianer 1909: 149–159.
[40] Cf. Burkitt 1921–1923.

Of the ten Haftarot readings from the Shabbatot of Rebuke and from the Shabbatot of Consolation, eight figure among the readings indicated in the Burkitt lectionary for the Lent and Easter season. The exceptions are Isa 1 and Isa 51:12ff. The following table illustrates the Christian lectionary parallels to the Haftarot readings of the Shabbatot of Rebuke and Consolation in the two primary witnesses (Pesiqta de-Rav Kahana and Pesiqta Rabbati). Gray indicates a missing or less close correspondence. Influence might have gone in either direction but I regard an influence from the Syriac to the Jewish material as more probable.

7 Conclusions

Toward the fifth/sixth centuries, Judaism and Christianity each had imbued the calendrical year with competing narratives that developed about the same time. The first attestations of a larger reading cycle in Judaism with the special Shabbatot before Pesach and Purim in the Mishnah and in the Tosefta puts the nostalgia of the sacrificial Temple cult very much at its center, much more than the mythological narratives from Genesis and/or Exodus. In the earliest sources, the single events are not well connected. Christians constructed a full-fledged liturgical year in the fourth century with the imperial construction boom in Jerusalem and the necessity and possibility to entertain a growing number of pilgrims in Jerusalem and analogously in other cities of the empire. Readings suited to time and place created an environment that enabled the individual to experience a re-enactment of Jesus' birth and Passion that filled half of the cyclical time. Lectionaries enable the export of this re-enactment to areas far away from Palestine.

In fifth-century Palestinian Judaism, a series of three plus seven Shabbatot before and after Tish'ah Be-'Av linked through a series of Haftarot creates a pre-sentiment for the season in Tishri. The theological interpretation of scripture has received a pedagogically well-chosen moment in the curriculum of the Jewish people. It takes place in a liturgical time that was relatively empty on the Christian side. The series of Shabbatot provides the participants with some answers to Christian claims and images of supersession. And it transforms the historical narrative into an exhortation for individual and collective rightful conduct and repentance that ensures divine benevolence and eventual redemption. Together with the special Shabbatot before Pesach, this series converts the Jewish Palestinian liturgical year originally built around the triennial cycle into an annual year just as for the Babylonian Jews and the Mediterranean Christians because only now, the majority of events are annual rather than triennial, even in Palestine. While the triennial

	Pesiq. Rab Kah.	Pesiq. Rab.	BL 14528 ("Burkitt")	Old Syriac Bible MSS BL 17105 (1) BL 12175 (2) BM 14432 (3)	Armenian Lectionary (Renoux)	Georgian Lectionary (Tarchnisvili)
Rebuke 1	Jer 1:1ff.	=	Monday in mid-Lent	Palm-Sunday (1)	Monday 2nd week of Lent	=
				Sun. after Easter (margins of 1)		
Rebuke 2	Jer 2:4ff.	=	Tuesday in mid-Lent	Friday after Easter (1)		
Rebuke 3	Lam 1:1ff./Isa 1:21–27	Jer 37:1ff.	Good Friday (Jer 37: 12–38:6)		1:16–20 Teach. of Catech.	Litanies
Consolation 1	Isa 40:1ff.	=	Wednesday in mid-Lent	= (2,3)	Friday 1st week of Lent	
			Thursday in week before Easter		=	
			Monday in week after Easter			
Consolation 2	Isa 49:14ff.	=	Ascension	= (2)		Litanies
Consolation 3	Isa 54:10ff.	=	Easter	= (2)		
Consolation 4	Isa 51:12–16	=	Commemoration of Bishops	Holy Saturday (2)		Litanies
Consolation 5	Isa 54:1ff.	Zech 9:9ff.	Easter	= (2)	Friday 4th week of Lent	
				Dedication of a Church (3)		
Consolation 6	Isa 60:1ff.	Zech 2:14ff.	Holy Saturday	= (2,3)	=	Holy Saturday
			Easter			Epiphany
			Tuesday after Easter			
Consolation 7	Isa 61:10ff.	=	Easter		Thursday after Easter	Thursday bef. E.

cycle continued to be practiced for centuries, it has become hybrid: the Haftarot system is already quite annual.

Finally, two crucial points may not yet have been emphasized enough. First, not only do the biblical texts create a perception of meaningful time, but the liturgical cycle in which these texts – which may originally have been quite unrelated to each other – have been inserted as bricks turns the unrelated or loosely connected texts into a plot, into continuous scenes of an overarching narrative. A moment in the year becomes recognizable, definable by reference to a part of this narrative. Time, or, more precisely, the year has become biblicized.

Furthermore, the relation of time and biblical texts goes both ways. Some of the biblical stories were already intrinsically linked to calendar dates by their content: the Exodus takes place in the first month and the revelation of the Torah in the third. Similarly, the crucifixion and Easter are clearly located on the calendar. There are, however, other narratives that had to be linked to a certain point in cyclical time through interpretation and here the association becomes completely dependent on the culture's "lectionary." On the Christian side, this concerns the events of Jesus' birth and youth, Christmas, Epiphany, the *hypapante*. On the Jewish side, this concerns Exodus 30 on Shabbat *Sheqalim*, Numbers 19 on Shabbat *Parah*, and even Lamentations on Tish'ah Be-'Av. The Bible has become seasoned.

Bibliography

Alexander, P. S. 2013. "Was the Ninth of Av Observed in the Second Temple Period?" In R. Boustan et al. (eds.), *Envisioning Judaism. Studies in Honor of Peter Schäfer on the Occasion of His Seventieth Birthday*. 2 vols. Tübingen: Mohr Siebeck, 1:23–38.

Amphoux, C. B. and J. P. Bouhot (eds.). 1996. *La lecture des Épitres catholiques dans l'Église ancienne*. Histoire du texte biblique 1. Lausanne: Éditions du Zebre.

Barth, L. 1982. "'The Three of Rebuke and Seven of Consolation': Sermons in Pesikta de-Rav Kahana." *Journal of Jewish Studies* 33: 503–515.

Bell, C. 2009. *Ritual: Perspectives and Dimensions*. Oxford: Oxford University Press.

Burkitt, F. 1921–1923. "The Early Syriac Lectionary System." *Proceedings of the British Academy* 11: 301–339.

Cohen, N. G. 2007. *Philo's Scriptures: Citations from the Prophets and Writings: Evidence for a Haftarah Cycle in Second Temple Judaism*. Journal for the Study of Judaism in the Persian, Hellenistic, and Roman Periods, Supplement Series 123. Leiden: Brill.

Danby, H. 1933. *The Mishnah: Translated from the Hebrew with Introduction and Brief Explanatory Notes*. Oxford: Oxford University Press.

Doering, L. 2013. "Sabbath." In R. Bagnall et al. (eds.), *Encyclopedia of Ancient History*. Chichester: Wiley-Blackwell, 5988–5990.

Fishbane, M. 2002. *Haftarot: The Traditional Hebrew Text with the New JPS Translation.* Philadelphia: Jewish Publication Society.

Fleischer, E. 1991. "Inquiries Concerning the Triennial Reading of the Torah in Ancient Eretz-Israel." *Hebrew Union College Annual* 62:43–61 (in Hebrew).

Gelardini, G. 2007. *Verhärtet eure Herzen nicht: Der Hebräer, eine Synagogenhomilie zu Tischa be-Aw.* Biblical Interpretation Series 83. Leiden: Brill.

Hezser, C. 1997. *The Social Structure of the Rabbinic Movement in Roman Palestine.* Texts and Studies in Ancient Judaism 66. Tübingen: Mohr Siebeck.

Knohl, I. and S. Naeh. 1992. "Milluim Ve-Kippurim." *Tarbiz* 62: 17–44 (in Hebrew).

Leonhard, C. 2006. *The Jewish Pesach and the Origins of the Christian Easter: Open Questions in Current Research.* Studia Judaica 35. Berlin: De Gruyter.

Levine, L. 2013. *Visual Judaism in Late Antiquity: Historical Contexts of Jewish Art.* New Haven: Yale University Press.

Lønning, I. 1972. *Kanon im Kanon: Zum dogmatischen Grundlagenproblem des neutestamentlichen Kanons.* Forschungen zur Geschichte und Lehre des Protestantismus 10. Reihe, Vol. 43. Oslo: Universitets Forlaget – München: Chr. Kaiser Verlag.

McClure, M. L. and C. Feltoe. 1919. *The Pilgrimage of Egeria.* Translations of Christian Literature Series III. Liturgical Texts. London: Society for Promoting Christian Knowledge.

Naeh, S. 1998. "The Torah Reading Cycle in Early Palestine: A Re-examination." *Tarbiz* 67: 167–187 (in Hebrew).

Renoux, A. (= C.) 1969. "Le codex arménien Jérusalem 121. 1. Introduction. Aux origines de la liturgie hiérosolymitaine. Lumières nouvelles." *Patrologia Orientalis* 35: 1–215.

Renoux, A. (= C.) 1971. "Le codex arménien Jérusalem 121. 2. Édition comparée du texte et de deux autres manuscripts, introduction, textes, traduction et notes." *Patrologia Orientalis* 36: 141–390.

Rotelle, J. 1990. *The Works of Saint Augustine: 3. Sermons (230 – 272B) on the Liturgical Seasons.* Hide Park, NY: New City Press.

Rouwhorst, G. 2001. "The Origins and Evolution of Early Christian Pentecost." In M. F. Wiles and E. F. Yarnold (eds.), *Studia Patristica XXXV: Papers presented at the Thirteenth International Conference on Patristic Studies held in Oxford 1999. Ascetica, Gnostica, Liturgica, Orientalia.* Leuven: Peeters, 309–322.

Schwartz, J. 1987. "The Encaenia of the Church of the Holy Sepulchre, the Temple of Solomon and the Jews." *Theologische Zeitschrift* 43: 265–281.

Schwartz, S. 2001. *Imperialism and Jewish Society, 200 BCE to 640 CE.* Princeton: Princeton University Press.

Stemberger, G. 1999. "Schriftlesung II. Judentum." *Theologische Realenzyklopädie* 30: 558–563.

Stern, E. 2004. *From Rebuke to Consolation: Bible, Exegesis and Ritual in the Literature of the Tisha b'Av Season.* Brown Judaic Studies 338. Providence, RI: Brown University.

Stern, S. 2001. *Calendar and Community: A History of the Jewish Calendar, 2nd Century BCE – 10th Century CE.* Oxford: Oxford University Press.

Stökl Ben Ezra, D. 2003. *The Impact of Yom Kippur on Ancient Christianity.* Wissenschaftliche Untersuchungen zum Neuen Testament 163. Tübingen: Mohr Siebeck.

Stökl Ben Ezra, D. 2007. "Parody and Polemics on Pentecost: Talmud Yerushalmi Pesahim on Acts 2?" In A. Gerhards and C. Leonhard (eds.), *Jewish and Christian Liturgy and Worship: New Insights into Its History and Interaction*. Jewish and Christian Perspectives 15. Leiden: Brill, 279–293.

Stökl Ben Ezra, D. 2011. "Fasting with Jews, Thinking with Scapegoats: Some Remarks on Yom Kippur in Early Judaism and Christianity, in Particular 4Q541, Barnabas 7, Matthew 27 and Acts 27." In T. Hieke and T. Nicklas (eds.), *The Day of Atonement: Its Interpretations in Early Jewish and Christian Traditions*. Themes in Biblical Narrative 18. Leiden: Brill, 165–188.

Stökl Ben Ezra, D. 2013. "Festivals (Jewish)." In R. Bagnall et al. (eds.), *Encyclopedia of Ancient History*. Chichester: Wiley-Blackwell, 2663–2666.

Stökl Ben Ezra, D. (ed.). 2015. *Thesaurus Antiquorum Lectionariorum Ecclesiae Synagogaeque*. www.lectionary.eu.

Stroumsa, G. 1999. "Mystical Jerusalems." In L. I. Levine (ed.), *Jerusalem – Its Sanctity and Centrality to Judaism, Christianity, and Islam*. New York: Continuum, 349–370.

Venetianer, L. 1909. "Ursprung und Bedeutung der Propheten-Lektionen." *Zeitschrift der Deutschen Morgenländischen Gesellschaft* 63: 103–170.

Verhelst, S. 1997. "Pesiqta de-Rav Kahana, chapitre 1, et la liturgie chrétienne." *Studii Biblici Franciscani Liber Annuus* 47: 129–138.

Werner, E. 1959. *The Sacred Bridge: The Interdependence of Liturgy and Music in Synagogue and Church during the First Millennium*. New York: Columbia University Press.

13

THE ROMAN EMBER DAYS OF SEPTEMBER AND THE JEWISH NEW YEAR

Robert Hayward

The traditional liturgical calendar of the Roman Church requires that each year four times (*quattuor tempora*) be set aside for special observance and devotion. These times, in English commonly called Ember Days, are made up of the Wednesday, Friday, and Saturday in single weeks that fall in the four seasons of the calendar year.[1] In the traditional Roman rite as celebrated today, these times are fixed: they occur in the first week of Lent (and thus in February–March); in the week following Pentecost (May–June); in September after Holy Cross Day, which occurs on the fourteenth of that month; and in December in the week following the third Sunday of Advent. These days are formally designated as days of fasting and penitence, and the liturgical arrangements for them certainly display some of the characteristic features associated with fast days.[2] Many of the texts used liturgically on these days, however, are by no means restricted to themes of penance and sorrow for sin; rather – and this is somewhat marked in the case of the Ember Days of September – an element of joyous celebration is often apparent.

Academic studies of the Roman liturgy agree, more or less unanimously, on several matters about these days. First, the observance of Ember Days was originally confined to the local Roman Church and was not adopted outside the city until the seventh century CE at the

[1] For modern scholarly discussion of the Ember Days, see especially Morin 1897: 337–346; Fischer 1914; Leclercq 1907–1953: 2014–2017; Daniélou 1956: 114–136; Willis 1964: 49–97; Talley 1990: 465–472; Verstrepen 1993: 339–365.

[2] Thus (with the exception of the Ember Days following Pentecost) the liturgical color of these days is violet, associated with mourning and sorrow; the *Gloria in excelsis* is not sung at Mass, nor the hymn *Te Deum laudamus* at Matins; the acclamation *Alleluia* before the reading of the Gospel is omitted, and a Tract substituted for it.

earliest.[3] Second, only three of these four times were observed from the beginning. There is clear evidence that the spring Ember Days were a later addition and that the original three sets of Ember days were related not only to fasting and penance but also to thanksgiving for harvests of the corn, oil, and wine.[4] This aspect of the Ember Days is still strongly represented in their liturgies and will concern us presently. Third, most of the Masses for these days betray signs of great antiquity, their most prominent pre-Constantinian feature being the provision of one, two, or more readings from the Old Testament before the Epistle and Gospel.[5] Finally, there is general agreement that these days were at first not fixed to any particular calendar date but were announced in advance by church authority. We are fortunate in still possessing ancient forms for the proclamation of these days.[6]

In a series of recent publications, Daniel Stökl Ben Ezra has argued that the Ember Days of September exhibit a curious relationship to the Jewish High Holy Days, which might often have coincided with them in time.[7] Especially arresting are Stökl Ben Ezra's observations on the liturgy of Ember Saturday in September, in which he detects several points of contact with the liturgy and customs of Yom Kippur. Indeed, he has argued that these points of contact provide evidence for a distinct but complex relationship between Jews and Christians in Rome, particularly in the time of Pope St. Leo the Great (reigned 440–461). This pope's remarks about the Jewish observances on Yom Kippur and the coincidental Christian fasting and penance on the Ember Days reveal a remarkably irenic attitude toward the Jews, and even a sympathy for them as they engage in their ancestral customs. Stökl Ben Ezra

[3] The Ember Days are unknown to the Eastern churches; and their reception in European dioceses outside Rome was gradual and sporadic. For the somewhat erratic and unsystematic adoption of the Ember Days in places outside Rome, see Willis 1964: 51.

[4] A convenient survey of the evidence, summarizing and analyzing earlier discussion, may be found in Willis 1964: 53–59.

[5] Justin Martyr's account of the Eucharist (*1 Apol.* 66–67), dating from the mid-second century CE, suggests that readings from the prophetic books and other texts were read as long as time permitted. The once widespread notion that the Ember Days owed their origin to the pagan observances at Rome of *feriae sementivae*, *feriae messis*, and *feriae vindemiales* (see e.g. Morin 1897; Chavasse 1965: 758–767; and, with careful qualification, Willis 1964: 53, 56–57 has been convincingly overturned by Talley, "The Origin of the Ember Days." The precise date of their origin is difficult to determine. Scholarly opinions range from the Apostolic period to the late fourth century CE, with the weight of the evidence (although not decisive) tending to support an earlier rather than a later date within this time frame: for detailed discussion, see Stökl Ben Ezra 2003a: 305–308.

[6] These, and a number of prayers for the Masses, are preserved in the Gelasian Sacramentary (MS Vat Reg Lat 316): see Mohlberg 1960: 101–104.

[7] See Stökl Ben Ezra 2001: 53–57; 2003a: 303–322; 2003b: 259–282.

notes that "the concepts and rites" of the Saturday Ember day of September in particular "parallel the concepts and rites of Yom Kippur," and that the Christian and Jewish observances have similar goals in terms of the piety they represent. The *obligatory* character of the rites is strongly emphasized by both religious groups.[8] Particularly impressive is the information he brings forward to illustrate the explicitly communal aspects of the observances undertaken by Jews and Christians at this time.[9] The evidence he adduces is convincing; this chapter attempts to broaden and corroborate Stökl Ben Ezra's arguments by considering the first in order of the Jewish Holy Days, Rosh Ha-Shanah, and its likely influence on the Ember Days of September. We shall discover that the Roman liturgy for the Wednesday Mass of the Ember week in September is replete with ideas, themes, and expressions recalling the concerns of Rosh Ha-Shanah. Psalms 81 and 33 play a prominent role in both the Jewish understanding of Rosh Ha-Shanah and in the Mass of Ember Wednesday. The Mass also evinces a keen interest in grapes and the vintage, emphasizing the joyful elements prominent in the prayers, chants, and readings for this Mass.[10]

We begin by rehearsing the results of Peter Lampe's researches into the geographical whereabouts of Jews and Christians in the city of Rome in the first centuries CE.[11] Lampe has assembled a mass of archaeological and literary evidence, whose cumulative effect demonstrates beyond reasonable

[8] See Stökl Ben Ezra 2003a: 312–313, with many examples drawn from Leo's sermons.

[9] See Stökl Ben Ezra 2003a: 313–314.

[10] The Jewish festivals of the month of Tishri cover an extended period of time, from Rosh Ha-Shanah (New Year's Day, 1–2 Tishri), through Yom Kippur (Day of Atonement, 10 Tishri), to Sukkot (Feast of Tabernacles, 15–21 Tishri), culminating in Shemini 'Atzeret (the eighth day following Tabernacles, and closing the cycle of celebration, 22 Tishri). Christians, however, appear to have "telescoped" these days, sometimes perceiving (e.g.) Yom Kippur and Sukkot as aspects of the same celebration. The sounding of the *shofar* both on New Year's Day and on Yom Kippur may have tended to obscure the differentiation of those days in the perception of non-Jews. For discussion of this phenomenon, see Stökl Ben Ezra 2003a: 69. Christian "condensation" of the three major sets of Jewish High Holy Days into the liturgies of the three Ember Days, therefore, is not quite so surprising as it might seem at first blush. It may also be noted that in late Second Temple times, Yom Kippur could be marked by an element of rejoicing and feasting, as well as confession of sin and sorrow and self-affliction for misdeeds: see Baumgarten 1999: 184–191, who notes (189) that the Mishnah speaks of the feast made by the High Priest on Yom Kippur after the Temple Service was concluded (m. Yoma 7:4). Earlier, Philo (*Spec.* 1:194) had explained the language of Lev 16:31, describing Yom Kippur, to demonstrate that this day was the greatest of all the festivals, and it is clear from what he says that the Day of Atonement could be regarded as having a joyous aspect: on this, see further Harlé and Pralon 1988: 155. Even in modern times, an element of rejoicing characterizes the Sephardi ritual for Yom Kippur as well as for Rosh Ha-Shanah: for details, see Idelsohn 1960: 223–248.

[11] See Lampe 2003; further information on this topic may be found in Green 2010.

doubt that Jews and Christians lived in the closest proximity to each other. The region known today as Trastevere was home to both communities, as was the area through which passed the *Via Appia*, a place of Jewish settlement from the third century at the latest, and home to Christians as well. Both groups also began to settle later in increasing numbers along the *Via Nomentana*.[12] Significantly, Lampe also draws attention to the role played in early Roman Christianity by what he calls "a broad stream of tradition from the Synagogue." As one of his principal witnesses to this "stream of tradition," he cites the First Epistle of Clement, listing parallels in these citations with what he describes as "post-canonical Jewish traditions."[13] This matter will require further attention presently.

The Jewish New Year, Rosh Ha-Shanah, is celebrated both in the Land of Israel and in the Diaspora on two consecutive days, the first and second of the month of Tishri.[14] It is marked by the ritual sounding of the *shofar*, a trumpet of ram's horn famed for its distinctive timbre. It is inconceivable that Christians living so close to the Jewish communities in Rome should have been unaware of this solemn ceremony, which necessarily required a fair amount of noise. The earliest strata of rabbinic literature indicate clearly that hearing the sound of the *shofar*, and hearing it properly, is essential if a Jew is to fulfill his obligations in observing the feast. Thus the Mishnah states that one who hears clearly the sound of the *shofar*, even if it is being blown in a pit, cellar, or a large barrel, has fulfilled his statutory obligation, whereas if the sound which he manages to hear is indistinct or unclear to him, then he has not fulfilled his religious duty.[15] The ceremony was popular, was observed in the provinces of the Land of Israel and in the Diaspora, and was regarded as providing the feast with its main characteristic, since in the Bible it is named as *zikron teru'ah* (Lev 23:24; literally, "a memorial of trumpet blast") and *yom teru'ah* (Num 29:1; literally, "a day of trumpet blast"). In both these verses,

[12] For the details, see Lampe 2003: 19–40.

[13] See Lampe 2003: 74–76. He attributes the presence of such traditions in Roman Christianity to "Christians from the sphere of influence of the Synagogue" (76) and has earlier (69–75) stressed the importance of "God-fearers" as channels of such traditions, but see further later.

[14] For the two-day celebration, see m. 'Erub. 3:7–9; y. 'Erub. 3 [21c] where the former Prophets are said to have instituted the observance of the two days; and b.'Erub. 39b; Roš Haš. 30b. Contrast, however, m. Meg. 3:5; t. Meg. 3:6; b. Meg. 31a, which legislate for one day's observance, and see discussion in Herr 1983: 142–143 and the opposing views of Fleischer 1984: 293–295 (both in Hebrew). For a detailed discussion and investigation of the festivals in the rabbinic period, see Tabory 2000 (in Hebrew), and the same author's essay 2006: 556–572 and bibliography there cited.

[15] See m. Roš Haš. 3:7, which stipulates that if a man were to pass behind a synagogue, or if his house were close to a synagogue, and he heard the *shofar*, he has fulfilled his obligation, providing that he had directed his heart to observe the precept.

teru'ah refers to a trumpet blast on the *shofar*. These designations almost inevitably ensured that the feast would be closely associated with Psalm 81, which includes the express command to sound the *shofar* at the new moon, a time further defined as "the day of our solemn feast" (Psalm 81:4). From the time of the Tosefta onward, the rabbis explicitly link this Psalm to the New Year.[16] Its significance for the month of Tishri in particular is underlined by the discussion in b. Roš Haš. 8a–b. In that discourse, Psalm 81:4 is brought forward to prove that divine judgment takes place at New Year in the month of Tishri, since the verse includes the words בכסה ליום חגנו, "on the full moon of the day of our festival": although in all likelihood the first of these Hebrew words originally refers to the full moon, the Sages note that it recalls the verb כסה, "to cover," and that the phrase may therefore signify "in the covered time of the day of our festival." The only festival day that occurs when the moon may be said to be "covered" is Rosh Ha-Shanah; so the Psalm must be referring to the month of Tishri.[17] The Psalm is then employed to demonstrate that both Israel and the nations are judged at this time: the "statute for Israel" (Psalm 81:6a) that the festival represents indicates that Israel, as king of the nations, is judged first; then the "ordinance of the God of Jacob" (Psalm 81:6b) refers to the judgment of the nations. Verses 4–5 of Psalm 81 in particular are also recited as part of the Qiddush ceremony after

[16] At t. Roš Haš. 1:10, Psalm 81 is cited with reference to the festival without further explanation. The citation follows quotation of Psalm 33:15, which is offered as a proof text for the famous statement that, on New Year's Day, all who enter the world pass before the Almighty like troops (see also m. Roš Haš. 1:2). It should be noted that the Mass employs Psalm 33:12, 6 as a Gradual chant: these verses refer to God's selection of Israel as his people, and to his creation of the world. This chant occupies a significant position in the Mass, following the reading from Nehemiah/2 Esd and immediately preceding the Gospel. Both the Mishnah and Tosefta refer here to the judgment of the world, whose inhabitants God inspects like troops passing before him in review. The manuscripts of the Mishnah (MS Budapest Akademie Kaufmann A50 and MS de Rossi 138, Parma Biblioteca Palatina 3173, consulted online at http://jnul.huji.ac.il/dl/talmud/mishna/selectmi.asp [May 4, 2017]) describe those passing before the Almighty as *kbny/w mrwn*, a difficult expression for which the Babylonian Talmud (b. Roš Haš. 18a) offers three explanations: "like a flock of sheep," "like the steps of Beth Maron," and "like troops of the house of David." The Jerusalem Talmud (y. Roš Haš. 1:3, 57b) has two explanations: "like these sheds" and "like this auxiliary troop" (partly corrupt in the MSS). For this phrase, the Tosefta (t. Roš Haš. 1:11) has *[kb]nwmrwn*, "like troops": we have here a loan-word from Latin (*numerus*), possibly transmitted via Greek νούμερον, signifying a body of troops. The military language may be of interest for the Mass, in view of the first Collect's petition that the frailty of the worshippers *subsistat* through the remedies of divine mercy. This verb is often used in a military context of resisting and standing against opponents in battle.

[17] For both the historical meaning of כסה as "full moon" and the rabbinic connection with Rosh Ha-Shanah see Kedar-Kopfstein 1989: 205–209 (208–209).

the Morning Service in synagogue; in the Ashkenazi rite, they are read before the recitation of Qaddish preceding the 'Amidah in the Evening Service.[18]

The Introit at Mass of the Wednesday in the Ember week in September consists of carefully arranged elements taken from Psalm 81:2–6, as follows:

> *Exsultate Deo adjutori nostro: jubilate Deo Jacob: sumite psalmum jucundum cum cithara: canite in initio mensis tuba, quia praeceptum in Israel est, et judicium Deo Jacob. Testimonium posuit illud, cum exiret de terra Aegypti.*

> Rejoice in God our helper: be joyful to the God of Jacob; take up the pleasant psalm with the lyre: sound the trumpet at the beginning of the month. For it is a precept in Israel and a statute for the God of Jacob. He set it as a testimony in Joseph, when he went forth from the land of Egypt.[19]

The selection of items from these Psalm verses is striking. The command to "rejoice" and "be glad" on what is supposed to be a day of fasting is remarkable enough; but then the congregation is told about sounding a trumpet, and that "at the beginning of the month," a calendar reference that will soon be reinforced and further defined by the second reading at this Mass. Heavily emphasized is the notion of legal ordinance: *praeceptum, iudicium*, and *testimonium* follow one another in quick succession, before the mention of Israel's redemption from Egypt. We should recall that it is Christians, not Jews, who have chosen these segments of the Psalm: it is they who are drawing attention to the *initium mensis* as a precept, and this they will continue to do. Thus meditation on, and reverence for, God's commandments will form the theme of the Offertory chant in this Mass, which consists of Psalm 119:47–48 almost in their entirety. As we have already noted, however, it is the second reading, taken from Nehemiah (Vulgate 2 Esd) 8: 1–10 that takes up and develops the notion of divine precept, with specific reference to the calendar.[20]

The theme of this reading is the Great Assembly in the days of Ezra, when all the Jews were assembled in Jerusalem to hear the reading of the Law *quam*

[18] For the latter, see Adler n.d.: 14. For Psalm 81 in the liturgy for New Year, see also Elbogen 1995: 147 (ET: Scheindlin 1993: 123).

[19] The text of the Introit is taken from *Missale Romanum ex Decreto Sacrosancti Concilii Tridentini* (Regensburg: Fred. Pustet, 1932). Translations are mine.

[20] Our earliest sources for the lections prescribed for the Ember Day Masses are the *Comes of Würzburg* and the *Comes of Alcuin*, whose manuscripts are dated to around 700 and to the ninth century respectively: both contain material of much earlier vintage. See the table of lections for the September Ember Days reproduced and discussed by Stökl Ben Ezra 2003a: 318, and the briefer comments of Willis 1964: 95, who remarks that these days, not coinciding with any of the major Christian festivals, have not absorbed the liturgical themes of other festival days.

praeceperat Dominus Israeli.[21] This took place on the first day of the seventh month, *in prima die mensis septem*, a nice detail that manages deftly to conjoin the Jewish calendar with the month of September in the Roman reckoning in which the Wednesday Ember Day is being observed. The Latin version of Neh 8:1–10 tends to emphasize certain aspects of the Hebrew base text.[22] Thus in verse 3, Ezra is said to have read the book *aperte*, "openly, plainly," a detail not included in the Hebrew at this point, but later (verse 8) spoken of with some insistence. Again, at the end of verse 3, we are told that the ears of all the people were "attentive," Latin *erectae*, to the book: this represents the clarification of the Hebrew words ואזני כל העם אל ספר התורה, which mean literally "and the ears of all the people were toward the book of the Torah." In various ways, the Latin version underscores the liturgical character of this assembly; but most noteworthy is its interpretation of verse 7, according to which the Levites who supported Ezra in his task "made silence among the people" (*levitae silentium faciebant in populo*) so that the Torah might be heard (*ad audiendam legem*). This language of hearing is not found in the Hebrew

[21] See Neh/2 Esd 8:1, in the Vulgate version. The book of Nehemiah (2 Esd), like the book of the Twelve Prophets, very rarely features as a source of readings in the Mass lectionary. Thus the use of Amos 9:13–15 as the first reading at the Ember Wednesday Mass (see note 36) is also highly unusual and concords with the selection of other passages from the Prophets and Writings typical of the Ember Day Masses, but rare on other occasions. This phenomenon is noted by Stökl Ben Ezra 2003a: 319–321, who catalogues three possible ways of explaining such use of Scriptural material otherwise rarely utilized in the lectionaries. The first of these, the adoption of Jewish ritual custom, is unlikely to account for the readings of Ember Wednesday: information available offers no extant Jewish source for readings either from Amos or from Nehemiah in association with the liturgy of Rosh Ha-Shanah. The second possibility, that the readings represent a polemic reaction to Jewish usage, seems unlikely given what we have noted thus far. This leaves the possibility that the readings were chosen by someone seeking in a more or less scholarly manner to focus on texts that spoke of the seventh month, the "bookish" approach to the liturgy, as Stökl Ben Ezra describes it. The same scholar, however, rightly notes (2003a: 319, note 130) that our knowledge of the Synagogue lectionary at Rome around the time of Pope Leo the Great is extremely limited, and our reliance on evidence provided by the classical rabbinic texts for what was read in the Synagogue Service may not, in fact, cast light on Roman practice. The possibility that the Roman synagogues read parts of Amos or Nehemiah, therefore, cannot be entirely discounted.

[22] The Latin text of this reading in the traditional Missals follows closely the received text of the Vulgate, albeit with certain omissions. The standard liturgical introduction to Old Testament lections, "*in diebus illis*," brings about the suppression of part of 8:1 (*et venerat mensis septimus filii autem Israhel erant in civitatibus suis*) so that the reading begins "*congregatus est omnis populus.*" In verse 4, the Missal omits the note *et steterunt iuxta eum* introducing the names of six Levites standing at Ezra's right hand, and the names of the corresponding seven Levites on his left. Similarly, in verse 7 the Missal again omits the names of the thirteen Levites listed there, and in verse 9 omits both the further description of Nehemiah as *ipse est Athersatha*, and the whole of the last segment of the verse, *flebat enim omnis populus cum audiret verba legis*.

base text, and it serves to strengthen the Latin translation of verse 8, which tells how the Levites made the people understand by reading out the Torah *distincte et aperte ad intelligendum*.

The stress on *hearing* and *precept* in this Ember Day Mass could hardly be greater. On the first day of the seventh month, the Jews are sounding the *shofar*, the hearing of which is a solemn religious obligation: on that self-same day, Ezra and the Levites had ensured that the people heard and understood the precepts of the Torah. The Mass of Ember Wednesday juxtaposes these items; it does not conclude the reading from Nehemiah until it has told how Ezra and his colleagues declared that this day (the first of the seventh month) is holy to the Lord and had therefore insisted that the people should not mourn or weep.[23] Rather, the Jews are to eat rich foods and drink sweet wine and send portions to those with nothing prepared, since this is a holy day to the Lord (Neh/2 Esd 8: 9–10). The people must on no account be sad. That this injunction is of some importance for the Ember Day is indicated by the use of Nehemiah (Vulgate 2 Esd) 8:10 as the Communion chant for the Mass at the close of the rite.[24]

The information presented to date seems to cohere in large measure with Daniel Stökl Ben Ezra's observations on Pope Leo I and his stance toward the coincidence of Yom Kippur and the Ember Days of September. Thus, while noting that Leo does (but with no great vigor) insist on a distinction between the Jewish and Christian observances at this time of the year, Stökl Ben Ezra can declare:

> Unlike Chrysostom, Leo does not complain about Christians actually participating in Jewish festivals and the fast, but he defends himself against accusations of Judaizing by exploiting the similarities as belonging to the apostolic Jewish heritage. Like the Ten Commandments, the Solemn Feasts are the valuable part of the Old Testament precepts, which have been adopted into the new covenant. Judaizing is orthodox, – if it is apostolic.[25]

Stökl Ben Ezra concludes by quoting a passage from Leo's sermon 92:1, in which the pope states that what had for a long time been Jewish custom could become Christian observances on the basis of Jesus' declaration that he had not come to destroy the Law, but to fulfill it.[26] His remarks have a particular

[23] The Missal seems to have taken this injunction to heart and omits from the text of Neh 8:9 the report that the people were in truth weeping when they heard the words of the Torah. See details in note 22.

[24] A translation of the verse yields: "Consume rich foods, and drink sweet wine, and send portions to these who had not made things ready for themselves: for the day is holy to the Lord. Do not be sad, for the joy of the Lord is our might."

[25] Stökl Ben Ezra 2003a: 315–316.

[26] Leo the Great, *Sermo* 92:1, cited by Stökl Ben Ezra 2003a: 316.

pertinence if we may regard all the Ember Days as having been viewed in antiquity as sharing common features and theological concerns. This seems, indeed, to have been the case: the evidence for such an opinion is convincingly set out in detail by Geoffrey Willis and need not be rehearsed here.[27] Thus the prayers for the Ember Days of the fourth month in the Gelasian Sacramentary (MS Vat. Reg. Lat. 316, dated to the earliest part of the eighth century) can properly be applied also to the Ember Days of September, when they declare the days to be *observationes antiquae* and petition the Almighty that the congregation be instructed by these days *legalibus institutis*, "by means of customs established by law."[28] This way of speaking of the Ember Days is, indeed, strongly reminiscent of the language and ideas presented in the sermons of Leo the Great.

Mass of Ember Wednesday in September displays further affinities with the Jewish New Year in its references to creation and the fruits of the earth. Immediately after the second reading, the Gradual chant juxtaposes the themes of a people (Israel and the Church) chosen by God, who is the creator of all things. Psalm 33:12, 6 are sung, declaring the blessedness of the elect people, in tandem with the proclamation "by the Word of the Lord were the heavens made firm." With these sentiments should be compared the distinctive references, in the *Mussaf* service of Rosh Ha-Shanah, to the Lord's creation of the world associated with one of the best-known elements of the Liturgy for this day. This is the solemn chanting of three groups of prayers severally designated *Malkhuyyot* (which speak of the Kingship of God), *Zikhronot* (referring to God's remembering of his creatures at this time), and *Shofarot* (speaking of the sounding of the *shofar*). These three groups of prayers were traditional already in the time of the Mishnah.[29] We may note especially the concern with God's creation voiced in the words that describe Rosh Ha-Shanah: "this day, on which was the beginning of Thy work, is a memorial of the first day; for it is statute for Israel and ordinance of the God of Jacob."[30] Indeed, the association of Rosh Ha-Shanah with the creation of

[27] See especially Willis 1964: 49–97.

[28] The prayer *Ad Populum* for the Wednesday Mass petitions God that the Church, by perpetually recounting the ancient observances, may profit in the future; and the Post-communion prayer for the Friday asks that the congregation be instructed *legalibus institutis*.

[29] See m. Roš Haš. 4:6, which names the prayers but does not quote them. It lays down the rules for them, which obtain to this day: ten Scriptural verses are allocated to each section, three from the Torah, three from the Prophets, and three from the writings, with a concluding verse from the Torah. For the Hebrew text, I have consulted Adler n.d.: 135–140.

[30] Translation as in Adler n.d.: 137. The closing words of this section are, of course, a quotation of Psalm 81:5.

the world and with God's choice of Israel as his people represents central themes in the Synagogue liturgy for the festival. Already in the fourth century CE, the Jerusalem Talmud records the words of a prayer for the New Year ascribed to Rav, a scholar associated with both the Land of Israel and with Babylonia.[31] The Jerusalem Talmud (y. Roš Haš. 1:3 [57a]) designates this prayer as *teqi'ata'* "blowing (sc. the *shofar*)," that is, specifically a prayer associated with the sounding of the *shofar*;[32] and it runs as follows:

> This day is the beginning of Thy works, a memorial of the first day: for it is a statute for Israel, a law of the God of Jacob. And concerning the provinces, on it [i.e., New Year's day] shall be declared which is destined for the sword and which for peace; which for famine and which for satiety; and the creatures shall be reviewed on it for life or for death.[33]

Striking here is the unmarked quotation of Psalm 81:5, referring to the statute for Israel and the law of the God of Jacob, now juxtaposed with God's activities on the first day of the creation of the universe. As we have seen, t. Roš Haš. 1:11 cites a verse from this very Psalm and sets it alongside a quotation from Psalm 33 with reference to God's judgment of the world, a theme also integral to Rav's prayer. It is thus of considerable interest to observe that the Mass for Ember Wednesday makes use of both these Psalms, adducing Psalm 33:6 to point to God's creation of the world, and retaining and indeed emphasizing in its use of Psalm 81 as the Introit the strong sense of precept and commandment involved for Israel.[34] The sense that the first day of the seventh month is the day of the creation of the world is further

[31] He died in 247 CE. He is sometimes referred to as Abba Arika (see e.g. b. Ḥul 137b), but his scholarly distinction was of such magnitude that he is usually named simply as Rav. Tradition associates him with the arrangement of the Synagogue Service for New Year's Day, especially the inclusion of the prayer *'Aleinu* preceding the *Malkuyyot* verses in the *Mussaf* service

[32] On the *teqi'ata'* and its use in the liturgy, see further Elbogen 1995: 141–142, and p. 143 for Rav's prayer in particular.

[33] The translation is mine. That the world was created in Tishri is the view of the Tanna R. Eliezer b. Hyrcanus given in a *baraita* at b. Roš Haš. 10b–11a, in dispute with another Tanna, R. Joshua b. Hananiah, who holds that it was created in Nisan. For the same dispute, see also Gen. Rab. 22:7.

[34] The Latin of the Introit (selections from Ps. 81:2–6) is not that of St. Jerome's translations of the Hebrew or the Old Greek version, but a form of the Vetus Latina (as is not infrequently the case in the proper chants of the Roman Missal). The text of this Introit is cited as representative of the Vetus Latina by Sabatier 1743: *ad loc.*, along with similar readings of the verses witnessed by the Psalters of the Mozarabic and Milanese Rites, and writings of Ambrose and Augustine.

represented in the Synagogue service, most notably in prayers of the *Mussaf* service proclaiming God's kingship and his remembrance of Israel.[35]

God's incomparability and sovereignty likewise feature in both Jewish and Christian liturgies at this time. Thus in the Evening Service of Rosh Ha-Shanah , the second benediction of the *'Amidah*, which in any case invariably twice asks the rhetorical question who is like God, requires on this day an additional third such question: "Who is like you, merciful Father, Who remembers his creatures for life in mercy?" It should be noted that the first Gradual chant of the Ember Wednesday Mass quotes Psalm 113:5, 7 and asks the same question, referring to God as the one who raises up the needy and the poor.

This Gradual chant follows the first reading at the Mass, about which we have as yet said nothing since it has, as far as I am aware, no counterpart in Jewish observance of Rosh Ha-Shanah. This reading is taken from Amos 9: 13–15, which prophesies abundant harvests of corn and grapes for the future.[36] Israel's restoration is proclaimed; ruined cities will be rebuilt; and their inhabitants will plant vineyards and gardens, drink wine, and eat fruits. The time of the year – the September grape harvest – accords well with the hopes expressed in this reading.[37] As we have seen, the Introit had spoken of the beginning of the month, and the second reading explicitly mentions the first day of the seventh month. The reading from Amos, placed between these two texts, highlights the abundance of produce, fruits of the earth, wines, and plentiful harvests that only the God of Israel can supply, and which the prophet confidently predicts that he will grant in the days to come.

It must be asked why a day that is officially proclaimed by the church as a time of fasting and penitence should be provided with chants and readings that speak of rich foods, wines, and celebrations and employ themes occurring also in the Jewish New Year festivities taking place at more or less the

[35] For *liturgical* purposes, the first day of Tishri counts as New Year's Day; the service of the Synagogue leaves no doubt of this – hence Rav's prayer quoted earlier. For other sources that present Tishri as "the first month" of the year, see (e.g.) Josephus, *A.J.* 1:80; T. Ps.-J. of Gen. 7:11; S. 'Olam Rab. 4:1; and Pirqe R. El. 23. The opinion of R. Joshua, that the world was created in Nisan, is based in part on the description in Exod 12:2 of the month of Israel's Exodus from Egypt as "the beginning of the months . . . the first of the months of the year." R. Joshua's concern to stresss the redemption of Israel is discussed by Urbach 1979: 1.671–672.

[36] The text of this prophecy in the traditional Missals agrees with that of the Vulgate, except that the liturgical lection is prefaced with the standard introduction, "*Haec dicit Dominus*," which replaces the *dicit Dominus* of 9:13, *ecce dies veniunt dicit Dominus*.

[37] In biblical times, the people of Israel would have been well aware of the New Year celebrations observed by their neighbors: the cultures of the ancient Near East generally seem to have associated New Year with the judgment and renewal of the world, the stability of the created order, and resolutions to avoid past mistakes and sins, which need to be acknowledged: see further Snaith 1974.

same time? This question is doubly insistent, inasmuch as Rosh Ha-Shanah is well known for its concern with God's judgment of the world. Might it not have been more appropriate for the church to mirror this aspect of the Jewish festival, as corresponding more closely to its own stated business of fasting and penance for sin, expressed (for example) in the Gelasian Sacramentary's *Denuntiatio* and the sermons of Pope Leo the Great? The association of judgment with Rosh Ha-Shanah is ancient, well known, and mentioned directly by the Mishnah without further explanation.[38] Now one might, in truth, find the matter of judgment implied in two out of the four prayers of this Mass, and these we consider presently; but in themselves these prayers do nothing to overturn the strong impression given by the chants and readings of an act of worship centered on rejoicing, particularly for the fruits of the earth and the precepts of God, their incomparable creator.

One strong reason for the character of this Mass and its close affinities to certain aspects of Rosh Ha-Shanah suggests itself when we recall that observance of the Ember Days was originally confined to the city of Rome. To that city, from the second century CE onward, flocked teachers and preachers such as Marcion, Valentinus, and Carpocrates, whose doctrines are often labeled somewhat generally as "gnostic."[39] All these, and others like them to a greater or lesser degree, either held the Jewish Scriptures in low esteem, or rejected them outright. They tended to regard the material

[38] The earliest rabbinic reference to New Year's Day as related to judgment is, of course, m. Roš Haš. 1:2, and the idea is taken up and developed by later texts. The association of judgment with this day is almost certainly older than the time of the Mishnah's redaction, being attested by the Liber Antiquitatum Biblicarum (LAB), a writing incorrectly attributed to Philo of Alexandria, which was composed in all probability in the first or early second centuries CE. This text is today represented only in a Latin translation, most likely made in the fourth century CE. At LAB 13:6, we read the Almighty's command to Moses: "On the festival of trumpets there will be an offering on behalf of your watchmen. Because on it I review creation, so as to take note of the entire world. At the beginning of the year, when you present yourselves, I will decide the number of those who are to die and who are to be born. On the fast of mercy you will fast for me for the sake of your souls, so that the promises made to your fathers may be fulfilled." For this translation, see Jacobson 1996: 1.113. The particular difficulties presented by the Latin of this section, along with problems of interpretation, are addressed by Jacobson in the same volume, 512–516; his comments on the date of LAB as a whole, and of its translation into Latin, may be found on 199–210 and 273–277, respectively. One can only speculate whether this Latin translation was known to Roman Christians, but the possibility is intriguing: see further James 1917, who remarks (54) on the extraordinary similarity in the Latinity of LAB and 4 Ezra, leading him to comment that "one is tempted to say that they are by the same hand." Christians certainly were responsible for the preservation and transmission of the latter text.

[39] For these and other teachers, and their relationships with the Catholics at Rome, see Lampe 2003: 241–256, 414–416 on Marcion and his students; 292–318 on the Valentinians; and 319–320 on the somewhat shadowy figure of Carpocrates and his followers.

creation, the work of the God of Israel, as of dubious value at best, at worst as the product of a low-grade, even malicious demiurge. The Catholic Church vigorously resisted these men and their teachings, which treated the Old Testament with such disdain and tended to downgrade and demean the status of the created universe. Faced with a common threat, Jews and Christians in Rome may have sensed that on matters of such fundamental importance as the divine origin and gift of Scripture and the goodness of what God had created, they should stand together. Of this there can be no certainty; but the sustained emphasis in the Mass of Ember Wednesday in September on the goodness of creation as expressed in the rightness and propriety of fruits and wine (and this on a day of penance), together with the statutory, legal obligation to proclaim such things, rather speaks in its favor.[40] As we now see, this observation will hold good in the face of what might, at first blush, seem to indicate a rather different interpretation of the Mass from the one advanced to date.

Daniel Stökl Ben Ezra's study of Yom Kippur offers a sophisticated model that calibrates the effects of Jewish rituals of Yom Kippur on Christian attitudes and practices in both positive and negative ways. Thus he can discern aspects of the September Ember Days that reflect, but at the same time significantly stand aside from, contemporary Jewish practices.[41] In the case of the Ember Wednesday of this month and its affinities with Rosh Ha-Shanah, we should be aware of Stökl Ben Ezra's insights, not least because the Gospel of that Mass may reflect a rather less than wholehearted acceptance of the Jewish festival than other parts of the Mass so far examined. This Gospel reading is taken from Mark 9:16–28 and tells of a child racked by an evil spirit, and the ineffective attempts of Jesus' disciples to exorcise it. Only late in the narrative do we learn that the evil spirit is deaf and dumb: it is expelled by a direct command of Jesus. The Latin is worth noting: *Surde et mute spiritus, ego tibi praecipio, exi ab eo* ("deaf and dumb spirit, I command you, Come out of him!"; Mark 9:25). The inability of the disciples to exorcise the spirit is explained in the final verse of the section: *Hoc genus in nullo potest exire, nisi in oratione et jejunio* ("This kind cannot go out by any means other than by fasting and prayer"; Mark 9:29). Here, at last, is a rationale for fasting and prayer on this Ember Wednesday, which in this text is presented as

[40] The September Ember Days might also correspond in time to the celebrations of the *ludi romani*, marking the dedication of the Capitoline temple, which was recalled on the Ides of the month (September 13). The feasting associated with these games seems not to have concerned Jews or Christians to any great degree.

[41] See especially his description of the theoretical model he has adopted for assessing the *nature* and *character* of the perceptions Jews and Christians might have had of each other and of each other's rituals, discussed in Stökl Ben Ezra 2003a: 4–10, 116–132.

a prerequisite for the ability to hear and to speak. Now those Christians who were so minded might discern in this Gospel the figure of the deaf and dumb child as representing the Jews, unable to hear clearly, distinctly, and plainly the words of their own Law, unless they are healed by Christ. This, however, is very far from being a necessary interpretation of the text, which could equally be applied to the Christian congregation, who are being reminded of the supernatural power associated with fasting and prayer. This view is reinforced by the fact that (to the best of my knowledge) there exists no Patristic comment on these Gospel verses that even remotely suggests that the dumb and deaf child in any way "represents" the Jews.[42] That the Gospel has in mind the Christian congregation at least primarily is, indeed, strongly suggested by the prayers of the Mass.

In the first of these prayers, a collect, the priest petitions God on behalf of the congregation to support the worshippers' frailty by means of the healing powers of God's mercy (*misericordiae tuae remediis*), so that what has been worn away by human circumstances may be restored by divine gentleness. This collect immediately precedes the reading from the prophet Amos. A second collect, preceding the reading from Nehemiah (2 Esd), asks God to grant that his household now supplicating him and abstaining from bodily foods may also fast from faults in their minds (*a vitiis quoque mente jejunent*). In these prayers, the Christian congregation is clearly in view as standing in need of divine mercy for its faults: implicit is the understanding that this congregation stands under divine judgment and requires God's remedies and kindness to escape it. The subtext becomes clear when the Gospel is read: fasting and prayer have a supernatural power to effect healing, liberation from vices, and deliverance from evil powers.

Granted all this, the Gospel nonetheless remains as a witness to the fact that this is a Christian liturgy, not a simple mirroring or general reconfiguring of Rosh Ha-Shanah in a non-Jewish setting. Stökl Ben Ezra is entirely justified in drawing attention to the complexities and nuances of these Christian *observationes antiquae*, which owe so much to Jewish thought and practice, and yet pursue their own course and agenda. Elsewhere, I have argued that the Roman Church had recourse to Jewish tradition in its fight against

[42] See Lapide 1639: 602. The homily on this Gospel at Matins on Ember Wednesday is taken from the writings of St. Bede the Venerable (Book 3.38): this, likewise, advances no anti-Jewish sentiments and offers no "symbolic" or "allegorical" interpretation of the deaf and dumb child. While the boy's dumbness is taken to refer to inability to confess the faith, and his deafness to an incapacity for hearing the words of truth, the exposition of the Gospel is otherwise quite unspecific on points of detail. The homily is concerned to emphasize the supernatural power of prayer and fasting as instruments for the defeat of demonic powers.

"gnosticism," especially as a counterblast to those who denigrated the created order.[43] It seems possible that the Mass of the Ember Wednesday in September might represent yet another example of such a procedure. This suggestion is strengthened when one considers the repeated references in the texts of the Mass to grapes and wine. They are found in the first reading from Amos, which speaks of the one who tramples the grapes (9:13) and predicts the future planting of vineyards by Jews restored to their homeland, who will drink the wine of those grapes (9:14). Wine makes another appearance in the second reading from Nehemiah. This has already been noted: Ezra solemnly commands the people to drink sweet wine on the first day of the seventh month (Neh 8:10), and this command is repeated as the Communion chant of the Mass, which constitutes one of the "last words" of the whole liturgical function.

Wine is an essential element of the Eucharist for the Roman Church, whose members were not unaware of groups that claimed to celebrate the Eucharist while using water instead of wine in the chalice. While it is not entirely clear whether Marcion and his followers substituted water for wine in their Eucharistic rites,[44] there were several other groups that adhered to such a practice, and apparently sought to promote it.[45] For the Roman Church, the use of wine in the Eucharist is nonnegotiable, a command of Christ himself followed by the Apostles: so much is clear from Justin Martyr's account of the Eucharist at Rome around the middle of the second century CE.[46] It need hardly be said that wine is also an essential element in Jewish religious observance; its careful avoidance by groups hostile to the teaching of the Hebrew Scriptures, and those disposed to regard the created order as the

43 See Hayward 2009: 101–123.

44 Research into this matter during the past few decades is summarized and discussed by Moll 2010: 121–123.

45 For a wide-ranging discussion and analysis of use of water instead of wine in the Eucharist, "agape," and other Christian celebrations, see McGowan 1999.

46 Justin, *1 Apol.* 66–67, gives the matter of the Eucharist as bread and wine. He notes specifically that the Apostles had handed down what Jesus had commanded the church to do, quoting the "dominical words." Justin's reference to water in this section of his *Apology* is an allusion to the practice of adding a certain amount of water to the wine, following the established custom of the Greco-Roman world, which regarded the drinking of unmixed wine as a barbarian excess. Irenaeus, writing c. 180 CE, regarded the Roman Church as a sure guardian of Christian orthodoxy (*Haer.* 3.4.1) and spoke of the Eucharistic chalice as "firstfruits" of creation (*Haer.* 4.18.2) in his opposition to groups claiming secret knowledge of other, allegedly more authentic Christian practices and beliefs. His reference is to the wine in the chalice: by speaking of the Eucharistic elements as "firstfruits," a sacral category description derived from the Hebrew Scriptures, he necessarily rules out water as the principal ingredient of the Eucharistic cup. Scripture nowhere refers to offerings of water as "firstfruits."

product of an "inferior" Jewish deity, is all too easy to comprehend. As a counterblast to the gnostics, the Ember Wednesday of September is fixed firmly in sacred time and sacred calendar, which the texts of this Mass underscore as God-given commandments, precepts, testimonies for both Jews and Christians, sacred time that celebrates the goodness and joy of the created order that the Almighty saw, "and behold, it was very good."

Bibliography

Adler, H. (ed.). n.d. *Service of the Synagogue: A New Edition of the Festival Prayers with an English Translation in Prose and Verse, New Year.* London: Routledge and Kegan Paul.

Baumgarten, J. M. 1999. "Yom Kippur in the Qumran Scrolls and Second Temple Sources." *Dead Sea Discoveries* 6: 184–191.

Chavasse, A. 1965. "Les Quatre-Temps." In A. G. Martimort (ed.), *L'Église en Prière. Introduction à la Liturgie.* Paris: Tournai, 758–767.

Daniélou, J. 1956. "Les Quatre-Temps de septembre et la fête des Tabernacles." *La Maison-Dieu* 46: 114–136.

Elbogen, I. 1995, *Der jüdische Gottesdienst in seiner geschichtlichen Entwicklung.* Hildesheim: Georg Olms (reprint of 1931 edition).

Fischer, L. 1914. *Die kirchlichen Quatember. Ihre Entstehung, Entwicklung und Bedeutungen liturgischer, rechtlicher und kulturhistorischer Hinsicht,* München: Veröffentlichungen aus dem kirchenhistorischen Seminar 4/3.

Fleischer, E. 1984. "Concerning the Celebration of *Rosh Ha-Shanah* in *'Eretz Israel.* " *Tarbiz* 53: 293–295 (in Hebrew).

Green, B. 2010. *Christianity in Ancient Rome: The First Three Centuries.* London: T&T Clark.

Harlé, P. and D. Pralon. 1988. *La Bible d'Alexandrie*, vol. 3: *Le Lévitique.* Paris: Cerf.

Hayward, R. 2009. "What Did Cain Do Wrong? Jewish and Christian Exegesis of Genesis 4: 3–6." In E. Grypeou and H. Spurling (eds.), *The Exegetical Encounter Between Jews and Christians in Late Antiquity.* Journal for the Study of the Pseudepigrapha, Supplement Series 18. Leiden: Brill, 101–123.

Herr, M. D. 1983. "More on Two Days of *Rosh Ha-Shanah* in *'Eretz Israel.*" *Tarbiz* 53: 142–143 (in Hebrew).

Idelsohn, A. Z. 1960. *Jewish Liturgy and Its Development.* New York: Schocken Books.

Jacobson, H. 1996. *A Commentary on Pseudo-Philo's Liber Antiquitatum Biblicarum, With Latin Text and English Translation.* 2 vols. Leiden: Brill.

James, M. R. 1917. *Biblical Antiquities of Philo.* London: SPCK.

Kedar-Kopfstein, B. 1989. "Glossen zur traditionellen biblischen Philologie." *Zeitschrift für Althebraistik* 2: 205–209.

Lampe, P. 2003. *From Paul to Valentinus: Christians at Rome in the First Three Centuries.* London: T&T Clark.

Lapide, C. 1639. *Commentarius in Quatuor Evangelia.* Antwerp: Heirs of M. Nutius.

Leclercq, H. 1907–1953. "Quatre-Temps." *Dictionnaire d'archéologie chrétienne et de liturgie* XIV/2: 2014–2017.

McGowan, A. 1999. *Ascetic Eucharists: Food and Drink in Early Christian Ritual Meals.* Oxford: Clarendon Press.

Mohlberg, L. C. 1960. *Liber Sacramentorum Romanae Ecclesiae Ordinis Anni circuli.* Rerum Ecclesiasticarum Documenta, Fontes 4, Rome: Herder.

Moll, S. 2010. *The Arch-Heretic Marcion.* WUNT 250, Tübingen: Mohr Siebeck.

Morin, G. 1897. "L'origine des quatre-temps." *Revue Bénédictine* 14: 337–346.

Sabatier, P. 1743. *Bibliorum Sacrorum Latinae Versiones Antiquae seu Vetus Italica, et ceterae quaecumque in codicibus Mss. et antiquorum libris reperiri potuerunt.* Reims: Florentain.

Scheindlin, R. P. 1993. *Jewish Liturgy: A Comprehensive History.* Philadelphia: Jewish Publication Society.

Snaith, N. H. 1974. *The Jewish New Year Festival: Its Origin and Development.* London: SPCK.

Stökl Ben Ezra, D. 2001. "'Christians' Celebrating 'Jewish' Festivals of Autumn." In P. J. Tomson and D. Lambers-Petry (eds.), *The Image of the Judaeo-Christians in Ancient Jewish and Christian Literature. Papers Delivered at the Colloquium of the Institutum Iudaicum, Brussels 18–19 November, 2001.* Wissenschaftliche Untersuchungen zum Neuen Testament 158. Tübingen: Mohr Siebeck, 53–57.

Stökl Ben Ezra, D. 2003a. *The Impact of Yom Kippur on Early Christianity.* Wissenschaftliche Untersuchungen zum Neuen Testament 163. Tübingen: Mohr Siebeck.

Stökl Ben Ezra, D. 2003b. "Whose Fast Is it? The Ember Day and Yom Kippur." In A. Becker and A. Reed (eds.), *The Ways That Never Parted: Jews and Christians in Antiquity and the Early Middle Ages.* Texts and Studies in Ancient Judaism 95. Tübingen: Mohr Siebeck, 259–282.

Tabory, J. 2000. *Jewish Festivals in the Time of the Mishnah and Talmud.* Jerusalem: Magnes Press (in Hebrew).

Tabory, J. 2006. "Jewish Festivals in Late Antiquity." In S. T. Katz (ed.), *The Cambridge History of Judaism, vol. 4: The Late Roman-Rabbinic Period.* Cambridge University Press, 556–572.

Talley, T. J. 1990. "The Origin of the Ember Days. An Inconclusive Postscript." In P. de Clerck and E. Palazzo (eds.), *Rituels: Mélanges offerts à Pierre-Marie Gy, O.P.* Paris: Cerf, 465–472.

Urbach, E. E. 1979. *The Sages: Their Concepts and Beliefs.* 2 vols. Trans. I. Abrahams. Jerusalem: Magnes Press.

Verstrepen, J.-L. 1993. "Origines et instauration des Quatre-Temps à Rome." *Revue Bénédictine* 103: 339–365.

Willis, G. G. 1964. "Ember Days." In G. G. Willis (ed.), *Essays in Early Roman Liturgy.* Alcuin Club Collections 46. London: SPCK, 49–97.

14

CELEBRATIONS AND THE ABSTENTION FROM CELEBRATIONS OF SACRED TIME IN EARLY CHRISTIANITY

Clemens Leonhard

It is widely assumed that Christians celebrated festivals throughout antiquity.[1] At first sight, this presupposition is not unwarranted. Fifth-century urban Christians had elaborate systems for the organization of sacred time at their disposal. Complicated astronomical calculations established a date for Easter, which was preceded and followed by structured sacred time. Epiphany and Christmas constituted another cluster that attracted holy days and structures of sacred time. In Jerusalem, many additional festivals and days of commemoration invited monks and pilgrims to live their lives embedded in times laden with meaning, purpose, structure, and ritual.

Similarly, modern scholars sometimes presume that festivals documented in the fifth century must somehow be traced back to first-century Judea. In this respect, van Goudoever's suggestion is not atypical: "De tous les éléments de la liturgie, les fêtes sont peut-être le plus permanent: il est pratiquement impossible de changer la date ou la forme des anciennes fêtes, et la création d'une nouvelle fête religieuse est presque inconcevable."[2] Such assumptions, made without any documentation on which they might rely, presuppose an extraordinary stability and continuity of practice within Jewish and Christian groups spanning the gap of centuries. By itself, this approach is not inadmissible because the mere lack of data about festivals need not reflect the fact that nobody celebrated them. The following observations suggest, however, that the burden of proof should be reversed: whoever wishes to claim continuous practice would have to adduce corroborating evidence,

[1] The topic of this chapter emerged from discussions with Israel Yuval and Daniel Stökl Ben Ezra at the Wissenschaftskolleg zu Berlin in 2011/12. I am grateful to the editors of this volume for a thorough critique and Inga Markert as well as Tobias Albers for general observations upon earlier versions of this text.

[2] Van Goudoever 1967: 213–214.

while the silence of the sources should otherwise be taken seriously, that is, as pointing to the absence of practice.

In the fourth century, Christians in different urban centers developed a tight network of the public construction of sacred time. Thus, the present chapter is concerned with evidence preceding that epoch, that is to say, from the first three centuries CE. This earlier period witnesses a slowly growing interest in the celebration of sacred time and a staunch opposition against that development at the same time. The chapter is designed to point out developments and interpretative problems of this epoch. Hence, it does not try to present a running history of the sanctification of time. Its argumentative structure is designed to challenge the assumption that Christians just continued to celebrate sacred time more or less unabated from the epoch of the Second Temple until their festivals, rituals, and customs surface again in the sources of the late fourth century. In hindsight, the epoch under consideration emerges as a period of experimentation that set the course of developments in some cases but failed to do so in others.

The discussion begins with the Hierapolis inscription that mentions Pentecost and Azyma (§ 1). It asks which generalizations are admissible from this third-century text regarding the performance of biblical and other festivals. The discussion then proceeds to the topic of the Sunday as a seemingly powerful Christian device to create a continuous performance of sacred time (§ 2). This chapter ends with the suggestion that the Sunday was not part of any kind of primordial repertoire of celebrations in those groups that would later be seen as nascent Christianity. This conclusion raises the question whether the first three centuries only constitute a big *argumentum e silentio* regarding festivals. Thus, §§ 3 and 4 present the two opposing stances of this epoch, partly endorsing, partly challenging the *silentium* of the argument. Next, § 3 collects some of Tertullian's testimonies about Christians as people who celebrate more festivals than anybody else. This in turn requires another retrospective look at the New Testament evidence for an explicit opposition against the celebration of festivals and against the performative and ritualized creation of sacred time in § 4. The conclusion exposes a basic principle that may have been operative behind the opposing stances presented here. It seems that every participant in these ancient discussions agreed to the presupposition that Christians – like everybody else, Jews, Greeks, and Romans – celebrate *the festivals of their cities*. They just disagree about the nature, place, character, and of course the *calendar* of that city.

1 Celebrating Festivals: Pentecost, Azyma, and Kalends

In this context, the performance of certain customs on the dates of Pentecost and a festival called "Azyma" (unleavened bread) that emerges from the famous inscription on the tomb of Publius Aelius Glykon Zeuxianos Aelianus in Hierapolis (normally dated to after 212 CE: Ameling 2004: 414–422) serves as a test case. The owner of the tomb had a copy of the inscription on the outside face of the tomb stored in the public archives (as claimed by the inscription). He deposited an endowment of 200 dinars for the association of the directorate of the purple-dyers and 150 dinars for the association of the carpet-weavers. These associations were expected to distribute the interest to their members "on the festival of the Kalends, in the fourth month on the eighth day," that is, on the Kalends of January, on "the festival of unleavened bread," and on "the festival of the fiftieth (that is, Pentecost)." In return for these benefits, the associations apparently agreed to decorate the tomb on these days.

The combination of the Kalends and the Jewish festivals is awkward from a modern perspective. It impedes the assessment of the donor's and the associations' relationships to what can be known about Judaism and its festivals in that epoch. On the one hand, the inscription does not prove that these associations celebrated the Jewish and the Roman festivals. On the other hand, it is not implausible that they did. However, if these two associations of local craftsmen and entrepreneurs celebrated those festivals, many questions would arise as to how they might have performed and understood such a celebration. It stands to reason that they held a banquet – which is not implied in the inscription. It is interesting to note that the donor expects these two associations to know the dates of two biblical festivals although their social context does not betray any relationship to rabbinic Judaism. Rabbinic texts understand the Kalends as utterly pagan.[3]

Arye Edrei and Doron Mendels suggested that Jews of the Western Diaspora did not have any access to the rabbinic traditions that were only handed down orally and preserved in languages that were not known to Jews in the Western parts of the Roman Empire.[4] The link of these Westerners to biblical Israel was the Septuagint, which was not expanded in the rabbinic way of building a legal and exegetical tradition besides it, but by means of adding pseudepigraphical writings. Language barriers were certainly not as

[3] Cf. m. 'Abod. Zar. 1:2. The question of compatibility with certain aspects of Jewish practice is, however, more complex; cf. Hadas-Lebel 1979: 428–430.

[4] Edrei and Mendels 2007/2008.

tight as Edrey and Mendels suggest. On the one hand, two associations in a Greek-speaking environment knew about *azyma* and *pentēcostē* respectively and may even have been used to celebrating them. On the other hand, they do not betray any knowledge of (or just ignore) rabbinic rules about those festivals. The inscription suggests at least the existence of different forms of Judaism.

At first glance, the centuries of silence in religious practice do not seem as dark as suggested in the introduction to this chapter. Decorating a tomb on the festival of unleavened bread as well as on the Kalends is, however, evidence for no more than the continuity of fragments of the biblical organization of time. What is more, the combination of Pentecost, Azyma, and the Kalends points to a successful integration of such fragments of biblical time into Greco-Roman concepts and social institutions. Modern readers of the inscription face Glykon, a citizen of ancient Hierapolis, who may or may not have regarded his house as distinguished from others because it kept two ancient Jewish festivals. As distinct from a rabbinic approach to this question, he regards the Kalends as being on a par with the Jewish festivals. Whatever Glykon celebrated during his life, after his death, sacred time is used as a means to show his integration into the larger society to which he belongs. The establishment of this sacred time was guaranteed by two *collegia*, one of which represented the most important industry of Hierapolis. To put it crudely, Glykon does not want to be remembered as a Jew, but as a respected citizen of Hierapolis.

2 First-Century Christian Sundays?

Before any discussion of yearly or monthly festivals that establish sacred time in Christianity, the Christian Sunday must be mentioned. The high frequency and continuity of this institution makes it a formidable manifestation of sacred time. The early development of Christian celebrations of the first day of the week is, however, fraught with difficulties of terminology, chronology, and (the usual) dearth of sources.

Two reconstructions of the early history of Sunday celebrations have been proposed: according to the first one, Christians have held meetings and even meals on Sundays since apostolic times; according to the second, the Sunday became a preferred day for holding assemblies only during the second century. The following observations endorse the latter alternative.

When Justin Martyr writes an Apology for Christianity (apparently not long after the middle of the second century), he takes it for granted that his group meets for banquets on the first day of the week. He cannot but refer to this day by its Roman name, "the Day of Helios," to make himself understood. Although he tries to explain several Christian concepts to his reader(s), he does not call this day "the Day of the Lord" or the like. Justin mentions creation as well as Jesus' resurrection and appearance in front of his disciples as reasons for his choice of the "day of Helios" for his Christian group of philosophers.[5] In other passages, he explains and defends the number Eight as laden with symbolism in support of the celebration of the eighth day.[6]

In the same century, the Letter of Barnabas (Barn. 15.9) also refers to Christ's resurrection, appearance(s), and ascension on the eighth day, as well as to the Sunday as a turning point in the history of the world (V. 8). Barnabas briefly explains the emergence of the eighth day of the week as a day for joy (*agomen ... eis euphrosynēn*)[7] in contrast to the Sabbath and hence as manifestation of the divine rejection of the Sabbath.[8] Barn. 15 does not indicate that Christian gatherings should be held on this day of the week.

Regarding Jesus' death, the Synoptic Gospels disagree with John about the day within the month of Nisan. John regards the daylight of Friday as the fourteenth of Nisan, whereas the Synoptics construct their narrative with Friday as the fifteenth of Nisan. In spite of this divergence, all Gospels agree on the fact that Jesus' empty tomb was discovered on a Sunday.

Emphasizing the eighth day, the Gospel of John mentions the appearances of the risen Christ in front of his disciples on the first day of the following week (20:19). Thomas is absent from that meeting but joins his fellows "eight" (20:26) days later. Unfortunately, it cannot be known how those eight days were counted.[9] If the Christian gathering on the first day of the week was instituted as a ritualized rejection of the Sabbath (as it can be seen in Barnabas),[10] these references to the "eighth" day point to a seven-day week.

5 Justin, *1 Apol.* 67.3, 67.8 translation: Minns and Parvis 2009: 259–263.

6 Leonhard 2006: 137.

7 The Greek term implies celebrating, keeping, or performing a holy day, date, festival, custom, and so on (cf. LSJ meaning IV), not necessarily an assembly; cf. Athenaeus, *Deipn.* 7.80.3 = 307–308 [LCL 224, 436–439], writing (ironically) about the (fast of) the middle day of the Thesmophoria.

8 Leonhard 2006: chap. 4, esp. 136–139.

9 The eight days may have been counted inclusively – i.e., pointing to a week that runs from Sunday to Sunday. Be that as it may, even much later traditions still kept wondering how one should count Jesus' "three" days in the grave; note Aphrahat's tortuous attempt to establish three "days" between Jesus' death and resurrection; *Demonstratio* 12.12. Buchinger 2010 shows that the idea of a *triduum paschale* is not yet operative in the centuries under discussion here.

10 Vinzent 2014b: 284–285 assumes that the Sunday was Marcion's innovation. Justin accepted but modified it. This assumption fits the notion of Barn. as a text that likewise rejects the Sabbath to recommend the following weekday to the Christians; cf. 287–289. In his

There is no precedent for singling out the day following the Sabbath as a day of assemblies in the context of first century Christianity beyond what scholars infer from John's narratives.

These observations evoke the association of the other texts of the New Testament that are sometimes quoted in support of a first-century Christian custom of holding meetings on Sundays. Thus, it is true that Acts 20:7 makes Paul celebrate the Eucharist on the first day of the week. The author of Acts does not call this day the "eighth day," but it follows a reference to "seven days" of the implied narrator's ("we") sojourn in Troas. This passage is embedded in a sequence of chapters (Acts 19:8–21:18) that seem to be conspicuously interested in dating events of Paul's journeys – much more than the rest of the book.[11] The reason for this phenomenon is not known. The sequence of dates in these chapters is not a fragment of a liturgical calendar. It seems that the author of Acts wants to feign authenticity of the narrative here. It is not warranted to conclude from this that the local congregation in Troas would have regularly gathered on Sundays.[12]

Even fewer echoes of liturgical celebrations can be detected in the narrative about the beginning of Revelation unless one presupposes that any combination of "lord" and "day" must point to the technical terminology of "the Lord's Day," that is, Sunday. John receives his visions on the "lordly day" (Rev 1:10).[13] The commentaries regularly refer to studies either by Willy Rordorf (1962) or Samuele Bacchiocchi (1977): the former defends,

opposition to Marcion, Justin connects the Sunday with the creation and remarks that the true Christians (his congregation) read the Old Testament prophets.

[11] Vinzent 2014b: 276–277; Young 2003: 117 and further references there, as well as Leonhard 2006: 123–124 n. 9.

[12] For a different approach, cf. Barrett 1998: 951–952; Fitzmyer 1998: 668–669; Pervo 2009: 510, 513. Dating the composition of Luke's Gospel into the second century, Markus Vincent (2014a) opens up new avenues for the interpretation of this passage. As a composition that would have been written in Justin's lifetime, the book of Acts could presuppose some knowledge of Sunday as a date for Christian meetings that was just emerging. On the one hand, a post-Marcionite Luke could attest to the fact that Sunday was still in the making as the characteristic day for Christian gatherings. On the other hand, this Luke could try to tell a seemingly authentic story of Paul's travels. In that case, he would most vaguely mention a meeting on a Sunday, because he knew that his readers would debunk his story as a rather primitive forgery if a first-century Paul should celebrate a day that emerged in their own lifetime.

[13] Cf. Bacchiocchi 1977: 111–131. *Kyriakē hēmera* does not occur elsewhere in the New Testament. The Septuagint does not render *yom YHWH* by means of this term. The adjective *kyriakos* appears in the New Testament only as a designation of the Lord's meal, *kyriakon deipnon* in 1 Cor 11:20. Paul does not indicate that the Lord's banquet should be held on each first day of the week. Didache 14:1 *kata kyriakēn de kyriou* is no less obscure than Rev 1:10 and has received wide-ranging interpretations.

and the latter attacks, the notion that followers of Jesus met on the first day of the week in the middle of the first century.[14] Bacchiocchi appropriately reads "lordly day" in the light of Old Testament prophecies about "the Lord's (= YHWH's) day" instead of the much younger terminology of "the Lord's Day" for Sunday. The rest of the book of Revelation that follows this introduction is replete with allusions to images pertaining to descriptions of the Day of the Lord according to the Old Testament.[15] It stands to reason that the "lordly day" of Rev 1:10 points to the "Lord's Day" representing a nontemporal entity, not a technical term for the liturgical organization of a Christian concept of the week. The term anticipates the following visions, which indicate neither that the world came to an end at the moment when John allegedly saw them nor that these visions were received on a special day (in terms of a liturgical calendar). Rev 1:10 does not, after all, hint at a congregational gathering. Sunday would not, furthermore, be a typical day for the reception of revelations. On the contrary, the author of Revelation does not want to disclose on which day of the week his hero received this revelation – perhaps because this information would be utterly unimportant for the understanding of the book.[16] Thus, even one of the youngest texts of the New Testament has not yet any interest in, or knowledge of, a Christian structure of sacred time based on the practice of regular, weekly meetings on the day after the Sabbath.

[14] Aune 1997: 84 rejects Bacchiocchi's (1977: 123–130) assumption that "the Lordly day" should echo the biblical notion of "the Lord's day." However, Aune bases his conclusions on Rordorf's 1962 study. Rordorf argues for an early date for regular Christian celebrations of the Eucharist on Sunday accompanied by the emergence of certain aspects of interpretation of this day. Aune does not adduce new evidence and also misrepresents Bauckham 1982 (in 1997: 83). Bauckham argues that "the Lord's" is virtually identical in meaning with "the Lordly" *viz.* day. It is, hence, inadmissible to quote Bauckham as a source for the semantic differentiation between the adjective and the noun in the genitive. There is no *linguistic* reason that precludes an identification of the two concepts – "the day of the Lord" and "the Lordly day." Following Rordorf's approach, one must read Rev 1:10 as the first literary attestation of the Sunday as the "Lord's day"; cf. Giesen 1997, 85–86; similarly Ford 1975: 382. If Rev was to be dated to the late *second* century (which would make sense in the context of Vinzent's [2014a] dating of the basic New Testament texts), it would indeed be difficult to claim that "the Lordly day" of Rev 1:10 should *not* have evoked associations of a kind of Christian Sunday. Regarding the *early* second century, one must however ask what the concept of "the Lordly day" should have comprised to classify as a precursor of the later Christian Sunday. In any case, Rev 1:10 does not provide evidence for a liturgical custom to celebrate the Eucharist on every first day of a seven-day week.

[15] Cf. Bacchiocchi 1977: 129–130.

[16] The notion that the literary John should have received his vision during a liturgical celebration is not supported by the text. Cf. Ford 1975: 382 for a different opinion.

The oldest document that allegedly attests Christian gatherings on Sundays is the First Letter to the Corinthians. Its last chapter begins with Paul's recommendation that each should put his share for the collection into a savings box "on the first day of the week (*kata mian sabbatou*)" (1 Cor 16:2). The letter does not give a reason for the choice of the day of the week. It only explains that it is useful to accumulate money week by week as against a collection *en bloc* when Paul would visit the group to take the money with him to Jerusalem.

Several reasons are normally given for Paul's choice of the first day of the week.[17] Paul could have wanted to avoid collection of money on Sabbaths. The text does not speak about, or hint at, a liturgical gathering as a background of the collection.[18] It may just envisage the Corinthian Christians performing small collections of savings in their homes to be able to give generously when Paul visits the town.[19] Paul's suggestions may just reflect his own strategy to maximize the sum that can be collected.

This does not, however, explain why Paul communicates this suggestion in an official letter to the whole congregation of Corinth and why he chooses a certain day of the week. Every Corinthian could follow his advice on any day during the week. The question evokes the association of a small group of famous documents and inscriptions that also speak about the collection of money at certain intervals within associations – more precisely: once a month, not weekly.[20]

[17] Schrage 2001: 428 collects reasons given in the secondary literature for the collection of money on Sundays. Cf. also Fitzmyer 2008: 614–615.

[18] Rouwhorst 2001: 251.

[19] Young 2003: 112–116 (arguing that *hekastos ... par' heautō* implies "each one ... at home"); cf. Klinghardt 1996: 326; Vinzent 2014b: 275–276. Schrage 2001: 429 emphasizes that the verse does not imply an official financial organization on the level of the congregation.

[20] Inscription of the *collegium* of Diana and Antinous in Lanuvium, col. I 10–13; translation by Bendlin 2011: 213. According to Bendlin, the reading also explains *Dig.* 47.22.1 *praef.*–1.1; originally from the Third Book of *Institutiones* composed by the jurist Aelius Marcianus between 218 and 235, Bendlin 2011: 223 and cf. the discussion there. Laubry and Zevi 2012 present two fragments of inscriptions from Ostia that refer to decisions of the Senate in a similar way as the Lanuvium inscription. They conclude that innovative legislation of the emperor Hadrian is implied in these texts, one of which refers to the date of 121. Hence, a general expectation of *collegia* to meet only monthly would not be expected for the first century. Weekly meetings according to 1 Cor 16 would not, however, interfere with imperial standards and customs regarding associations because a Christian group in first-century Corinth could never attain to (and was also not interested in) Senatorial recognition. Thus, it is assumed here that Christian customs to hold business meetings were influenced by normal ways of how groups and associations would structure their organization without strict adherence to a time interval. Downs 2008: chap. 3 assumes that the collection was done during the liturgical meetings of the Corinthian congregation, 128. He does not yet envisage

One of those inscriptions, however, mentions that the *collegium*, which had the inscription cut in stone, used to meet twice in August (col. II 11–13), which is clearly a case of inconsistency if only monthly meetings should have been allowed. Andreas Bendlin restores and explains another text, namely a roughly contemporaneous decree of the Roman Senate regarding a maximum of monthly gatherings as referring to the business meeting of the *collegium* only. Its members would collect five *asses* on the occasion of their monthly meetings, which were not identical with the occasions to hold a banquet.[21]

These documents indicate that groups did not only meet for dinners. Furthermore, the collection and management of the group's finances need not have taken place at the same occasion. The *conventus* of the association as well as the meeting of the Christian group in Corinth (1 Cor 16:2) was not regarded as a proper occasion for the celebration of a festive dinner. Thus the Corinthians may have been used to meeting on the first day of each week, on which occasion Paul suggested performing a public collection of money.[22] They would not dine, that is, celebrate the Eucharist, at that time. The First Letter to the Corinthians does not, therefore, support the reconstruction of the regular celebration of Christian meals on Sundays in New Testament times.

The three New Testament passages just mentioned refer to "the Lord's day" (or the like). They could be read in the light of texts that emerged in the second century beginning with Barnabas and including Justin's *First Apology* as well as Ignatius's *Letter to the Magnesians* (9.1; whose date is debatable). However, apart from the Gospel narratives, which set the discovery of the empty tomb on a first day of the week, there is no hint at an understanding of this day as an occasion for the remembrance of Christ's resurrection. The notion of a "weekly Pascha" is a phenomenon of the early fourth century that presupposes the Dominical Pascha after it had

a business meeting of the congregation as different from its meeting for banquets. Groups such as Pauline churches did not, of course, distinguish between Eucharistic and non-Eucharistic meals (McGowan 1997). However, they may have been used to distinguishing between gatherings for banquets (= Eucharists) and other meetings without banqueting.

21 Bendlin 2011: 240.

22 As Thraede 2004 has shown, Pliny knows next to nothing about the Christian groups according to his letter to Trajan. It must be mentioned in the present context that Pliny refers to two meetings of the group of Christians "on the determined day," only one of which is an afternoon meal. This piece of evidence must not be read into 1 Cor 16:2. It indicates what Pliny thought to be a plausible description in the eyes of the emperor. Neither Pliny, nor Ignatius (*Magn.* 9:1), nor the Didache (Did. 14:1; cf. n. 26) know anything about Sundays as dates for regular Christian celebrations. These texts have been discussed in Leonhard 2006: 122–135; cf. also Vinzent 2014b: 270, 278–279, 289.

prevailed over the Quartodeciman celebration. The agreement of the four Gospels on the first day of the week as the day of the resurrection does not legitimize first century customs to celebrate Sundays. It points to a *literary*, not a *liturgical*, tradition. First-century groups of followers of Jesus did not celebrate Sundays.

3 Tertullian: Celebrating Twice

In the preceding sections, it has been claimed that nascent Christianity was made up of groups that were quite loath to celebrate festivals or to keep appointed times, even to perform banquets on recurring days. Tertullian (155/160–220/240 CE) emphasizes that Christians celebrate lots of festivals. He vilifies the customs and performances of the Roman festivals but continues (*Apol.* 35.4 *CCL* 1.145; transl. *ANF*):

> Poor we, worthy of all condemnation! For why do we keep/perform the votive days and high rejoicings in honor of the Caesars with chastity, sobriety, and virtue? Why, on the day of gladness, do we neither cover our door-posts with laurels,[23] nor intrude upon the day with lamps?

All Christians who are included in Tertullian's "we" keep the festival days of the Roman calendar.[24] They only refuse to join the performance of some customs within festive practice. Beyond the pagan festivals, Christians continue to celebrate their own ones. Tertullian claims that the fifty days of Pentecost alone surpass any number of pagan festivals with regard to the quantity of sacred time.[25]

Tertullian claims for Christianity the legitimacy and political status of a *curia*; that is to say, he describes Christianity as an official and public section of the Roman city of Carthage.[26] In general, his understanding of the role of the churches in their city is modeled on public

[23] Cf. however his statement in *Idol.* 15.1 *CCL* 2.1115 transl. *ANF* "But 'let your works shine,' says He; but now all our shops and gates shine! You will now-a-days find more doors of heathens without lamps and laurel-wreaths than of Christians." Although Tertullian disapproves of it, Christians do not strictly refrain from decorations of their gates; cf. Schöllgen 1982: 19–20.

[24] Waltzing/Severyns 1961: 75 note ad loc.: Christians did not celebrate imperial festivals. Yet, they celebrated the Emperor "dans leur cœur" (*conscientia* 35.1).

[25] In a similar way, Tertullian discusses the problem of Christians celebrating pagan holidays and festivals in *Idol.* 14.6–7 *CCL* 2.1115.

[26] *Apol.* 38.1 *CCL* 1.149. He goes on to ponder the Roman fears of associations that interfere in politics and votes. Cf. Bendlin 2005: esp. 78–79, 102–104. Tertullian invented the term *religio licita* (*Apol.* 21.1 *CCL* 1.122) for his own rhetorical purposes. He does not avail himself of a coined juridical expression. I am grateful to Benedikt Eckhardt for sharing with me his

institutions.[27] Tertullian does not compare Christianity with clubs or associations. He speaks about *tribus* (as a traditional division of the city of Rome) *et curiae et decuriae*. Only at the beginning of the chapter does he call Christianity a (political and hence potentially dangerous) "party" (*factio*) and a *corpus* that makes a *coetus* and a *congregatio*.[28] At the end of the chapter, he ventures a pseudo-legal definition saying that one should refrain from referring to Christianity as a party at all. Christianity should be compared with a *curia*, not a *factio*. The term *collegium*[29] does not appear in the *Apology*. Tertullian also gives a list of festivals of the respective "others" whom Christianity surpasses in frugality and decency. He only refers to municipal, public celebrations of certain places within the Roman Empire, not to festivals that are typical of *collegia*, which did not have the same status as *curiae*. Tertullian does not discuss technical characteristics of Christianity in an accurate way. His reference to Christianity as a *curia* lays a resounding and massively exaggerated claim to a place in the grand, public sphere of Carthage.

Tertullian assumes that Christians celebrate Pesach and Pentecost along with the festivals of their city, such as "*Saturnalia*, the festivals of New Year and Midwinter and the *Matronalia*."[30] Romans expect their fellow citizens to celebrate them. Tertullian acknowledges the basic correctness of that expectation. Apparently, Christians who perform aspects of festivals of their city and the empire (such as associations and other public institutions) increase their legitimacy as a social body. Even Tertullian claims for himself that he

observations and bibliographies about Tertullian's reference to *curiae* and about the comparison of Christian groups with Roman voluntary associations.

27 "Why, the very air is soured with the eructations of so many tribes, and curiae, and decuriae. The Salii cannot have their feast without going into debt; [...] the smoke from the banquet of Serapis will call out the firemen" (*Apol.* 39.15 *CCL* 1.152, transl. *ANF*). Dawson 2014 reviews studies of an inscription from the West Tunisian town Chemtou (ancient Simitthus), *CIL* VIII 14683, that summarizes the bylaws of the town's *Curia Iovis* from 185 CE – roughly Tertullian's time. Dawson refers to Schmidt's (1890) theses as an authoritative interpretation of this inscription even against more recent approaches. Schmidt argues convincingly that the inscription was not erected by a *collegium* but by a larger and more public part of the ancient municipality, a *curia*.

28 Tertullian does not talk about a Christian *concilium*, which would be the technical term for the convention of a *curia*; cf. Schmidt 1890: 605–606.

29 Tertullian knows the concept. Cf. his use in *Spect.* 7.3 *CCL* 1.233 l. 10; 11 *CCL* 1.238 l. 7 and 9 (*collegium artium musicarum*, etc.). Tadeusz Kotula 1968: 116–118 (cf. 101–102) reads Tertullian's testimony as evidence in support of the thesis that the North African *curiae* had lost most of their former political functions and had become groups like mere associations. Regarding Tertullian's testimony, this is a bit of circular reasoning, for Tertullian thinks that *curiae* are powerful sections of the society of Carthage.

30 *Idol.* 14 CCL 2.1115.

would buy and set up flowers on a festival day of the society of Carthage. He would just refrain from wearing them as a crown. These claims emphasize that the Christians' structuring of sacred time is well integrated within their city. Although they may mark their identity by means of tiny adaptations within socially expected and acceptable practice, they abide by and large by the general rules.

4 "The Jerusalem Above Is Free – This Is Our Mother"

Tertullian's church of early third-century Carthage claimed an official position within its city by integration of public sacred time into the lives of its members. This helped them integrate into the society of *their city*. It is obvious that Carthage was *their city*. This evokes a question of ancient theology: what is the Christians' city? Prior to Tertullian's time, several New Testament texts had given a different answer to that question. Having considered Tertullian's approach, the position of these New Testament texts comes to the fore more prominently.

Christians were not required to sever their connections to the rest of their societies. The extent to which individuals sought integration into, or dissociation from, these societies varied greatly. Some Corinthian Christians, for example, did not only taste a bit of meat from "idol worship" in a *triclinium* of a non-Christian friend or business partner but participated in public banquets in temples, "in a place of idols."[31] They would not only just be seen by coincidence. They would on purpose celebrate publicly to demonstrate their integration into the social network of the city and its important subgroups. In a first round of arguments, Paul permits that practice claiming that the pagan deities are just nothing(s). He locates the problem in the eye of the "weak" Christian beholder, not in the practice of the "strong" Christian diner. In a second round of arguments, he implicitly admits that he failed in his attempt to reinterpret idol-food as harmless. In the end, he adopts a more or less "weak" approach and advises his readers that everyone should just totally avoid the practice. Paul's inconsistence may be the result of his just not being bold enough to stick to his initial position. However, it seems more plausible that he was actually more sympathetic to the weak Christians than to the strong ones.

[31] 1 Cor 8:10. Paul regards eating as problematic if a weak person observes a strong one "reclining in a pagan temple" – *eidōleiō katakeimenon*. Tertullian would not recline in public at the feast of Bacchus but he would also not reject food that originated from that context.

In Rom 14, Paul associates the consumption of meat and wine (v. 21; without variant readings) with the distinction of days. He suggests to his readers that they should adopt a position that "selects/singles out *all* days," that is, a position that does not single out days at all. Andrew McGowan has shown that the position of the weak in this context is neither exclusively nor typically Jewish.[32] It is not, likewise, concerned with fast days. These weaker members of the group opt for a more comprehensive disengagement from Roman society than the stronger ones. The principles of their practice include rules about the consumption of meat (in general, not specifically meat sacrificed in temples) and wine, and the performance or disregard of festivals. In Rom 14:2, the acceptance of the consumption of meat is mentioned first. Verse 5 likewise mentions the distinction of certain days first. The weak ones reject the consumption of meat together with the distinction of days. Their strategy of disentanglement from the surrounding society includes the rejection of the typical sacrificial cuisine (meat and wine) and the performance of sacred time – without any interest in a distinction between more Jewish or more Roman ways of "singling out days." In this letter, Paul seems to hold a more distanced position with regard to the details of religious practice in the field.

Regarding festivals, Paul assumes a weak position in the Letter to the Galatians.[33] Thus, he tries to dissuade the implied readers from organizing their time according to the "weak and inferior elements (*stoicheia*)" such as "days, appointed times (*kairoi*), months, years."[34] A strong position would have implied that celebrating or not celebrating festivals would be indifferent. The metaphor of the "weak elements" invites associations of astral bodies or gods of the Greek and Roman pantheon as well as the four elements.[35] The Galatians are warned not to return to the celebration of festivals and timekeeping of their former status – the festivals of their society, city, or their formerly favorite sanctuary.[36]

32 McGowan 1999: 226–231.

33 In analogy to 1 Cor 8:4–5, Paul could have emphasized that keeping festivals is as useless as it is innocuous. He also refers to his weakness in this context; Gal 4:13–20. Among other things, Paul wants to divert the Galatians' material support for his competitors to himself; cf. Gal 4:17–8 and 6:6.

34 Gal 4:8–11. Vinzent 2014b: 273, 282 n. 80 refers to Philo, *Spec.* 2:42 (that every day is a festival [*heortē*] for people who follow the laws of nature) and writings of Christians who reject the celebration of Jewish festivals. He also observes that Paul does not speak about "weeks" here.

35 Mayer-Haas 2003: 88–89 n. 32 quotes Philo, *Spec.* 2:225 for the opinion that Greeks worship the elements. For Philo, cf. Witulski 2000, 133–141.

36 Cf. Witulski 2000, esp. 152–168; Nanos 2002: 267–269. Hardin 2008: 116–147 argues in favor of an interpretation of Gal 4:10 in terms of the Roman imperial cult. He (repeats and)

Which festivals could the Galatians have been tempted to celebrate? The answer can be sought in several passages of the Letter to the Galatians. In the present context, Gal 4:26 emerges as a significant signpost: "By contrast, the Jerusalem above is free – this is our mother."[37] Neither Paul nor the redactors of the New Testament as a collection explain the concept of the "heavenly Jerusalem" here. As members of their society, readers of Galatians were supposed to celebrate the festivals of their cities. Such a performance would be a real "return" to their former state. As readers of the Torah, they could have felt compelled to observe biblical festivals. Paul would not have approved of this more than he approved a "return" to the festivals of their city: as true citizens of the heavenly Jerusalem, they are free from both.

Thus, Paul acknowledges the principle that citizenship determines one's participation in festivals.[38] He uses the principle to declare absolute freedom from any such performance in the social reality of the church because the Christians' "mother" city is the heavenly Jerusalem, where no festivals are kept.[39] For citizens of the heavenly Jerusalem, the differences between Jews

accepts Witulski's (2000) similar assumptions about the reference of Gal 4:10 but tries to integrate the text into the argument of Gal as a whole without Witulski's solution to interpret this verse as part of a long interpolation. Hardin also observes that those passages that have often been interpreted as traces of the Galatian's orientation toward the Torah are sufficiently ambiguous to be read in a context of the Roman imperial cult, too; 138–142.

37 Betz 1979: 238 and cf. 246–248.

38 In Tertullian's writings, this observation is obvious. It also corresponds to the participation in the sacred time of their cities in voluntary associations: cf. Suys 2005; Ausbüttel 1982: esp. 49–59. Christian groups cannot be compared directly with Roman voluntary associations. The parallels and differences have been discussed in recent publications, cf. Harland 2003. With respect to sacred time, it stands to reason that Christian groups adhered to a principle that one celebrates the festivals of one's city. Paul's remark about the Christians' mother city as a commonplace and obvious assumption thus fits perfectly with his opposition against festivals. In this mind-set, a declaration of adherence to a city entails a clarification of the festivals that one celebrates. Tertullian suggests a strong position – for the participation in public life – against Paul's weak one. Cf. Meiser 2007: 201–202.

39 See the following paragraph and cf. Schwemer 2000: 224–225. In her description of the consequences that a Christian is a citizen of the heavenly Jerusalem, Schwemer does not take into consideration Paul's statement about the festivals. Paul also refers to this notion in Phil (3:20). Christians claim that their *politeuma* is in heaven. Younger texts of the NT expand this notion: on earth, Christians are (like) strangers: Eph 2:19; 1 Pet 1:17; cf. John 18:36. Theobald 1988: 20 claims that Phil 3:20 was Paul's reuse of a notion of his adversaries. In the context of the present discussion, it is more plausible to read this statement as Paul's own opinion from the beginning; cf. Schwemer 2000: 228–236. Likewise, Gal 4:26 and Rom 11:26 imply that the Christians' citizenship is in the future/heavenly city. This does not prevent the literary figure of Paul in Acts 25 to use his Roman citizenship to improve his no less literary situation.

and Greeks, and so on, as well as between circumcised and uncircumcised have been obliterated.

Some of the first Christians – the strong ones – celebrated the festivals of their earthly cities, participated in public banquets in temples, and had no qualms regarding the consumption of (sacrificial) meat and wine. Several of the younger New Testament texts reflect a strictly analogous position of the weak ones. They also celebrate the festivals of their city and the place of their (spiritual) citizenship – (heavenly) Jerusalem. Now, one might expect Christian groups to celebrate festivals of earthly Jerusalem, such as Sukkot, Pesach, Shavuot, or Yom Kippur.[40] Before or after the destruction of the Temple, they could have aspired to affiliate themselves to a more or less ideal Israel as living in Jerusalem and performing rituals prescribed in the Torah. This idea did not, however, occur to them. The New Testament is absolutely silent about the performance of such celebrations of Christian groups. On the contrary, younger New Testament texts are quite explicit about the duties of citizens of the heavenly Jerusalem with regard to their performance of festivals.

The Letter to the Hebrews (13:14; cf. 12:22) urges its readers to leave their earthly cities to reach the heavenly one to which they actually belong and in which their names are recorded in the lists of citizenship. "Leaving the cities" means following Christ with all consequences.[41] For Hebrews, the Christians' city is essentially part of the heavenly or (remotely) Platonic sphere of ideas. For the notion of a heavenly city, the author of Hebrews could have referred directly to Plato's *Republic* (9.592 a–b). Hebrews teaches that God had instructed Moses how to build the Temple showing him a heavenly model of the earthly sanctuary (Exod 25:9, 40; Heb 8:5). If Christians celebrate the festivals of their place of citizenship, it must be asked which festivals were celebrated in their heavenly city – not in the earthly Temple of Jerusalem before its destruction.

Like the Letter to the Hebrews, the Book of Revelation advises its readers to leave a city (18:4)[42] and to join the inhabitants of the heavenly Jerusalem.

[40] Col 2:16 states what would be the obvious question from later times on (until today) and reformulates Gal 4:10 saying that one should not be concerned about "festivals, new moons, Sabbaths" (cf. Isa 1:13). It just adds that Christians should also reject the days, times, and festivals of the Bible, which opens up another debate than Gal.

[41] If one asks which city should be left and abandoned by Christians, the answer could be generic – the very society/city where they happen to be living. It could also point to Rome, which is called "the eternal city" by Ovid (*Fast.* 3.72); Koester 2001: 571. The texts quoted in the following paragraphs were collected and interpreted by Wenger 1954 and Theobald 1988.

[42] Backhaus 2009: 473. Stökl Ben Ezra 2003: 192 suggests that Heb 13:14 advises the addressees of the letter, the community in Jerusalem to leave the earthly city in concrete terms.

They are supposed to celebrate the festivals of their city to maintain and increase their integration into it as its citizens. What did they celebrate? Rev 21 does not leave any doubt that nobody should start celebrating Yom Kippur or Pesach or even the Sabbath, which is not mentioned in this book.[43] On the contrary, Rev 21:22–26; 22:5, 17 not only emphasizes that there is definitely no temple in the heavenly Jerusalem. The text also adds that neither sun nor moon is shining there because God is its light (echoing Isa 60:1–2, 19–20 but against Isa 66:23), and that its gates are never closed because there is no night. Hence, neither time nor calendar is even possible in the heavenly city. The Christians who follow the customs of their (true) hometown are not provided with a means to use sacred time to establish, shape, display, and negotiate the relationship to their ancestral city, the heavenly Jerusalem.[44]

Hebrews and Revelation urge their readers to abide by certain ethical standards (cf. Heb 12:4–13:19) or to "keep the words of the prophecy of this book" (Rev 22:7). They do not forbid the performance of festivals or the construction of sacred time. However, it is obvious that their imagery can be exploited much more easily by opponents than by supporters of festivals.

Against the backdrop of Tertullian's discussion of festivals in his apology for Christianity, these New Testament texts voice a resounding "no" to the celebration of festivals. Before Tertullian could boast Christianity's rich, almost exuberant number of festivals, to promote his group's integration into Roman society, the New Testament texts had repudiated festivals. Even though this is not the only voice that can be heard speaking in the New Testament, the texts quoted here emphasize their groups' rejection of Roman society, its customs, and the integration of its members into that

[43] Heb mentions the Sabbath in a close analogy to 1 Cor 5:7. The idea of Christ as the Christians' Pesach animal does not mean that any Christians at Corinth would have celebrated Pesach; rather, it implies that they must cast out a sinner from their group. After the watershed event of Jesus' death, his followers live in an age of unleavened bread. Similarly Heb 4:1–13 implies that anyone who reached "God's rest" would not need or want to keep the Sabbath. One must overcome the temptation to construct a category of "Jewish Christians," "Jewish believers in Jesus," or the like here. This terminology only betrays the modern interpreter's conviction that he or she should know the borders between Jews and Christians with sufficient accuracy to identify and describe hybrids. The terminology of "Jews" and "Christians" is at best approximative and always tainted with judgments of reception history. One must describe groups that *keep* certain festivals and *perform* certain customs, not groups (let alone individuals) that *are* Jews, Christians, or even hybrids.

[44] The idea that those Christians who were most advanced in Christian philosophy – the *gnōstikoi* – had left the more inferior status behind, a status wherein one celebrates "festivals" (*heortais*) and "appointed days" (*hēmerai apotetagmenai*), is still supported by Clement of Alexandria; *Strom.* 7.7.35.1–7 (esp. 3 and 6; *ANF*).

society. As all other groups, Christians met for banquets and other purposes. They did not begin to establish a Christian system of sacred time by means of scheduled meetings until the middle of the second century,[45] when experiments in that direction started with Pesach and Pascha.

5 Conclusions

This chapter exposed texts of the literary history of Christianity to the thesis that inhabitants of the ancient world celebrated the festivals of their city. Even if they did not do that in practice, they at least knew that they were expected to do it. Celebrating and not celebrating, as well as partly celebrating in any kind of nuance, were hence a means of establishing and expressing one's inclusion in, or exclusion from, the surrounding society. The chapter concludes that even groups whose members fiercely rejected any celebration at all would nevertheless claim that they acted in accordance with the rules and customs of celebrating the festivals of their city – just not the earthly city in which they happened to actually live. In other words, texts that offer a form of self-conception on the basis of the idea that Christians hold a citizenship in a heavenly city reject the establishment of festivals on the same grounds as texts that recommend Christianity as part and parcel of their municipal organization.

This principle emerges as operative from the Hierapolis tomb inscription. Instead of reading the Jewish, Christian, pagan, pagano-Jewish, or even Judeo-Christian inclinations of the owner of the tomb into this inscription, one must admit that he just uses sacred time to express his and his family's integration into the city of Hierapolis.

Any argument that early Christian groups did not celebrate festivals at all faces the objection that these groups may in any case have celebrated the Lord's Day or Sunday; § 2 argues that the establishment of the Sunday as a typical day for regular meetings of Christians is a phenomenon of the second century. It was not conceivable in apostolic times.

The principle that one ought to celebrate with one's city comes to the fore again in § 3. Tertullian emphasizes that Christians celebrate all kinds of festivals of Carthage and the Roman Empire in addition to their own lavish festival calendar. Tertullian himself might introduce some subtle changes into

45 The meal customs of a group are regarded as an important characteristic for the assessment of its quality. Thus, Justin briefly describes the Christian Eucharistic meetings on the "day of Helios" to his Roman readers (*1 Apol.* 67; see § 2). Following earlier studies by Gerard Rouwhorst, Leonhard 2006 argues that the Quartodeciman Pascha emerged as a reaction against Pesach toward the middle of the second century.

the performance to make the celebrations acceptable for himself as a Christian. His apologetic explanations exhibit Christianity as a group that fiercely fights for its total integration into the social web of Carthage – including the (almost) full acceptance of its neighbors' festivals.

Against the backdrop of Tertullian's ideas, § 4 goes back to New Testament texts that oppose the celebration of festivals and the observation of appointed times. These show an approach diametrically opposite to Tertullian's opinions. The groups that present their self-portraits in these texts shun pagan (as well as Jewish) celebrations of festivals. They construct themselves as citizens of a heavenly Jerusalem that has lost any means to even distinguish, let alone celebrate, appointed times and festivals. Against other positions that emerge in the margins of these texts, the groups discussed here aspire to the disentanglement from the societies of their cities.

Thus, Christians who emphasized their adherence to a heavenly Jerusalem (without temple and calendar) were free to meet at any date and at any time, but also to refrain from any celebration. The advantage of freedom from obligations to celebrate at appointed times is matched by a disadvantage of arbitrariness and the lack of any means to respond to the sacred time of others in one's vicinity. Christians who opted for social integration would join in their fellow citizens' celebrations of festivals even before Tertullian.

These observations have three consequences for the future assessment of the history of Christian religious festivals as performances of sacred time. First, there is no basis for any reconstruction of a continuous celebration of festivals between first-century Jerusalem and the cycles of Christian festivals from the late fourth century onward. Someone who wants to claim the opposite must bring reliable evidence for the continuous performance of festivals. Continuity is not the rule but the theoretical exception.

Second, any similarity or conspicuous dissimilarity between rabbinic and Christian liturgical practice stems either from a secondary and independent ritualization of biblical texts or from the parallel development of rituals in conflict from the second century onward. Jews and Christians may also share many customs and beliefs that they adopted independently from their Greek and Roman neighbors.

Third, theses about liturgical developments must dispense with the simplistic cliché that festivals and the construction of sacred time would slowly and continuously build up the identity of groups and individuals by shaping their collective memory. Attempts to use rituals to manipulate public memory are as ubiquitous in antiquity as they are today. Both the performance of certain customs and the resistance to the public pressure to participate in their performance belong to the repertoire of means to negotiate and legitimize power relations in the society. The earliest Christian texts reworked and discussed their biblical heritage. This did not require them to put into practice ancient notions about rituals such as the performance of certain festivals at certain times. For Tertullian, publicly visible practice, not ideology or memory, is at stake in the polemic about the creation of sacred time. The creation or acceptance of sacred time is not an objective but rather a side effect of certain strategies to strengthen one's own group. Furthermore, commemoration may be involved in discourses that legitimate performances that create sacred time. It need not be their cause. Christian groups that reject joining or establishing performances that create sacred time indicate that groups could dispense with that concept in theory and practice. Reconstructions must be receptive to historical records and constellations of data that do not point to a group's celebration of appointed times. There were as many good reasons for celebrations as there were reasons against them.

Bibliography

Ameling, W. (ed.). 2004. *Inscriptiones Judaicae Orientis, vol. 2: Kleinasien*. Texts and Studies in Ancient Judaism 99. Tübingen: Mohr Siebeck.

Aune, D. E. 1997. *Revelation 1–5*. Word Biblical Commentary 52. Nashville: Thomas Nelson.

Ausbüttel, F. 1982. *Untersuchungen zu den Vereinen im Westen des Römischen Reiches*. Frankfurter Althistorische Studien 11. Kallmünz: Michael Lassleben.

Bacchiocchi, S. 1977. *From Sabbath to Sunday: A Historical Investigation of the Rise of Sunday Observance in Early Christianity*. Rome: The Pontifical Gregorian University Press.

Backhaus, K. 2009. *Der Hebräerbrief*. Regensburger Neues Testament. Regensburg: Pustet.

Barrett, C. K. 1998. *A Critical and Exegetical Commentary on the Acts of the Apostles, vol. II:* Introduction and Commentary on Acts XV–XXVIII. International Critical Commentary. Edinburgh: T&T Clark.

Bauckham, R. J. 1982. "The Lord's Day." In D. A. Carson (ed.), *From Sabbath to Lord's Day: A Biblical, Historical, and Theological Investigation*. Grand Rapids, MI: Zondervan, 221–298.

Bendlin, A. 2005. "'Eine Zusammenkunft um der *religio* willen ist erlaubt...'? zu den politischen und rechtlichen Konstruktionen von (religiöser) Vergemeinschaftung in der römischen Kaiserzeit." In H. G. Kippenberg and G. F. Schuppert (eds.), *Die verrechtlichte Religion: Der Öffentlichkeitsstatus von Religionsgemeinschaften*. Tübingen: Mohr Siebeck, 65–107.

Bendlin, A. 2011. "Associations, Funerals, Sociality, and Roman Law: The *collegium* of Diana and Antinous in Lanuvium (CIL 14.2112) Reconsidered." In M. Öhler (ed.), *Aposteldekret und antikes Vereinswesen: Gemeinschaft und ihre Ordnung*. Wissenschaftliche Untersuchungen zum Neuen Testament 280. Tübingen: Mohr Siebeck, 207–296.

Betz, H. D. 1979. *Galatians: A Commentary on Paul's Letter to the Churches in Galatia*. Hermeneia. Philadelphia: Fortress Press.

Buchinger, H. 2010. "Was There Ever a Liturgical Triduum in Antiquity? Theological Idea and Liturgical Reality." *Ecclesia orans* 27: 257–270.

Dawson, C. 2014. "CIL VIII 14683 and the North African Curiae." Unpublished paper read at the 145th annual meeting of the Society for Classical Studies in Chicago; abstract no. 51.5: https://classicalstudies.org/annual-meeting/145/abstracts/chris-dawson (accessed May 4, 2017).

Downs, D. J. 2008. *The Offering of the Gentiles*. Wissenschaftliche Untersuchungen zum Neuen Testament 2/248. Tübingen: Mohr Siebeck.

Edrei, A. and D. Mendels. 2007/2008. "A Split Jewish Diaspora: Its Dramatic Consequences." *Journal for the Study of the Pseudepigrapha* 16: 91–137 and 17: 163–187.

Fitzmyer, J. A. 1998. *The Acts of Apostles: A New Translation with Introduction and Commentary*. Anchor Bible 31. New York, NY: Doubleday.

Fitzmyer, J. A. 2008. *First Corinthians: A New Translation with Introduction and Commentary*. Anchor Bible 32. New Haven: Yale University Press.

Ford, J. M. 1975. *Revelation: Introduction, Translation and Commentary*. Anchor Bible 38. Garden City, NY: Doubleday.

Giesen, H. 1997. *Die Offenbarung des Johannes*. Regensburger Neues Testament. Regensburg: Pustet.

Hadas-Lebel, M. 1979. "Le paganisme à travers les sources rabbiniques des IIe et IIIe siècles : Contribution à l'étude du syncrétisme dans l'empire romain." In W. Haase (ed.), *Aufstieg und Niedergang der römischen Welt* 2.19.2. Berlin: De Gruyter, 397–485.

Hardin, J. K. 2008. *Galatians and the Imperial Cult: A Critical Analysis of the First-Century Social Context of Paul's Letter*. Wissenschaftliche Untersuchungen zum Neuen Testament 2/237. Tübingen: Mohr Siebeck.

Harland, P. A. 2003. *Associations, Synagogues, and Congregations: Claiming a Place in Ancient Mediterranean Society*. Minneapolis, MN: Fortress.

Klinghardt, M. 1996. *Gemeinschaftsmahl und Mahlgemeinschaft: Soziologie und Liturgie frühchristlicher Mahlfeiern*. Texte und Arbeiten zum neutestamentlichen Zeitalter 13. Tübingen: Francke.

Koester, C. R. 2001. *Hebrews: A New Translation with Introduction and Commentary*. Anchor Bible 36. New Haven: Yale University Press.

Kotula, T. 1968. *Les Curies municipales en Afrique romaine*. Prace Wrocławskiego Towarzystwa naukowego, Seria A 128. Wrocław: Zakład narodowy im. Ossolińskich.

Laubry, N. and F. Zevi. 2012. "Inscriptions d'Ostie et phénomène associatif dans l'Empire Romain: Nouveaux documents et nouvelles considération." *Archeologia Classica NS* 63: 297–343.

Leonhard, C. 2006. *The Jewish Pesach and the Origins of the Christian Easter: Open Questions in Current Research*. Studia Judaica 35. Berlin: De Gruyter.

Mayer-Haas, A. J. 2003. *"Geschenk aus Gottes Schatzkammer" (bSchab 10b): Jesus und der Sabbat im Spiegel der neutestamentlichen Schriften*. Neutestamentliche Abhandlungen, New Series 43. Münster: Aschendorff.

McGowan, A. 1997. "Naming the Feast: The *Agape* and the Diversity of Early Christian Meals." In E. A. Livingstone (ed.), *Papers Presented at the Twelfth International Conference on Patristic Studies Held in Oxford 1995: Biblica et Apocrypha, Ascetica, Liturgica*. Studia Patristica 30. Leuven: Peeters, 314–318.

McGowan, A. 1999. *Ascetic Eucharists: Food and Drink in Early Christian Ritual Meals*. Oxford Early Christian Studies. Oxford: Clarendon Press.

Meiser, M. 2007. *Galater*. Novum Testamentum Patristicum 9. Göttingen: Vandenhoeck & Ruprecht.

Minns, D. and P. Parvis. 2009. *Justin, Philosopher and Martyr, Apologies*. Oxford Early Christian Texts. Oxford: Oxford University Press.

Nanos, M. D. 2002. *The Irony of Galatians: Paul's Letter in First-Century Context*. Minneapolis, MN: Fortress Press.

Pervo, R. I. 2009. *Acts. A Commentary*. Hermeneia. Minneapolis, MN: Fortress Press.

Rordorf, W. 1962. *Der Sonntag: Geschichte des Ruhe- und Gottesdiensttages im ältesten Christentum*. Abhandlungen zur Theologie des Alten und Neuen Testaments 43. Zurich: Zwingli Verlag.

Rouwhorst, G. 2001. "The Reception of the Jewish Sabbath in Early Christianity." In P. Post et al. (eds.), *Christian Feast and Festival: The Dynamics of Western Liturgy and Culture*. Liturgia condenda 12. Leuven: Peeters, 223–266.

Schmidt, J. 1890. "Statut einer Municipalcurie in Africa." *Rheinisches Museum für Philologie, New Series* 45: 599–611.

Schöllgen, G. 1982. "Die Teilnahme der Christen am städtischen Leben in vorkonstantinischer Zeit." *Römische Quartalschrift für christliche Altertumskunde* 77: 1–29.

Schrage, W. 2001. *Der erste Brief an die Korinther: 1 Kor 15,1–16,24*. Evangelisch-Katholischer Kommentar zum Neuen Testament 7,4. Zurich: Benziger – Neukirchen-Vluyn: Neukirchener Verlag.

Schwemer, A. M. 2000. "Himmlische Stadt und himmlisches Bürgerrecht bei Paulus (Gal 4,26 und Phil 3,20)." In M. Hengel, S. Mittmann, and A. M. Schwemer (eds.), *La Cité de Dieu. Die Stadt Gottes: 3. Symposium Strasbourg, Tübingen, Uppsala. 19.–23. September 1998 in Tübingen*. Wissenschaftliche Untersuchungen zum Neuen Testament 129. Tübingen: Mohr Siebeck, 195–243.

Stökl Ben Ezra, D. 2003. *The Impact of Yom Kippur on Early Christianity: The Day of Atonement from Second Temple Judaism to the Fifth Century*. Wissenschaftliche Untersuchungen zum Neuen Testament 163. Tübingen: Mohr Siebeck.

Suys, V. 2005. "Les associations cultuelles dans la cité aux époques hellénistique et impériale." In V. Dasen and M. Piérart (eds.), *Idia kai dèmosia: Les cadres "privés" et "publics" de la religion grecque antique: Actes du IXe colloque du Centre International d'Étude de la Religion Grecque Antique (CIERGA), tenu à Fribourg du 8 au 10 septembre 2003*. Kernos Supplements 15. Liège: CIERGA, 203–218.

Theobald, M. 1988. "'Wir haben hier keine bleibende Stadt, sondern suchen die zukünftige' (Hebr 13,14): Die Stadt als Ort der frühen christlichen Gemeinde." *Theologie und Glaube* 78: 16–40.

Thraede, K. 2004. "Noch einmal: Plinius d. J. und die Christen." *Zeitschrift für die neutestamentliche Wissenschaft* 95: 102–128.

van Goudoever, J. 1967. *Fêtes et calendriers bibliques*. 3rd rev. ed. Paris: Beauchesne [1st ed. 1961].

Vinzent, M. 2014a. *Marcion and the Dating of the Synoptic Gospels*. Studia Patristica Supplements 2. Leuven: Peeters.

Vinzent, M. 2014b. *Die Auferstehung Christi im frühen Christentum*. Freiburg i.Br.: Herder [revised German translation of idem 2011. *Christ's Resurrection in Early Christianity and the Making of the New Testament*. Farnham: Ashgate].

Waltzing, J.-P. and A. Severyns. 1961. *Tertullien, Apologétique*. Budé. 2nd ed. Paris: Les Belles Lettres.

Wenger, L. 1954. "Bürgerrecht." *Reallexikon für Antike und Christentum* 2: 778–786.

Witulski, T. 2000. *Die Adressaten des Galaterbriefs: Untersuchungen zur Gemeinde von Antiochia in Pisidien*. Forschungen zur Religion und Literatur des Alten und Neuen Testaments 193. Göttingen: Vandenhoeck & Ruprecht.

Young, N. H. 2003. "'The Use of Sunday for Meetings of Believers in the New Testament': A Response." *Novum Testamentum* 45: 111–122.

AUTHOR INDEX

SUBJECT INDEX

Lightning Source UK Ltd.
Milton Keynes UK
UKHW02n0127060718
325316UK00006B/27/P

9 781107 108967